Electronic Measurement Systems

To *Ria*
Michel
Maurice
Antoinette
Pascal

with love

Only knowledge in harmony is progress

AFPVP

Electronic Measurement Systems

Anton F. P. van Putten
University of Technology, Eindhoven

PRENTICE HALL
New York London Toronto Sydney Tokyo

First published 1988 by
Prentice Hall International (UK) Ltd,
66 Wood Lane End, Hemel Hempstead,
Hertfordshire, HP2 4RG
A division of
Simon & Schuster International Group

Printed and bound in Great Britain at
The University Press, Cambridge

Library of Congress Cataloging-in-Publication Data

Putten, Anton F.P. van, 1939–
Electronic measurement systems.

Bibliography: p.
1. Electronic measurements. I. Title.
TK7878.P87 1988 681′.2 87-29251
ISBN 0-13-251893-7

1 2 3 4 5 92 91 90 89 88

ISBN 0-13-251893-7
ISBN 0-13-251885-6 PBK

CONTENTS

Preface xi

1 Introduction to Electronic Measurement Systems 1

1.1 Information 1
1.2 Types of measurements 3
1.3 Why electronic measurement systems? 4
1.4 Future trends 6
1.5 Problems 7

2 Types of Measuring Systems 8

2.1 Introduction 8
2.2 Functional structures 9
2.2.1 Measuring structure for an active physical quantity 9
2.2.2 Measuring structure for a passive physical quantity 11
2.3 Signal structures 13
2.3.1 Signals with a unique character 14
2.3.1.1 Time analysis 14
2.3.1.2 Response analysis 14
2.3.2 Periodic signals 17
2.3.3 Sampled signals 19
2.3.4 Stochastic signals 21
2.4 Spatial structures 23
2.4.1 A multiple input/output configuration 24
2.4.2 A centralized data-acquisition system 25
2.4.3 A decentralized data-acquisition system with a digital multiplexer 28
2.4.4 A distributed multiplexer data-acquisition system 29
2.4.5 Telemetry 30
2.5 Automated electronic structures 31
2.6 Problems 33

3 Implementation of Functions 34

3.1 Introduction 34
3.2 Tools 34
3.3 Analog functions 44
3.3.1 The operational amplifier 44
3.3.2 Preamplifier with adjustable gain 55
3.3.3 The non-inverting operational amplifier 55
3.3.4 The differential amplifier 56
3.3.5 The comparator 57
3.3.6 The integrator 57
3.3.7 A constant current source 58
3.3.8 The logarithmic amplifier 59
3.3.9 An analog multiplier 60
3.3.10 A peak detector 61
3.4 Analog-to-digital and digital-to analog conversion 62
3.4.1 Coding 62
3.4.2 Digital-to-analog converter 65
3.4.3 Analog-to-digital converter 67
3.4.3.1 The dual-slope integration ADC 67
3.4.3.2 Autozeroing and AD conversion 69
3.5 Digital functions 70
3.5.1 A decoder 70
3.5.2 An encoder 71
3.5.3 A multiplexer 72
3.6 Miscellaneous 73
3.6.1 The conventional Wheatstone bridge configuration 73
3.6.2 An integrated Wheatstone bridge 75
3.6.3 The principle of the constant voltage, constant current bridge configuration 76
3.7 Problems 78

4 System Specifications 80

4.1 Introduction 80
4.2 Technical specifications 81
4.2.1 Application area 81
4.2.2 Resolution 81
4.2.3 Accuracy and inaccuracy 82
4.2.4 Sensitivity 84
4.2.5 Linearity 85
4.2.6 Offset and drift 86
4.2.7 Rejection factor 87
4.2.8 Dynamic range 91
4.2.9 The number of measurements per unit time 94
4.2.10 Environmental conditions 96
4.3 Types of errors 96

4.3.1 Systematic errors 96
4.3.2 Conditional errors 96
4.3.3 Stochastic errors 97
4.4 Improving measures for electronic systems 98
4.4.1 The feedback principle 98
4.4.2 Feedforward coupling 101
4.4.3 The influence of feedback on input and output impedance 103
4.5 A microprocessor-based measuring system 107
4.6 Problems 109

5 Reliability 112

5.1 Introduction 112
5.2 Reliability concepts 113
5.2.1 Definitions 113
5.2.2 Stochastic variables, probability distributions, functions and reliability 114
5.2.3 Reliability parameters 116
5.2.4 Mean time between failure (MTBF) 118
5.3 Families of lifetime distributions 120
5.4 Reliability of parts, components, equipment and systems 122
5.5 Availability, repairable and non-repairable systems 123
5.6 Boolean algebra and reliability 126
5.7 The reliability and MTBF of systems in series 129
5.8 The reliability and MTBF of systems in parallel 130
5.8.1 The active on-line standby system 131
5.8.2 The cold on-line standby system 134
5.8.3 A parallel k out of n system 136
5.9 Methods to determine the reliability of systems 137
5.9.1 The network reduction method 137
5.9.2 The path-tracing technique 139
5.9.3 The decomposition technique 140
5.9.4 The minimal cut set technique 142
5.9.5 Markov techniques 143
5.10 Design aspects 150
5.11 Causes of failure 155
5.12 Problems 157

6 Transducers 160

6.1 Introduction 160
6.2 Definitions and terminology 161
6.2.1 Types of energy forms 161
6.2.2 Types of energy sources 162
6.2.3 Modulating and self-generating transducers 163
6.3 State descriptions 169

6.3.1 The steady-state description 169
6.3.2 The dynamic state description 172
6.4 Tables of energy conversions 173
6.4.1 The radiant energy domain 173
6.4.2 The mechanical energy domain 174
6.4.3 The thermal energy domain 175
6.4.4 The magnetic energy domain 176
6.4.5 The chemical energy domain 177
6.5 Integrated semiconductors and transducers 178
6.6 Characteristics of silicon technology 179
6.6.1 Size 179
6.6.2 Complexity 179
6.6.3 Power consumption 181
6.6.4 Reliability 181
6.6.5 Price performance ratio 182
6.7 Physical effects in silicon 183
6.8 Radiant energy in silicon microtransducers 185
6.8.1 Radiant energy in compatible technology 186
6.9 Mechanical energy in silicon microtransducers 188
6.9.1 Mechanical energy in compatible technology 189
6.10 Thermal energy in silicon microtransducers 190
6.10.1 Thermal energy in compatible technology 192
6.11 Magnetic energy in silicon microtransducers 192
6.11.1 Magnetic energy in compatible technology 194
6.12 Chemical energy in silicon microtransducers 194
6.13 Interfacing to sensors 196
6.13.1 Sensors in bridge configurations 196
6.13.2 Biasing circuits 200
6.13.3 Bridge amplifier configurations 201
6.14 Future trends 208
6.14.1 Technological trends 208
6.14.2 Fundamental developments 209
6.15 Problems 210

7 Offset and Drift 212

7.1 Introduction 212
7.2 Principles of offset voltage and current 212
7.3 Calculation of the offset voltage and current 214
7.3.1 The bipolar transistor 214
7.3.2 The field-effect transistor 215
7.4 Offset behavior of differential amplifier stages 217
7.4.1 The bipolar differential input stage 217
7.4.2 The junction-FET differential input stage 218
7.4.3 The MOSFET differential input stage 219
7.5 Drift behavior of offset current and voltage 220

7.5.1 Stochastic variations in electrical conductivity 220
7.5.2 Aging, humidity and vibrations 220
7.5.3 Power supply variations 221
7.6 The influence of temperature 222
7.7 Temperature drift in bipolar transistors 223
7.8 Temperature drift in FETs 225
7.8.1 The junction field-effect transistor 225
7.8.2 The MOS field-effect transistor 226
7.9 Problems 228

8 Guarding and shielding 229

8.1 Introduction 229
8.2 The nature and causes of interference 229
8.2.1 Sources of interference 234
8.3 Aspects of guarding 235
8.3.1 Ground and earth 236
8.3.2 A grounded system 237
8.4 Aspects of shielding 239
8.4.1 A shielding cabinet 239
8.4.2 A shielding cabinet as part of an electronic circuit 240
8.4.3 The use of more than one shielding cabinet 241
8.4.4 The connection of a common in a multiple cabinet configuration 243
8.5 Interference caused by a magnetic field 246
8.6 Types of coupling mechanisms 249
8.6.1 Direct coupling 249
8.6.2 Magnetic coupling 251
8.7 Interference caused by power supply transformers 252
8.8 The differential amplifier and guarding 257
8.8.1 The common-mode rejection ratio 259
8.8.2 The CMRR in shielding circumstances 261
8.9 Improvement of the common-mode rejection ratio 263
8.9.1 The application of a floating guard 263
8.9.2 Active guarding or the application of double-shielded cables 267
8.10 Worked examples 267
8.11 Problems 272

9 Noise calculations 275

9.1 Introduction 275
9.2 Noise voltage, noise current and noise figure 276
9.3 The relationship between the noise voltage, noise current and noise figure 278
9.4 Calculating total noise voltage 279
9.5 Calculating the noise figure and the signal-to-noise ratio 284

9.6 The noise figure myth 286
9.6.1 The optimum source resistance 287
9.6.2 The optimum signal-to-noise ratio 289
9.7 Other noise calculating techniques 295
9.8 Problems 299

10 Physics of Noise 300

10.1 Introduction 300
10.2 Thermal noise or Johnson noise in conductors 300
10.3 Shot noise 301
10.4 Flicker noise 303
10.5 Noise in semiconductor components 304
10.5.1 Noise behavior of the bipolar transistor 304
10.5.2 Noise behavior of the field-effect transistor 306
10.5.3 Noise behavior of the diode 309

11 Ergonomics or Human Engineering 313

11.1 Introduction 313
11.2 Biological equipment of the human being 314
11.2.1 The auditory window 314
11.2.2 The visual window 315
11.3 Some other aspects of human perception of information 316
11.3.1 Transmission of information 316
11.3.2 Retrieval of information 316
11.3.3 Processing of information 317
11.4 Perception 317
11.5 Trends in the development of measuring systems 319
11.6 The observer–display interaction 320
11.7 Coding of information on displays 322
11.8 Problems 327

Appendix 1 Performance Specifications for Converters 328

Appendix 2 Standard Symbols for Logic Elements 335

Bibliography 336

Index 339

PREFACE

Electronic measurement systems have become complicated machines able to perform complex tasks in a large variety of environments. Moreover, such systems are found in every area of science and technology.

The large impact of digital electronics has changed the world of electronic measurement systems dramatically, both inside and outside the system. Multipurpose, multifunctional measurement systems which are capable of performing tasks that were impossible ten years ago are common place now. Another important aspect of these systems is society's increasing dependence on them; the reliability of systems has become of vital importance.

Measuring systems are realizations of technical concepts able to gather, handle, modify, store, retrieve and present information. The design of a measurement system not only requires a great deal of knowledge about components and their possibilities: aspects related to the underlying philosophy of a system and its specific architecture are increasingly important. For example, a large shift to digital control and microprocessor-driven measurement systems has taken place in a large variety of applications.

The design of a measurement system is a complicated process and often the designer of a system is highly specialized in a few areas, but in practice, the designer is confronted with numerous questions concerning the (technical) specifications, for instance:

1 What degree of accuracy is required of the system?
2 What are the environmental conditions, e.g., temperature range, shock, vibration and electromagnetic interference?
3 What is the required reliability or the expected mean time between failures?
4 Which type of transducer is needed for the physical quantity to be measured and what possibilities are available?
5 Which type of architecture should be applied?
6 What is the required sensitivity of the measurand?
7 What about noise behavior?
8 What is the required common-mode rejection ratio?
9 Which kind of ergonomic layout is best?

This book covers a large number of topics encountered in the design, application and verification of electronic measurement systems. The system is considered to be composed of three parts: the *input* stage, the *modifier* stage and the *output* stage. Each stage can be treated separately.

In the first four chapters a general picture is presented of measurement architectures, structures, the tools used to describe systems and implementation of components. Further system specifications are discussed in detail and illustrated with many examples. The remaining chapters discuss topics of major importance to the proper functioning of a system.

Not all subjects could be covered in the same detail here, mainly due to limitations of space. As will be recognized, attention is devoted to subjects that are usually not found together in one book. Often it is these aspects that ultimately determine if a system can operate within its stated specifications.

For several years, the material on which this textbook is based has been used for graduate and undergraduate students in electrical and electronic engineering with a basic knowledge of electronics, but of course some topics discussed have a much broader application than to electronics systems alone. For the sake of completeness, and to make an introduction more easy, some topics which may be assumed to be well known are also discussed here.

The contents of the chapters may be summarized as follows:

Chapter 1 is devoted to basic ideas of measurement systems, such as what information is and why it is preferred to handle information electronically.

Chapter 2 considers measurement architectures and structures. This includes functional structures, signal, spatial and automated structures.

Chapter 3 is devoted to the available tools, such as the Bode plot, the polar plot and the pole-zero plot and the relationships between them. Further implementations of frequently applied functions are presented. These implementations cover analog functions, analog-to-digital and digital-to-analog conversion and some miscellaneous functions. Under this last heading a new integrated bridge is discussed.

Chapter 4 is devoted to general and more specific technical system specifications. All the usually met technical specifications are presented with their meanings and consequences. This includes a discussion of the common-mode behavior, linear amplification and the implications of feedback. Error definitions, classifications and calculations are also presented. After some error causes are introduced, some frequently applied improving measures are reviewed. The chapter concludes with a discussion of a microprocessor-based measuring system.

Chapter 5 provides a comprehensive introduction to reliability concepts, including the basic concepts and calculation techniques used in this area. Markov techniques are included as the most generally applied and powerful techniques for calculating reliability.

Chapter 6 is devoted to transducers or sensors and actuators for energy conversions to and from the electrical energy domain. Basic ideas, concepts and definitions are discussed in detail and a general review of most known energy conversion mechanisms is presented. The possibility of silicon and compatible technologies for the manufacture of transducers is also discussed.

Chapter 7 discusses aspects of offset and drift in basic electronic active components. The influence of temperature on bipolar transistors, FETs and differential input stages is treated in detail. Offset and drift are treated separately.

Chapter 8 is devoted to the problems concerning guarding and shielding systems from electromagnetic noise. With the increasing trend to handling all kinds of information electronically, the environment is charged with an ever-increasing electromagnetic pollution. Moreover, electronic measurement systems are highly sensitive to electromagnetic interference and effective precautions should be taken to avoid these influences. All kinds of preventative remedies are discussed in detail. The influence on the common-mode rejection ratio of sources of interference is also discussed.

Chapter 9 presents a discussion of noise calculations. It is not necessary to have a detailed knowledge of the physics of noise for this, and it suffices to use only the specifications of the manufacturer. The theoretical discussion is left to the following chapter.

Chapter 10 treats the physics of noise in three important noise mechanisms and the noise behavior of the active components to be met in electronics today, such as the bipolar transistor and the field-effect transistor.

Finally, **Chapter 11** is devoted to human engineering or ergonomics. Some interesting aspects concerning man–machine relations are presented, in particular the ergonomics of the visual and auditory organs for the presentation of information.

Almost every chapter can be read independently from the others; hence a flexible presentation of subjects can be realized. The consequence of this approach is that some details are repeated in different contexts, but this can only improve understanding. Besides this, every chapter is provided with examples and exercise material.

It will be clear that writing a book is a tremendous task and cannot be realized without the help of others. I therefore would like to thank all the persons who have given advice directly and indirectly in the preparation of this book.

Special thanks are due to the following persons: Jim Sherwin of National Semiconductor Corp. for his permission to use material about noise calculations. I am also indebted to A. W. Wijkman, P. J. van den Akker and K. E. Kuijk from Philips Corp. for their permission to use relevant material concerning shielding and guarding. I also want to thank Paul A. M. Maas of Delft University Press; Paul P. L. Regtien and Prof. Simon Middelhoek, both of Delft University of Technology, for providing material about transducers and related topics. Special thanks are due to Dr Mark Browne of University of Manchester Institute of Science and Technology for his careful reading of the manuscript and his valuable advice.

Last but not least, I express my gratitude for the everlasting patience of my wife Ria and the influence of my sons Michel and Maurice on this book.

A. F. P. van Putten
Eindhoven
1988

Electronic Measurement Systems

One

INTRODUCTION TO ELECTRONIC MEASUREMENT SYSTEMS

Ever since human beings began to think, we have exchanged information and have given measures to quantities in an attempt to understand our surroundings. We appear to have a basic curiosity about and a need to investigate our environment. We recognize this with our senses, but our natural capacities are limited, so we have developed tools to help us fulfill our measurement needs.

In a continual process towards perfection, we are still refining our tools to improve our understanding of all kinds of mechanisms. Often models are shaped which describe more or less accurately the real world. This requires collecting information about the environment, a system or a process. Knowledge is related to information and, with collected information concerning a system or a process, we can increase our knowledge. This is a continuous story of interaction: model-making and performing measurements and vice versa.

1.1 INFORMATION

The word 'information' is interpreted as a type of arrangement or regularity which can be recognized in the environment, for instance, the letters on a page show a certain regularity and the message involved is recognized by the sequence in which the letters have been printed.

Information content is implemented in the order in which the letters appear on the paper, and a limited quantity of information is conveyed. If the same letters appear at random, however, all information is lost and interpretation is no longer possible. The amount of information has become zero. This example explains why it is said that information always involves a certain amount of regularity. Put in a reverse way, every regularity contains a quantity of information. Here order and disorder or information and chaos are each other's opposites.

This interrelationship, between information and chaos, was first shown by Shannon, who connected the idea of information to entropy. In accordance with the second law of thermodynamics, entropy increases in a closed system towards a maximum which corresponds with the minimum amount of energy in that system. Hence, entropy is defined as a measure of chaos.

Information can be quantified, which is what information theory is about. In this discipline, even the smallest amount of information is defined. Information can be represented in different forms, but it is always bound to a certain type of energy carrier or mass.

Five different types of energy carriers can be distinguished: radiation, electrical or magnetic energy, heat, chemical energy and mechanical energy.

Today everyone is familiar with the fact that information can be measured, handled, stored, retrieved and represented, depending on its application and requirements demanded of it. Besides this, information can be converted from one energy form into another by so-called transducers.

Our world is a very complex one with a tremendous amount of information available, coming from all kinds of processes and physical mechanisms, often so complex that it is not possible to understand all collected information directly. A good example of such a process is the climate, which is very complex indeed. If statements have to be made about the weather, a lot of data are collected concerning temperature, pressure, humidity, wind velocities and directions at different heights and so on. Finally, we make a weather forecast, which always involves a certain amount of uncertainty. This uncertainty is a basic characteristic of every statement which is based on data collected by measurement techniques. The reason for this may be that the process is too complicated, or the way the data are collected is imperfect, or the amount of data collected is insufficient.

In the above example, this uncertainty is quite obvious, but errors are inherent in every measurement technique. The result is that we create an image of part of the world with a limited amount of information. However, despite this implicit imperfection, measurement remains the heart of science and technology and without it no progress can be made.

The first thing to do when performing measurements is to characterize the type of process needed to investigate and determine the physical parameter(s) which are to be measured. This means criteria must be formulated for every measurement concerning type, response time, accuracy, etc. After characterizing the process, decisions can be made concerning which type of measurement will be applied, and then collection of data can begin. After the collection of the data we have to analyze, synthesize and evaluate the measurements. Finally an interpretation and check of the results must follow.

Collecting data corresponds with receiving information about the system or process. As more information is received, uncertainty concerning the system is decreased.

Two different kinds of information have to be distinguished. Firstly, in the process under consideration, the type of information must be recognized. This concerns the characteristics of the process. It is a qualitative description of the information involved and describes what sort of energy is present. This is called the *structural* information of the process. Secondly, the magnitude or intensity of the relevant characteristics must be discovered, for instance, scaling factors concerning how much, how fast, etc. This is called *metric* information. The process by which this information is collected is a quantitative process.

It will be clear from this that a qualitative process must precede a quantitative process. Ultimately, the measurement process can be interpreted as making assignment statements in accordance with predefined procedures by technical means. The result is a collection of numbers assigned to the parameters under consideration by a well-chosen relationship.

1.2 TYPES OF MEASUREMENTS

In practice, measurements can be performed in five different ways with five different levels of accuracy and degrees of freedom.

1 Given two unknown quantities of the same kind, if only a statement is required of whether they are equal, a *nominal* measurement is involved. Examples are a color impression of two different objects or the acidity of two liquids.
2 If additional information is collected about relative sizes, an *ordinal* type of measurement is performed. For instance, in a sample of boys and girls, the respective height and weight can be fixed and classified according to this kind of ordering.
3 When more specific information is required such as the magnitude of the respective intervals between the quantities of heights and weights in the above example, without making any statement about a kind of zero-reference, a type of *interval* measurement is performed. There is an increase in information content received.

   Interval measurements can take the form of a weight measurement in kilogrammes or miligrammes, a temperature measurement in °C, or a time measurement in hours or seconds. There is no zero or floating reference point.
4 We can choose a certain point as a reference of zero level and then perform ratio measurements by dividing every measurand by an arbitrarily chosen reference value, and determining their relative magnitudes. For example, all measurands are divided by the maximum measured value and then a so-called normalized measurement is performed; the largest value will be 1; all the others will be less than 1. This type of measurement is very often applied in physical experiments and, again, provides more specific information.
5 Finally, statements can be made about the magnitude of the reference compared with a standard reference. When a standard system is available, every physical parameter can be measured compared with its respective standard reference quantity with no degree of freedom and no possible ambiguity. These types of measurements are called *cardinal* measurements. The SI system is just such a well accepted reference system, although in aircraft industries other reference systems are still in use, for instance mph, knots, gallons, pounds, feet, etc.

It is interesting to note that in the real world it is often not possible to perform cardinal measurements. For instance, in biological and economic systems, the processes are so complex that the available tools fail to provide accurate predictions and statements. Here, time and the number of interacting parameters are the most disturbing and complicating factors and, last but not least, every measurement disturbs the process under consideration.

1.3 WHY ELECTRONIC MEASUREMENT SYSTEMS?

So far in this introduction, basic concepts have been discussed which you can find in most textbooks about measurement techniques and systems, but the next question to answer is why it is such a great advantage to handle information by electronic means using electric charges as information carriers.

Information can also be handled by pneumatic, hydraulic or mechanical means based on the same principles as in electronics. Here are the main reasons why it is preferable to handle information by assigning it to electrical carriers:

1 A very large dynamic power range can be covered by electronic means. The available power range is: 10^{-9} to 10^{+9} watts, which means a dynamic range of 10^{18}.

2 For electric charges, very high speed and acceleration is available due to the high ratio of charge to mass of the smallest possible electrical charge q, the electron:

$$\begin{aligned} q/m &= 1.6 \times 10^{-19}/9.1 \times 10^{-31} \\ &= 0.18 \times 10^{+12} \text{ C/kg} \end{aligned}$$

The acceleration equals

$$a = (q/m)E \qquad (\text{ms}^{-2})$$

where q again is the specific electronic charge, m the specific mass and E is the electric field strength to which the charge is subjected. If, for instance, the field strength is 1 V/m, we find

$$a = 18 \times 10^{10} \qquad \text{ms}^{-2}$$

which is an exceedingly large acceleration.

3 A very large time domain can be controlled from picoseconds (10^{-12} s) to hours, which means a time range of 10^{15}. No other energy carrier known can handle information in such an easy way with such high speeds and accelerations and with such huge dynamic ranges.

4 Information can be transported very efficiently by means of copper or aluminum cables, and via satellite or radio channels. When fiber optics are used for transport purposes, the highest specific transport–information density over long distances is available.

5 Information can be amplified by electronic means with very large gain factors from very low levels of signal amplitude (nanovolts) to hundreds of volts, which is the process ocurring most in electronic information handling. Gain factors of 10^8 are possible with one single device.

6 Finally, information can be modulated in other shapes, stored, retrieved and modified, and all known mathematical operations can be performed using either analog or digital methods at very high speeds and in very small volumes.

With these striking advantages compared with other techniques it is not surprising that we prefer to handle information electronically. The consequences are a tremendous growth in applications of electronic means at all levels and in all disciplines. It is the strength and simultaneously the weakness that our society has become completely dependent on electronic means and tools.

You can encounter the influence of electronics in the field of medicine, throughout industry, in research and development, in banking and insurance and in business and private life. Our society has really become an 'information society' as Toffler describes in his book *The Third Wave*. This dependency is the main reason there is a chapter on reliability in this book – there is an increasing need for statements about the reliability of the systems we depend on so heavily.

When considering the development of specific electronic tools or components for electronic measurement, the birth of the 4-bit microprocessor in 1971 cannot be overlooked. This component, now considered to be obsolete, was a simple 4-bit p-channel metal-oxide semiconductor (MOS) chip, initially manufactured as a custom-designed chip for a Japanese calculator manufacturer.

The impact of the new 32-bit microprocessors on a single piece of silicon will be equally enormous and cannot be foreseen. In the meantime, the microprocessor has already earned its place in measurement applications and technology due to its great flexibility in collecting and processing data.

One may well ask if there are any disadvantages to microprocessor technology. There are a few. The first is that microprocessors can function only within a limited temperature range. The lower limit is −50°C and the upper limit is about 175°C. They are also not suitable for applications in situations demanding very high degrees of intrinsic safety, for example in explosive environments. Intrinsic safety means that installed electric power may not exceed 200 mW.

Electronic measurement systems also fail when very large forces are required as, for example, in offshore industry.

Electronic systems can be best described as types of information channel consisting of three parts. The first is the input part, where the information is picked up. This information is very often a nonelectrical physical quantity to be converted into an electrical quantity by a type of transducer. Secondly there is a processing part where the information is processed and modified into the desired shape and form. Thirdly there is an output part where the information leaves the channel in order to be perceived by the observer.

Measured data should be presented in a form that is easy to recognize, which is primarily visual. Many physical principles can be employed for display purposes, the choice being dependent on different criteria such as price, power consumption, response time and ergonomic aspects. The latter is of particular importance in aircraft where a lot of data must be recognized and interpreted, sometimes in a split second.

1.4 FUTURE TRENDS

It is useful to discuss some future technological trends resulting from the great technological push of electronic tools and systems.

A lot of attention is paid to speech processing and this has already resulted in a completely automatic talking dashboard in an automobile. In the near future it may be expected that the way typewriting is executed will be replaced by speech commands being given to a type of processor able to recognize the human voice.

These are examples of systems made possible by the developments in technology one of which is the further decrease in line width in integrated circuit (IC) technology by means of electron- or X-ray photolithographic techniques. These techniques make it easily possible to reduce the line width below 1 μm. Another major development is the progress made in wafer and die size; wafer sizes of 6 inch diameter and chip areas of 100 mm^2 are being used. We will come to these aspects again in Chapter 6.

Memory size and speed will further increase, especially in NMOS and CMOS technology. Up to 4 megabits dynamic and static RAMs are already being manufactured. The compact disk is not only useful for recording music but is used for storage applications as well, where it is able to provide a storage capacity of 100 Mbyte on a single disk of 4 inch diameter.

For communication purposes, fiber optics will replace copper cables in most cases, due to cost and bandwidth capacity advantages.

Rapid progress is being made in gallium-arsenide (GaAs) technology and the question has been put forward already if GaAs will replace silicon. Cost aspects play a dominant role here. For digital electronics, GaAs appears promising, especially with respect to high-frequency behavior and speed. Multipliers and (small) RAM chips are already available.

In electronic measurement systems an increasing shift to digital data processing is taking place. Digital processing happens more and more directly after the data is collected from the outside analog world. Every physical process is analog in itself but these quantities are converted into an electrical *digital* quantity as soon as the minimum required signal level is acquired.

These developments will strongly influence all kind of systems in all kinds of industry and organization. With this in mind we can concentrate on important aspects related to the design of measurement instruments. If a measuring instrument has to be constructed it is not a matter of merely connecting the inputs and outputs of a set of subsystems. To start with, a careful and precise description of the quantity to be measured must be made. Important questions to be answered are:

- What quantity has to be measured?
- What is the measuring object?
- What are the environmental conditions?
- What accuracy is required?
- What reliability is required?
- What is the required display?

We can conclude the introduction with the following statement: 'no system is better than its weakest link.' Thus a check and comparison for every subsystem must be executed on all the relevant aspects mentioned above. The same questions will hold for the user also.

1.5 PROBLEMS

1.1 Give a definition of information and compare this concept with the concept of entropy.

1.2 To which kinds of carriers can information become bound?

1.3 What is meant by the characterization of the kind of information?

1.4 Discuss the relationships between information, chaos and uncertainty.

1.5 What is meant by structural information?

1.6 What is meant by metric information?

1.7 Discuss the main advantages of assigning information to electronic charge carriers.

1.8 Give a definition of analog information and digital information.

1.9 Which five types of measurement can be distinguished?

1.10 Why has reliability become an increasing factor of importance in the design of measurement systems?

Two

TYPES OF MEASURING SYSTEMS

2.1 INTRODUCTION

In a given process in a certain arrangement, a limited amount of information is available, generally as a function of time and place. Examples are the temperature distribution in a radiator, the flow velocity in a tube, the radiant intensity of a surface, mechanical stress in structures, and so on. Information retrieved from a process is obtained by taking a measurement. It can be said that measuring is making a kind of image on a well defined scale of symbols. More precisely, measuring is creating a representation of a physical quantity, which can be described as

$$y(t, s) = f[x(t, s)] \tag{2.1}$$

where t and s denote time and place respectively and f represents the functional behavior of the applied instrument.

The physical parameter of which information has to be retrieved is always bound to a certain energy carrier. Of the different types of energy, the electrical form is the preferred one, due to its flexible and easy way of handling. In many cases the quantity to be measured is of a character other than an electrical one. Therefore an energy conversion is required. This conversion is realized by the appropriate transducer or sensor. At the output of the measuring chain again a type of transducer can be used to make the measurand perceivable or suitable for registration. The center of the measuring chain between input and output is characterized by an electrical process.

Measurement systems have become a complex mixture of analog and digital electronics. To be able to discuss the several aspects and impacts of electronic measurement systems we have to introduce many (theoretical) tools.

In this chapter we will present a discussion of the classification of measuring structures. In the first instance, measuring instruments can be divided into four classes with respect to their structures.

1 The functional structure

The functional structure describes the way the measurement is performed and depends on the number of unknown related physical quantities to be measured and whether there is an active or passive quantity involved.

2 The signal structure
Measurement systems can be distinguished according to the type of signal to be measured or processed.

3 Spatial structure
In this case the spatial layout of the input quantities determines the structure and architecture. Data-acquisition systems, for instance, can lead to very complex spatial structured systems.

4 Automated structures
Here the digital structure of the measurement system is emphasized. This arises from the present trend in performing manual measurements from semi-automatic to fully automatic and computer-controlled systems.

2.2 FUNCTIONAL STRUCTURES

There are two basic types of functional structure. This distinction depends on the measurand's ability to carry and transport power. The measurand can be an active or a passive quantity. This characteristic determines the way the measuring system is composed and prescribes the functional structure of the system.

2.2.1 Measuring structure for an active physical quantity

A physical quantity is defined as *active* or *intensive* if it can carry and transport power. Examples of active quantities are force, electrical field, pressure, light intensity, and so on. As will be seen, for such signal sources, no auxiliary energy is required.

A set-up for measuring an active physical quantity is composed of a limited, well-defined number of functional blocks and can be represented as shown in Fig. 2.1. If a nonelectrical physical quantity has to be measured then, in the first functional block, an energy conversion to the electrical domain must be realized. This first functional block is often called a sensor or transducer. Transducers will be discussed in Chapter 6. For the moment it suffices to say that the sensor is able to produce an electrical signal which is a good representation for the physical quantity to be measured.

The second functional block is the measuring system assigned to the process, in which the electrical signal is modified to the desired shape. This electrical modification process can accomplish many different functions. A collection of possible functions, sometimes denoted as signal conditioning functions, is summarized in Table 2.1.

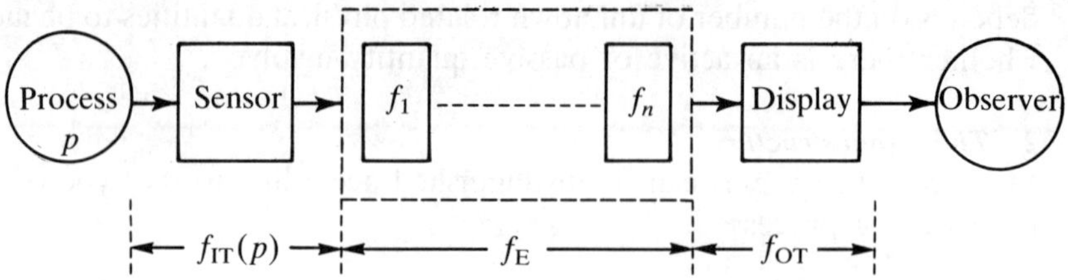

Figure 2.1 General arrangement for a measuring set-up

We will discuss several of these functions in Chapter 3. The electrical process can be written as

$$f_E = f_1 f_2 \ldots f_n \tag{2.2}$$

where $f_1, \ldots, f_n$ describe the different electrical functions to be distinguished in the second functional block.

The third functional block is composed of a type of display able to represent the measurand in a suitable form which can be perceived by the human senses, or in a shape suitable for storage. Generally, the function of a system measuring an active physical quantity can be written as

$$M = f_{OT}[f_E\{f_{IT}(p)\}] \tag{2.3}$$

where:

M	is the desired measurand
$f_{IT}(p)$	represents the input transducer's function of process p
f_E	represents the electrical process
f_{OT}	represents the output transfer function

When more than one unknown related physical quantity must be measured the situation is more complex. We can use the circuit as shown in Fig. 2.2. Assume a pressure and a temperature must be measured simultaneously, where it will be clear that the pressure is related to temperature. Then the measurement set-up should provide two signals and hence two equations must be solved to obtain required information.

Table 2.1 Review of some electronic functions

Amplification (of the small source signal)
A/D and D/A conversion
Modulation and demodulation
Determination of a maximum or minimum (peak detection)
Calculation of the average or RMS value
Derivative or integral of a signal
Sampling
Filtering
Measuring zero crossing, or timing intervals
Summing, multiplying
Counting

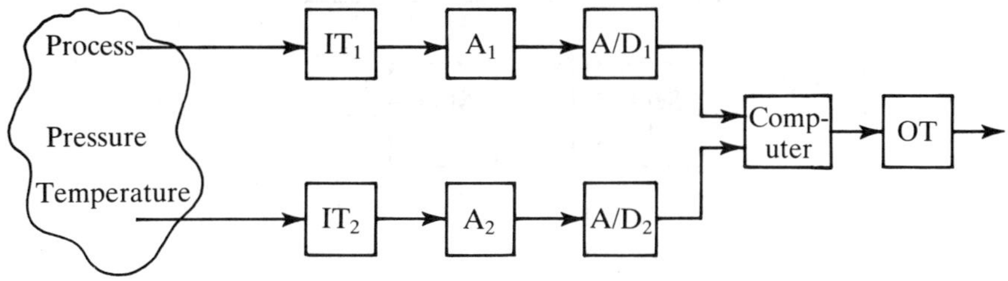

Figure 2.2 Measurement set-up for two unknown signals. The computer will calculate both quantities from two digital signals

In Fig. 2.2 two parallel measuring chains are shown. At the end of the chain a computer will solve both equations, providing the requested information. In this figure IT_1 represents a pressure transducer delivering a temperature-dependent magnitude. Transducer IT_2 represents a second temperature-dependent pressure transducer, delivering another and different dependency between pressure and temperature. So two independent equations can be formed. Often amplifiers are needed to amplify the very weak transducer signals. They are denoted by A_1 and A_2. The analog-digital conversion is executed by A/D_1 and A/D_2. Finally the output transducer OT will display the magnitude of pressure and temperature calculated by the computer.

2.2.2 Measuring structure for a passive physical quantity

A *passive* or *extensive* physical quantity is not able to transfer energy to a sensor directly. The relevant information is implied in the structure of that matter itself and can only be retreived by measuring the response gained when the object is subjected to a suitable stimulus.

Examples of these are mass, the elasticity of a spring, the strain in a bridge girder, the color-pattern of an image, an impedance, and so on. If such a physical quantity has to be measured, an auxiliary power source must be attached to that object in order to force the information through the measuring system. This configuration is illustrated in Fig. 2.3 and shows a symmetric layout.

A signal generator drives an actuator, which stimulates the object to generate a response from which the required physical quantity can be retrieved. A generator producing mechanical vibrations in a steel girder is a good example. In this measuring set-up both the stimulus and the response are measured and the ratio of the two values produces the result.

This configuration has several advantages. If the circuit shows a good symmetry and both the stimulus and the response have the same dimensions, the ratio delivered by the divider will cancel several errors. The influence of temperature may well be canceled if the symmetry is guaranteed. The measurand can be expressed in the following function:

$$M = f_{\mathrm{OT}} \frac{[f_{\mathrm{E1}}\{f_{\mathrm{IT1}}(P)\}]_{\mathrm{response}}}{[f_{\mathrm{E2}}\{f_{\mathrm{IT2}}(\mathrm{P})\}]_{\mathrm{stimulus}}} \tag{2.4}$$

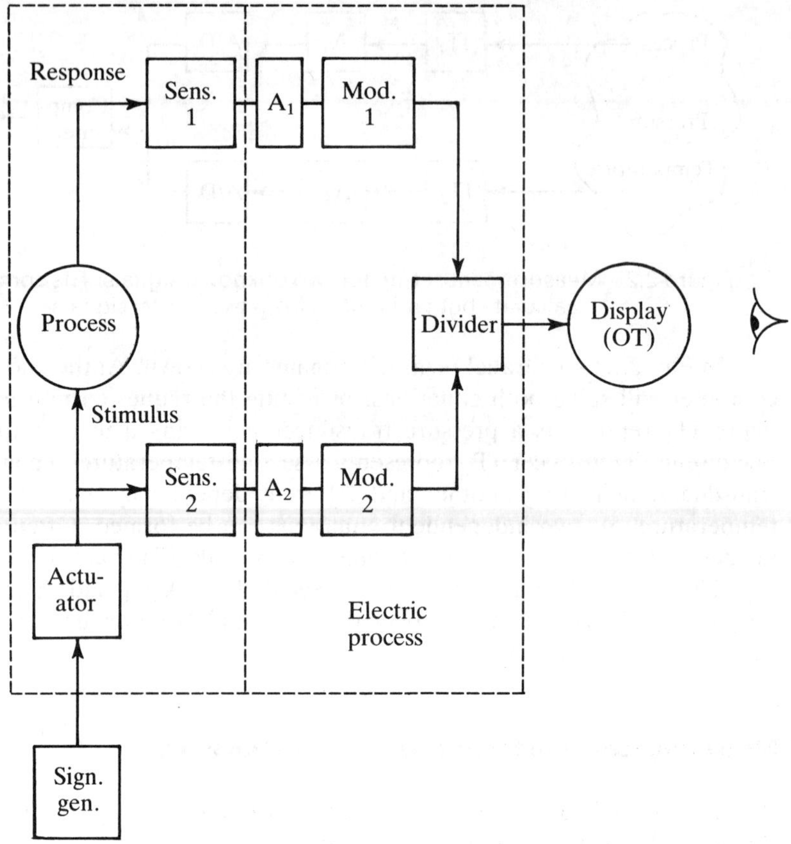

Figure 2.3 Measuring structure for a passive quantity

where the different functional blocks are:

- M measurand
- f_{IT1} input transduction by sensor 1
- f_{IT2} input transduction by sensor 2
- f_{E1} electronic modifier including gain in response branch 1
- f_{E2} electronic modifier including gain in stimulus branch 2
- f_{OT} output transducer

A very simple example can illustrate the concept although here the division is performed manually.

* * *

Example 2.1

Suppose an unknown impedance, Z_2, must be determined; the circuit in Fig. 2.4 can be used. An impedance is a passive quantity and cannot carry or transport

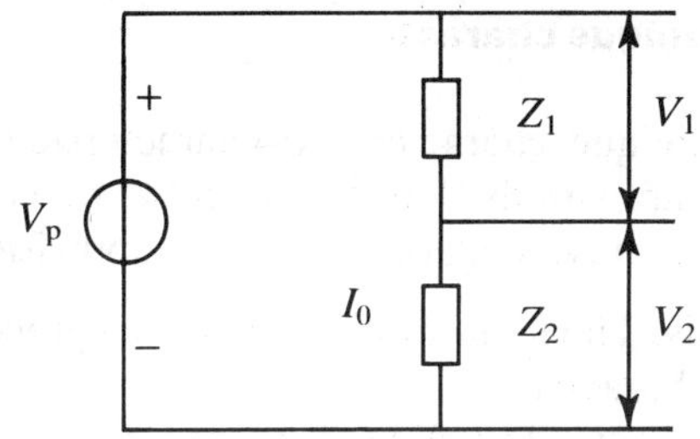

Figure 2.4 Measuring impedance as an example of a passive quantity

power, so an external power source, V_p, must be applied. If a second and known impedance, Z_1, is connected to the unknown impedance, a ratio measurement can be executed.

The external power source will force a current I_0 to flow through both impedances generating voltages across Z_1 and Z_2 and the result can be derived from the ratio of the two voltages V_1 and V_2:

$$Z_2 = Z_1 V_2 / V_1 \tag{2.5}$$

where

$$V_1 = I_0 Z_1$$
$$V_2 = I_0 Z_2$$

Note that I_0 and V_p are not implied in the expression and are canceled, hence any related interference is absent. The same holds for the influence of the ambient temperature on the resistive part of the impedances. It is said that Z_1 acts as a reference impedance.

* * *

2.3 SIGNAL STRUCTURES

The second possible distinction classifying electronic measurements is related to the type of signals they must process. The types of signals to be processed can have quite a different character and of course can have a great impact on the structure of the measurement system itself. Four different types of signals can be identified.

1 Signals with a unique character
2 Periodic signals
3 Sampled signals
4 Stochastic signals

Each of these will be discussed in turn.

2.3.1 Signals with a unique character

Signals with a unique character are characterized by their unique occurrence. Sometimes they are strictly dependent on the process or system under consideration. Considering these signals, two main subdivisions can be made:

(a) Time analysis. Time analysis is related to a process which shows a unique time dependent character.
(b) Response analysis. As will be discussed, three other subdivisions can be made here.

2.3.1.1 Time analysis

The behavior of the weather with respect to time shows a unique character concerning pressure, temperature and humidity. Measurements are performed analyzing the behavior of these features with respect to time. Commonly applied and characteristic signal operations are summarized in Table 2.2. They will be discussed in more detail in Chapter 3.

Table 2.2 Review of commonly applied signal operations executed on unique time-dependent signals

Zero crossing detection with respect to time
Peak detection with respect to time
Taking the average or RMS (root mean square) value of the signal
Dynamic logarithmic compression
Dynamic exponential expansion
Integration or taking the derivative
Summing, multiplying, division, etc.
A/D and D/A conversion

2.3.1.2 Response analysis

To make calculations possible in electronic systems it is well established that every electronic system is considered as a linear system. In fact this is not true, but the assumption is valid if we state that all signal changes in a system are small enough with respect to their biasing levels. For small signal changes, the system behavior is said to be linear.

By definition a system is called *linear* if it can be described by a linear differential equation, or

$$a_0 S_i + a_1 \frac{dS_i}{dt} + a_2 \frac{d^2 S_i}{dt^2} + \ldots = b_0 S_o + b_1 \frac{dS_o}{dt} + b_2 \frac{d^2 S_o}{dt^2} + \ldots \qquad (2.6)$$

where it will be assumed that all coefficients in the differential equation are constants and S_i is the input signal and S_o is the output signal. The consequences are:

(a) A proportional change of the input signal results in a proportional change of the output signal, or

$$aS_i \Rightarrow aS_o \tag{2.7}$$

where a is an arbitrary number.

(b) The superposition theorem holds, or

$$a_1 S_i 1 + a_2 S_i 2 \Rightarrow a_1 S_o 1 + a_2 S_o 2 \tag{2.8}$$

(c) An input signal, $S_i = \exp(\alpha t)\sin(\omega t + \phi)$, results in the same shape of output signal apart from a possible time shift.

1 The impulse function: $\delta(t)$

It is well established that every linear system can be characterized by its *transient*, or *impulse response* $h(t)$ on an *impulse function* $\delta(t)$. The function $\delta(t)$ is also known as the *Dirac-pulse* stimulus. The impulse function $\delta(t)$ is defined as a signal of unit area vanishing everywhere except in the origin, or

$$\int_{-\infty}^{\infty} \delta(t)\,dt = 1 \text{ with } \delta(t) = 0 \text{ for } t \neq 0 \tag{2.9}$$

Of course, the Dirac-pulse function is only a theoretical function, but appears to be applicable in many cases.

If the transient response $h(t)$ of a linear analog system is known, it can be shown that a response $y(t)$ to an arbitrary input $x(t)$ can be calculated by applying the convolution integral of the input signal $x(t)$ and the transient response $h(t)$. Convolution implies a binary operation between the functions $x(t)$ and $h(t)$. It finds its basis in the previously mentioned consequences for linear systems. Apart from this the system is considered as time-invariant, or a delay at the input results in an equal delay at the output. This is expressed in the following equation for analog systems. The same kind of statement can be made for digital systems, but this will not be discussed here.

$$y(t) = \int_{-\infty}^{\infty} h(\tau)x(t - \tau)\,d\tau \tag{2.10}$$

Equation (2.10) is also denoted as $y(t) = h(t)*x(t)$.

* * *

Example 2.2

Assume the impulse response function $h(t)$ of a system is given by

$$h(t) = \exp(-\alpha t)\varepsilon(t) \tag{2.11}$$

where $\varepsilon(t)$ is defined as the step function (see below) starting at time $t = 0$. Then the response $y(t)$ on an arbitrary function $x(t)$ may be written as

$$y(t) = \int_{0}^{t} x(\tau)h(t - \tau)\,d\tau \tag{2.12}$$

$$y(t) = \int_0^t x(\tau)\exp[-\alpha(t-\tau)]\,d\tau \tag{2.13}$$

* * *

Note that the boundary conditions of the integral change in accordance with $\varepsilon(t)$. To calculate convolution it is often more convenient to apply Laplace transforms. It can be shown that a convolution of two functions in the time domain corresponds with the product of their Laplace transforms, or

$$h(t) * x(t) \Leftrightarrow H(s)X(s)$$

Remember that the Laplace transform of a function is defined as in the following equation. More precisely, this is called the unilateral Laplace transform.

$$F(s) = \int_0^\infty f(t)\exp(-st)\,dt \tag{2.14}$$

The output signal can be interpreted as the sum of the transient responses on a continuously input series of the amplitude of the absolute value of the input signal $|x(t)|$. Another interpretation is that the zero state response $y(t)$ of any linear system equals the convolution of the input $x(t)$ with the impulse response $h(t)$ of the system. Hence, to determine $y(t)$ it suffices to find $h(t)$ and to take its convolution with the input $x(t)$.

Measuring transient responses has a great disadvantage due to the fact that a large amount of energy must be supplied to the system in a very short time. Infringement of the allowed amplitudes can occur easily, which often results in jamming of the system.

2 The step function: $\varepsilon(t)$

A more easily applicable stimulus is the *step stimulus*, $\varepsilon(t)$, defined as

$$\varepsilon(t) \stackrel{\text{def}}{=} \begin{cases} 0 & t < 0 \\ 1 & t \geq 0 \end{cases} \tag{2.15}$$

From the system, a step response is obtained, which offers information about the speed at which the system achieves a stable state. Note that the step function, $\varepsilon(t)$, is the integral of the impulse function, or

$$\varepsilon(t) = \int_{-\infty}^{t} \delta(t)\,dt \tag{2.16}$$

3 The ramp function: $t\varepsilon(t)$

A third and often applied function is the *ramp function*, by which a system or process can be characterized. A ramp function is defined as

$$x(t) \stackrel{\text{def}}{\Rightarrow} t\varepsilon(t) \tag{2.17}$$

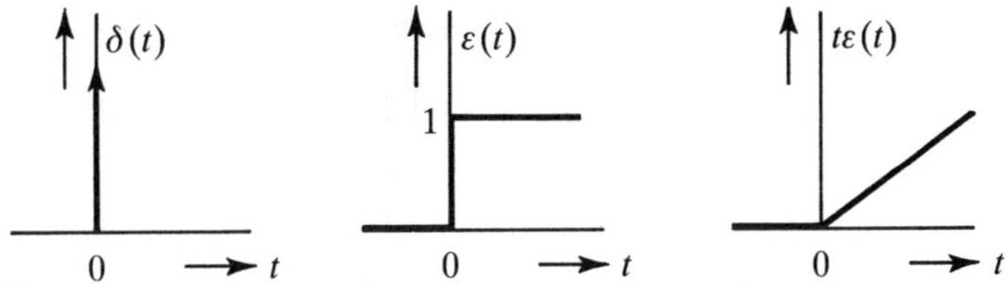

Figure 2.5 (a) The impulse function $\delta(t)$; (b) the step function $\varepsilon(t)$; (c) the ramp function $t\varepsilon(t)$

The response on a ramp function provides information about the linearity or non-linearity of a system.

The different types of test functions are depicted in Fig. 2.5.

2.3.2 Periodic signals

Periodic signals are one of the most frequently occurring and applied signals in electronic systems. They are characterized by their periodic and often harmonic behavior and frequently occur in natural environments. For instance, the strings of a violin or mechanical rafters can produce periodic harmonic vibrations. Periodic signals can be described in terms of their amplitude, frequency and phase.

It can be shown that every periodic function with period T can be expanded into an infinite series of *harmonics*. This infinite sum of terms is called a *Fourier series* and is expressed as

$$f(t) = a_0 + \sum_{n=1}^{\infty} [a_n \cos(n\omega_0 t) + b_n \sin(n\omega_0 t)] \tag{2.18}$$

where a_0 is a constant defined by

$$a_0 = \frac{1}{T} \int_{-T/2}^{T/2} f(t)\,\mathrm{d}t \tag{2.19}$$

and

$$a_n = \frac{2}{T} \int_{-T/2}^{T/2} f(t) \cos(n\omega_0 t)\,\mathrm{d}t \tag{2.20}$$

and

$$b_n = \frac{2}{T} \int_{-T/2}^{T/2} f(t) \sin(n\omega_0 t)\,\mathrm{d}t \tag{2.21}$$

The first harmonic is obtained by substituting $n = 1$ in (2.18). ω_0 is called the fundamental frequency and equals $2\pi/T$. For purposes of analysis it suffices to determine the period T and the coefficients a_n and b_n of the Fourier series

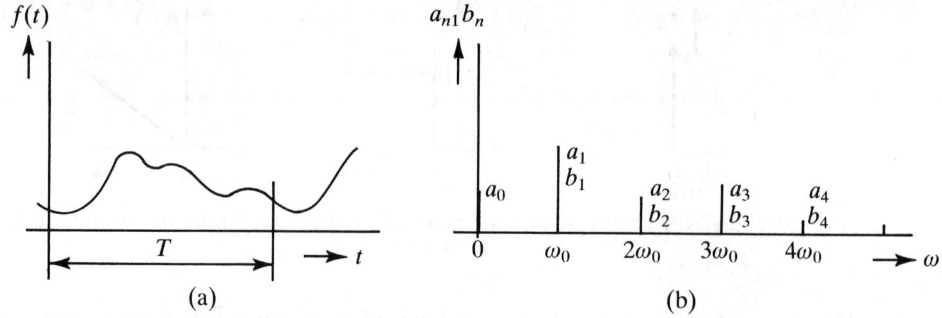

Figure 2.6 (a) Periodic signal *f(t)* with period *T*; *(b) amplitude spectrum of f(t)*

expansion. This is illustrated in Fig. 2.6, in which the coefficients represent the different amplitudes of the different frequency components.

The coefficients of the amplitudes of the different frequencies can be measured either by analog integration or by digital computation.

There is another reason for applying (co)sine functions to a system. (Co)sine functions are preferred for the measurement of the transfer functions of a system, because, inherent to their character of continuity, there is less risk for jamming. Furthermore sine-shaped test signals can reveal the transfer function of a system. Ultimately we can find the complex frequency response $H(j\omega)$ of a system, if sine-shaped input signals are chosen as test (input) signals.

It can be shown that the complex frequency response $H(j\omega)$ equals the Fourier-transformed impulse response $h(t)$ of the system, or

$$H(j\omega) = \int_{-\infty}^{\infty} h(t)\exp(-j\omega t)\,\mathrm{d}t \tag{2.22}$$

and conversely

$$g(t) = \frac{1}{2\pi}\int_{-\infty}^{\infty} H(j\omega)\exp(j\omega t)\,\mathrm{d}\omega \tag{2.23}$$

where

$$\begin{aligned} \exp(-j\omega t) &= \cos(\omega t) - j\sin(\omega t) \\ \exp(j\omega t) &= \cos(\omega t) + j\sin(\omega t) \end{aligned} \tag{2.24}$$

If the input signal $x(t)$ equals $\exp(j\omega t)$ then it can be shown that the output signal $y(t)$ is given by equation (2.25). This is illustrated in Fig. 2.7, where $X(j\omega)$ and $Y(j\omega)$ represent the Fourier-transformed input and output signal respectively.

$$Y(j\omega) = H(j\omega)X(j\omega) \tag{2.25}$$

By definition, the real part of $X(j\omega)$ and $Y(j\omega)$ are, respectively

$$\begin{aligned} \mathrm{Re}\{X(j\omega)\} &= \cos(\omega t) \\ \mathrm{Re}\{Y(j\omega)\} &= \mathrm{Re}\{\mathrm{H}(j\omega)\}\cos(\omega t) - \mathrm{Im}\{H(j\omega)\}\sin(\omega t) \end{aligned}$$

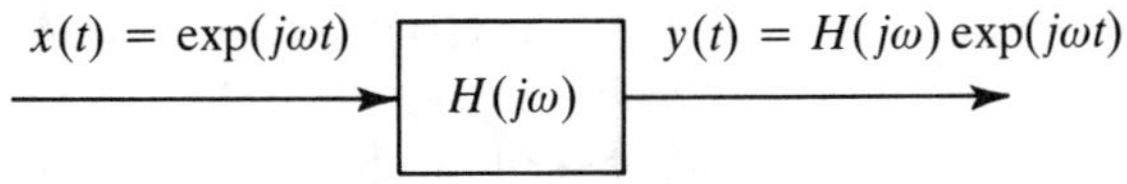

Figure 2.7 The complex frequency response $H(j\omega)$ represents the transfer function of a linear system

Multiplying the real part of the response, $\mathrm{Re}\{Y(j\omega)\}$, by $\cos(\omega t)$ and taking the time average, we find the real part of $H(j\omega)$. Conversely, if we multiply the real part of the response, $\mathrm{Re}\{Y(j\omega)\}$, by $\sin(\omega t)$, and execute a time average, the imaginary part of $H(j\omega)$ is found. This process is found in so-called synchronized detectors.

It is often convenient to depict a transfer function in a polar plot or in Bode plots. This will be discussed in Section 3.2. A sine-shaped input signal will not be distorted when it passes a linear system; only amplitude and phase changes may occur. This is a unique characteristic for sine-shaped input functions.

There is a third reason why applying sine-shaped signals is preferred. They can be manipulated easily for modulation purposes in order to *transport* measurement signals. Modulation can be executed in one of the parameters of the sine function – the frequency, the phase, or the amplitude. For instance, multiplying two sine functions in the time domain results in an amplitude which is proportional with the amplitudes of both signals. The generated frequencies are the sum and the difference of both source frequencies. In frequency modulation, the zero crossings are the only parameter of importance. Very accurate conversion of amplitude into the time domain can be realized by frequency modulation.

2.3.3 Sampled signals

Sampled signals are characterized by the values of the primary signal at certain equidistant times and places, allowing, for instance, the transmission of more than one signal through one transmission channel only. This is called time-sharing or time-multiplexing of signals and can be realized with different types of modulation.

Pulse amplitude modulation (PAM), pulse width modulation (PWM), pulse position modulation (PPM) and analog-to-digital (A/D) conversion are different possibilities.

With the increasing importance of processing signals in digital form, the A/D and the D/A converter have both become extremely important components which can be found in a large variety of instruments and systems. We will discuss the A/D and D/A conversion in more detail in Section 3.4 but we first have to explain the principle of *sampling*.

The signals involved in conversion by an A/D converter are:

1 the continuous-time signal $x(t)$;
2 the samples or discrete-time analog signal $x(nT)$; and
3 the output signal $x(n)$, which can be characterized as a discrete-time digital signal.

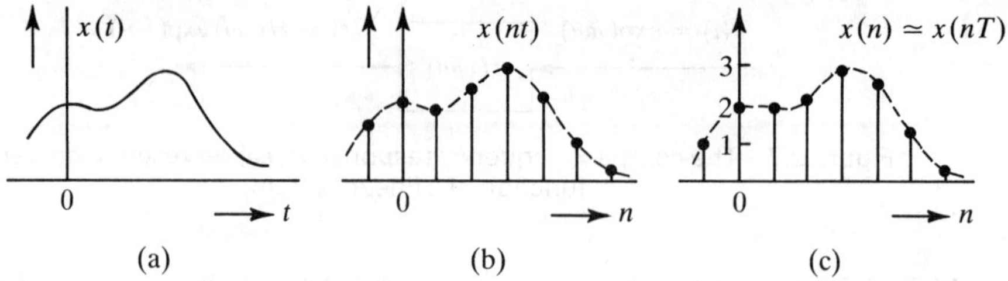

Figure 2.8 (a) Continuous time signal $x(t)$; (b) discrete time analog signal $x(nT)$; (c) discrete time digital signal $x(n)$

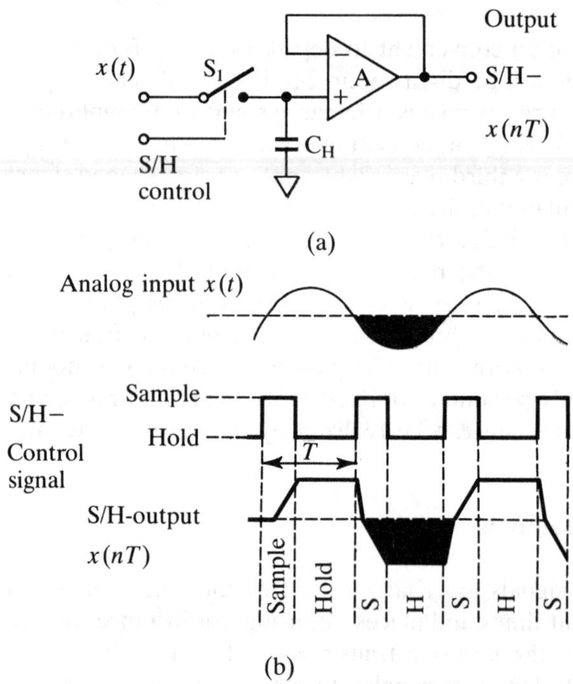

Figure 2.9 Sample and hold circuit for digitizing an analog signal $x(t)$

Ultimately the output of the A/D converter will represent a rounded-off sequence of numbers. The values are multiples of the smallest available digit. Thus, in an A/D converter a sampling and a quantization process are executed. This conversion process is illustrated in Fig. 2.8.

The quantization process implies a kind of *binary coding*. Binary coding is assigning a binary number to samples of the discrete-time analog signal $x(nT)$. Samples of the analog signal $x(t)$ are obtained by sampling the function $x(t)$. Sampling is a kind of modulation and can be realized applying the basic Sample and Hold circuit depicted in Fig. 2.9.

S_1 represents an electronic switch, controlled by the digital S/H signal. During the time the switch S_1 is closed, a buffered signal from the storage capacitor C_H, containing the input signal $x(t)$, is obtained. The output signal follows the input signal. When the switch is opened the input signal is stored in C_H and the output signal will be Hold, until the next Sample period starts. This illustrates the ideal situation; in practice, possible errors are: transfer errors in Sample to Hold state and vice versa, time-varying sampling intervals, time-varying hold intervals. Furthermore, an ideal amplifier and capacitor, and no external interference are assumed.

Now assume $x_1(t)$ is a periodic signal which is modulated (i.e. sampled) with a periodic Dirac impulse $x_2(t) = \Sigma\delta(t - nT_0)$, then this will result in an output signal $y(t)$, which is the product of $x_1(t)$ and $x_2(t)$,
or

$$y(t) = x_1(t)\left[\sum \delta(t - nT_0)\right] \quad \text{for } n = 0, 1, 2, 3, \ldots \tag{2.26}$$

This result can be transformed in the frequency domain. Applying the Fourier series expansion for $x_1(t)$ we get $x_1(j\omega)$. Reasoning in the same way, but now applying the Fourier integral transformation to the periodic Dirac pulse, yields

$$X_2(j\omega) = \frac{2\pi}{T_0}\sum_{-\infty}^{\infty} \delta(\omega + n\omega_0) \tag{2.27}$$

where $\omega_0 = 2\pi/T_0$.

The output signal in the frequency domain $Y(j\omega)$ can be obtained by applying the frequency convolution theorem to (2.26). This yields

$$Y(j\omega) = \frac{1}{2\pi}X_1(j\omega)*X_2(j\omega) \tag{2.28}$$

A graphical representation of this process is depicted in Fig. 2.10, where it can be seen that the frequency spectrum of $X_1(j\omega)$ mirrors with respect to the middle of the sample frequency $1/T_0 = f_0$. All the periodic frequency spectra of $X_1(j\omega)$ are equal and contain the same information. Every periodic frequency spectrum of $X_1(j\omega)$ is called an *alias*.

It can be deduced from the same figure that if the sample frequency f_0 is too low, an aliasing error will occur. The periodic spectrum of $X_1(j\omega)$ will show overlaps in the middle of the frequency spectrum. This is called an *aliasing error* and must be avoided to prevent loss of information.

The sampling theorem says that if the sampling frequency is high enough, the frequency spectrum will show no overlap and no aliasing errors will occur. This can be realized by confining the frequency band B of $x_1(t)$ to $f_0/2$ with the aid of a lowpass filter. To prevent loss of information in practical circumstances, the sampling frequency will always be more than twice the highest occurring signal frequency in $x_1(t)$, or $f_0 > 2B$. This theorem was first formulated by Shannon and Nyquist and is known as the *Nyquist frequency*.

2.3.4 Stochastic signals

In this case, the instantaneous value of a signal is described by its specific probability density function with respect to place and time. It is called a 'stochastic'

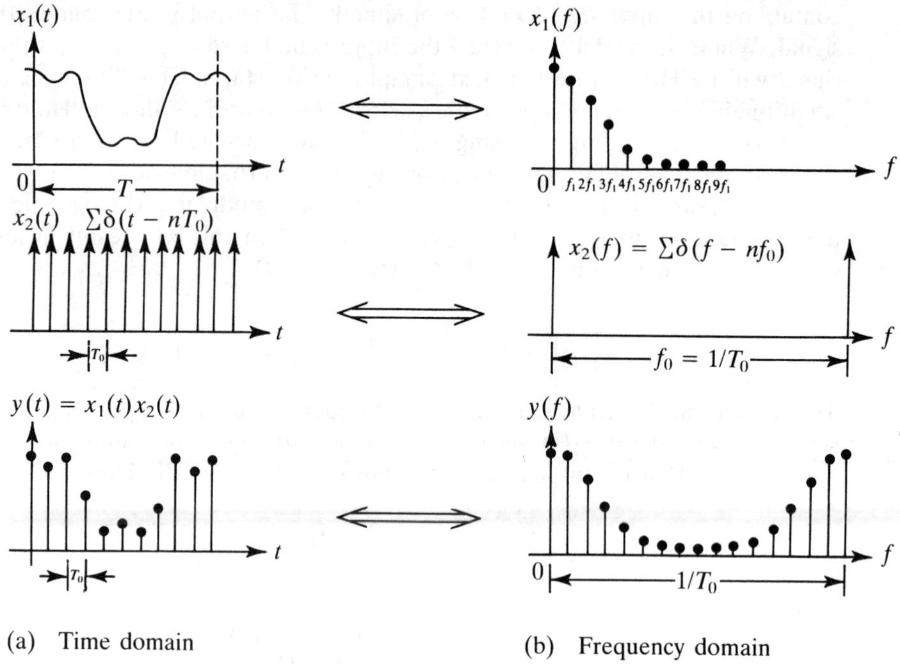

Figure 2.10 Illustration of the sampling theorem. The time domain representations are shown in the left-hand figures, and the frequency domain on the right. Note that $f = \omega/2\pi$

signal for the reason that only a statistical description of that signal can be given. For instance, noise is a representation of this kind of signal. Due to its very nature, a noise signal is only predictable in terms of amplitude probability density functions of place and time, $p(x, t)$.

The signal or noise generators are investigated by measuring the amplitude distribution $p(x, t)$ and the power spectrum $S(j\omega)$. The signal is fully described by these parameters. A detailed and comprehensive discussion concerning noise calculations in electronic systems and components is given in Chapter 9. Here we shall discuss two applications of noise.

Noise can be used as a test signal for response measurements of linear systems. If so-called *white noise* is used as the test signal $x(t)$, it can be shown that the *correlation function* R_{xy} of $x(t)$ and the response $y(t)$ of that system, equals the previously defined impulse response function $h(t)$, or

$$h(\tau) = R_{xy}(\tau) = \lim_{T\to\infty} \frac{1}{2T} \int_{-T}^{+T} x(t)y(t + \tau)\mathrm{d}t \tag{2.29}$$

White noise can be defined as noise with a constant power per unit bandwidth. Correlation is a measure for the dependency between two signals. The correlation is zero if the signals are independent. An advantage of measuring the impulse response by noise is the natural phenomenon of noise. Adding noise to a system

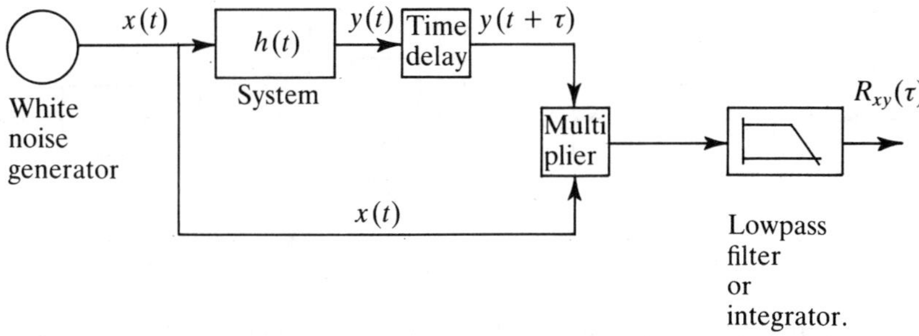

Figure 2.11 Measuring the impulse response $h(t)$ by applying white noise as a test signal

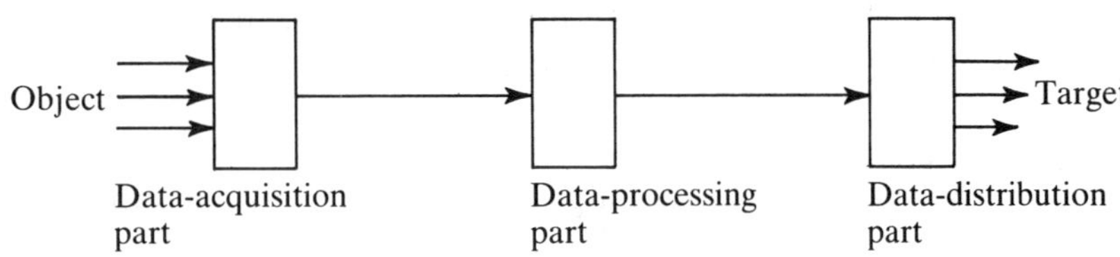

Figure 2.12 The measuring system considered as an information transport channel

will not disturb the system's process under consideration. A block diagram of an analog correlator is shown in Fig. 2.11.

There is another interesting application in which noise is involved. When a periodic signal with unknown form is buried in additional noise, it is possible to detect the periodic component by applying the *autocorrelation function* $R_{xx}(\tau)$, which is defined as

$$R_{xx}(\tau) = \lim_{T\to\infty} \frac{1}{2T} \int_{-T}^{T} x(t)x(t+\tau)\,\mathrm{d}t \tag{2.30}$$

At large values of the delay time τ, the autocorrelation function of the noise components is zero, but that of the periodic signal will be nonzero.

This application can be used for instance in long-distance data channels, and in circuits in which very weak signal sources are involved.

2.4 SPATIAL STRUCTURES

In this approach electronic measurement systems, we can interpret a system as a transport channel transporting information. Three functions can be distinguished here – a *data-acquisition* part, a *data-processing* part and a *data-distribution* part. This is illustrated in Fig. 2.12.

In the data-acquisition part, information is acquired from the measuring object

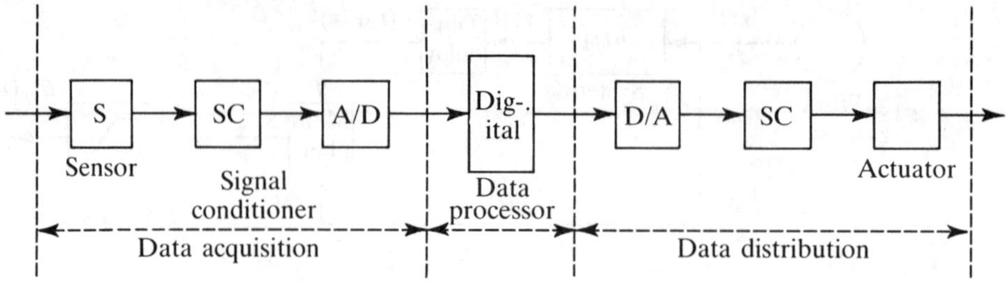

Figure 2.13 Expansion of the data-acquisition and data-distribution parts

and converted into data. Often more than one quantity must be measured, or, at different places, a multiple of the same physical quantity must be monitored, for instance, the temperature, strain or pressure in an oil refinery plant. The data-processing part contains all the required parts to bring the signal to the demanded form and condition. This part often starts with filtering and amplifying. The possible operations are summarized in Tables 2.1 and 2.2. The third part performs data distribution towards the target objects, where the information is required. Often a multiple output is involved.

The first and the last parts can be expanded as illustrated in Fig. 2.13. The data-acquisition part can be subdivided into a sensor, a signal conditioner and an A/D converter. The sensor produces a signal which is a continuous representation of the measurand and converts it into an analog electrical signal. The signal conditioner conditions the signal and makes it suitable for conversion into a digital signal by the A/D converter. The operations involved can be found in Tables 2.1 and 2.2.

In many cases, digital processing of the signal is preferred. The consequence of this is that after the data is processed, we have to convert the digital signal back into an analog signal. Finally, the analog signal is delivered to an actuator or output transducer. In the output transducer, the information is often converted into a nonelectrical form, able to activate a switch, a valve, an indicator, a type of display, etc. In the following section a multiple input and output configuration is discussed.

2.4.1 A multiple input/output configuration

In Fig. 2.14 a multiple input, multiple output system is shown. At the input an analog *multiplexer* connects the different input signals to the digital processor in a time-programmed serial sequence. A multiplexer can be defined as an electronically controlled switch with multiple inputs and one output. The switch is controlled in time and therefore the operation can be called time-multiplexing. At the output an analog *demultiplexer* is installed to provide multiple output signals. Also the demultiplexer is a type of electronic switch, time-programmed in a serial sequence of switching operations. In this configuration, only one A/D and one D/A converter are installed, which is cost effective, but a disadvantage might be that little time is available for conversion. Conversion has to take place during some

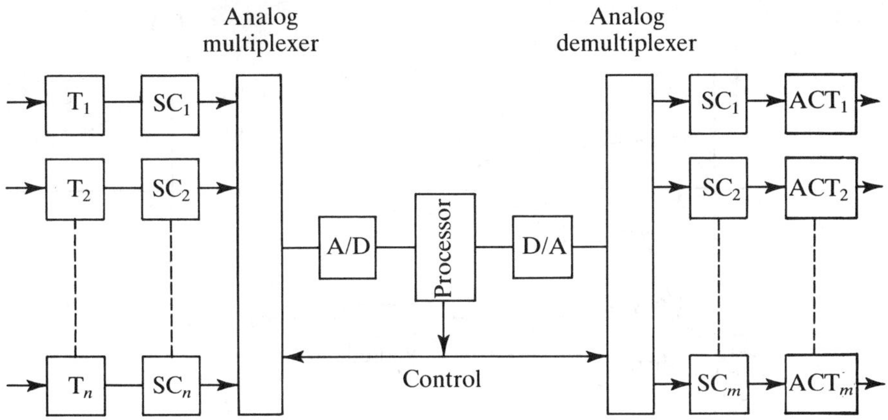

Figure 2.14 A multiple-input, multiple-output configuration

clock cycles of the processor. This can be avoided by giving every input and output channel its own A/D and D/A converters. Note that the number of input and output channels might be different.

We will discuss several other configurations of the concepts outlined above. Starting with a centralized data-acquisition system, presenting an example in which many aspects of such a system are characterized.

2.4.2 A centralized data-acquisition system

* * *

Example 2.3

Assume we want to measure the mechanical strain in a wing structure of an aircraft with the aid of strain gauges. A strain gauge is a resistive film-element consisting of thin wire. It changes it resistance when a mechanical load is applied to the gauge.

The strain in the wing can be measured at the surface of the structure and will be a function of load and position. Before the strain gauges can be attached, a qualitative measurement is executed. This is to discover the best locations to attach the strain gauges. This qualitative measurement can be performed by painting the structure with a type of lacquer whose reflection coefficient is highly sensitive to strain variations. A measurement set-up with which this effect can be revealed is depicted in Fig. 2.15. It consists of a photodetector, a rotating mirror and a lens system.

If the mirror rotates with a speed of n revolutions per minute, the system will convert the space-dependent strain function into a time-dependent current function $I(t)$. The photocurrent will be a function of the incident light falling on the photodetector reflected from the surface. This is expressed in the following relationship between the photocurrent and strain:

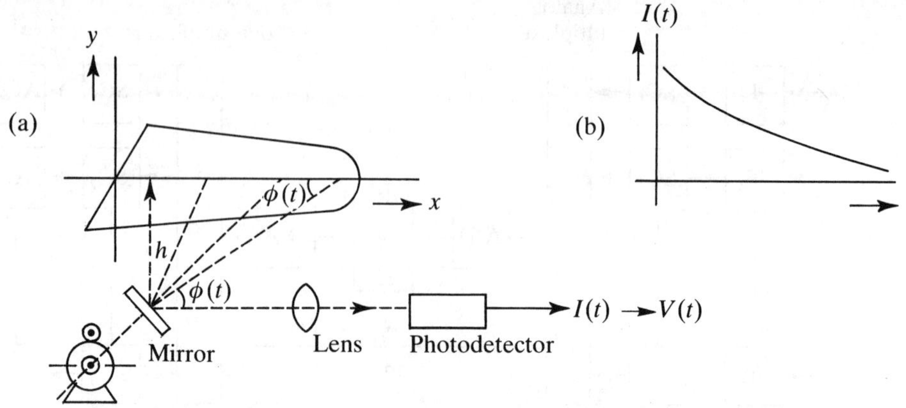

Figure 2.15 (a) Measurement and conversion of a space function into a time-dependent current; (b) its graphical representation

$$I(t) = f[S(x)] \tag{2.31}$$

where

$I(t)$ is the current function versus time
$S(x)$ is the strain function versus position

From the figure it can be derived that the x-coordinate function equals

$$x = h \tan\{\phi(t)\}^{-1} \tag{2.32}$$

where

x is the distance with respect to a certain reference point
h is the perpendicular distance from mirror to object
$\phi(t)$ is the angle between mirror and axis of the photodetector,

and $\phi(t)$ is given by

$$\phi(t) = \omega t + \phi(0) \tag{2.33}$$

where

ω $= 2\pi n/60$
n = revolutions per minute
$\phi(0)$ = phase shift

The result is an electrical signal $V(t)$ representing the strain as a function of intensity in one dimension, here the x-axis.

The measuring set-up is depicted in Fig. 2.16. The first block represents the mechanical analog multiplexer NE-AMX (rotating mirror) and is placed before the input transducer, IT. The photodetector is the input transducer. An amplifier stage, A, will deliver the required gain. The signal conditioner, SC, executes the appropriate electronic modification before A/D conversion can be performed. An interface, I, connects the system to the digital processor. Apart from this, the

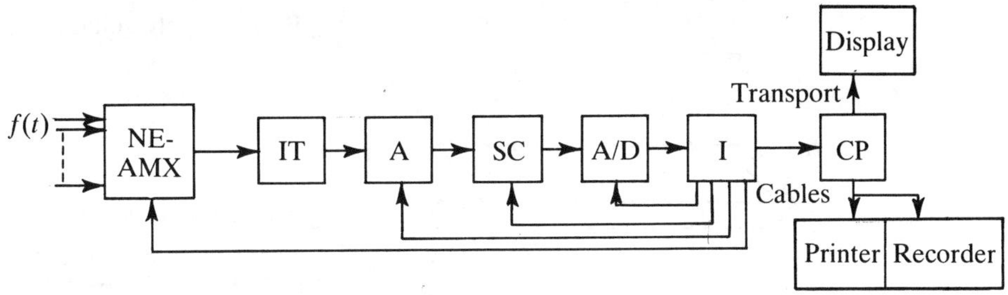

Figure 2.16 Block diagram of the qualitative measuring set-up for a non-electrical quantity

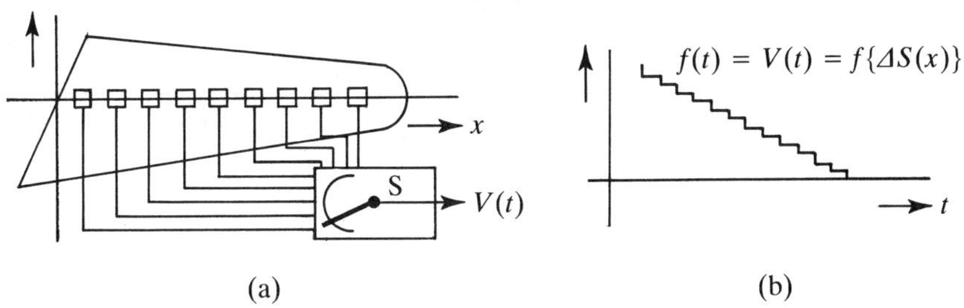

Figure 2.17 (a) The strain gauge measuring set-up for a wing structure; (b) graphical representation of the output signal

interface controls the different mentioned functions. An *interface* can be defined as a piece of hardware between two digital systems that provides the necessary (digital) conversions to make communication between both systems possible.

Finally a central processor (CP) with appropriate peripherals, for instance, a printer, a recorder or a display are depicted.

Once a qualitative measurement is executed to determine where the highest strains might be expected, an electrical quantitive measurement can be performed. The strain gauges are attached to the surface of the wing. Sometimes hundreds of these gauges are mounted on one structural part. This is depicted in Fig. 2.17.

All these strain gauges can be connected with two or three wires into a Wheatstone bridge configuration. Due to the very weak signals for strain gauges, this is the preferred configuration. Finally, the strain gauges could be connected to a central processing unit as shown in Fig. 2.18.

The A-MUX connects the bridge outputs sequentially to the input of the amplifier A. The result will be a time-multiplexed signal of the strain as a function of position along one axis in the structure.

* * *

The above given description is an example of a complete *centralized data-acquisition system*. Such a system has several disadvantages, e.g.:

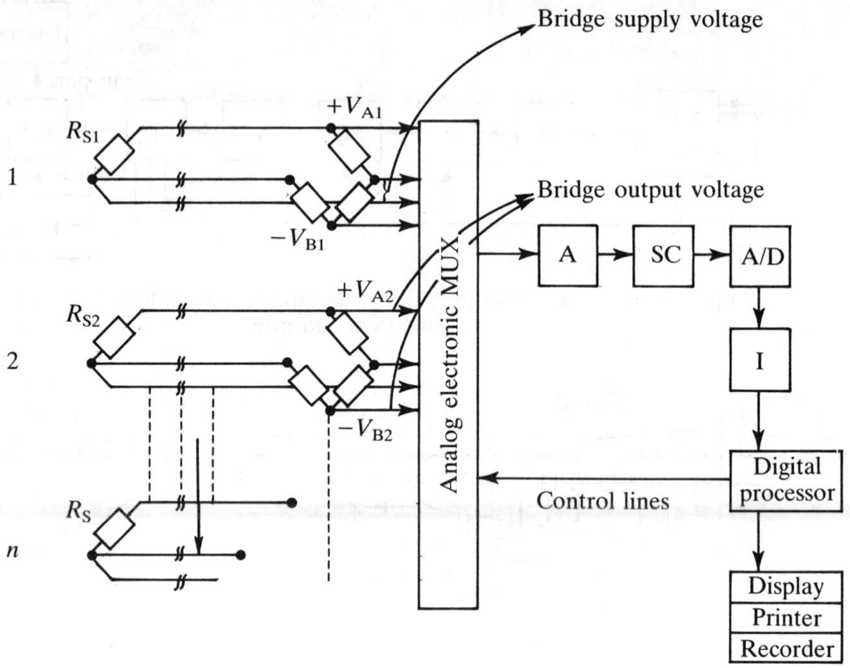

Figure 2.18 Centralized data-acquisition system with analog electronic multiplexer (A-MUX)

1 The transport of a weak transducer signal over long distances makes the system highly sensitive to interference, hence the result can be a bad signal/noise ratio.
2 Long transducer cables and the analog multiplexer decrease in general the result's accuracy. Especially when high demands are put forward with respect to switching speed of the MUX, a considerable reduction of the overall accuracy might occur.
3 If a large number of long cables is required the system will become expensive.

Another configuration, a decentralized data-acquisition system, can improve the system specifications to a large extent. This is discussed below.

2.4.3 A decentralized data-acquisition system with a digital multiplexer

Assume each transducer is given its own signal processing part. The signals are then converted again into a digital signal to decrease sensitivity to interference. Cables run between the respective A/D converters and multiplexer. In this configuration, no switching at the transducer's side is required. This will improve the overall accuracy because all bridges are active, avoiding any time-dependent settling effects. The multiplexer is a digital one controlled by the processor. A block diagram is shown in Fig. 2.19.

This configuration also has some disadvantages, for instance:

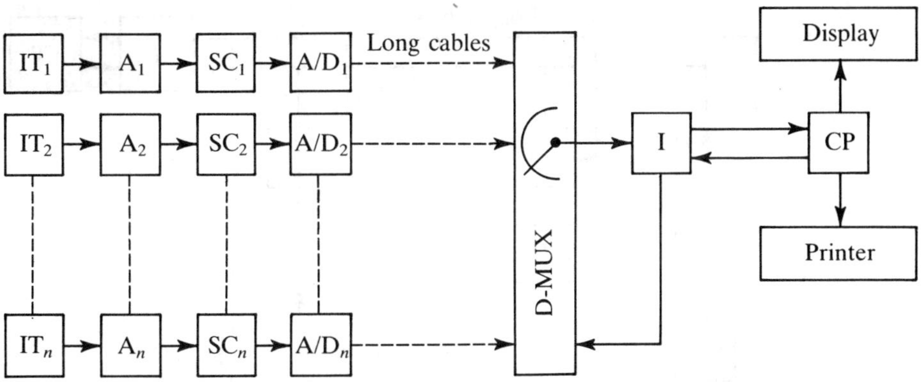

Figure 2.19 Decentralized data-acquisition system with a digital multiplexer (D-MUX)

1 The system is very expensive because each transducer has its own processing part. Depending on the application, this is sometimes justified and can be worthwile.
2 A large amount of cabling is involved.

A recent development shows that the first disadvantage may be overcome by the use of silicon integration technology. Integration technology makes it possible to integrate the sensor part and the signal processing part on to one chip. This will be discussed in more detail in Chapter 6.

In the following section, a distributed multiplexer system configuration is discussed.

2.4.4 A distributed multiplexer data-acquisition system

In this configuration, all the measuring units are switched sequentially in parallel on one long coaxial cable. Every data-acquisition part contains its own multiplexer switch operated by the processor's interface. The system can operate either with digital or with FM modulated signals. This concept is illustrated in Fig. 2.20.

The disadvantage of a large number of cables experienced with the configuration described in 2.4.3 has disappeared. Only the large amount of decentralized mounted electronic components and subsystems might remain as a disadvantage because this increases costs.

Here the same remarks can be made concerning integration technology. In the meantime 'smart' sensors have been developed and are already available. All the required electronics can be integrated on to one piece of silicon. This implies that operations such as gain, filtering, A/D conversion, digital processing, multiplexing, etc., are integrated on to one piece of silicon. This reduces cost and volume considerably.

There is one other important configuration worth mentioning. In some

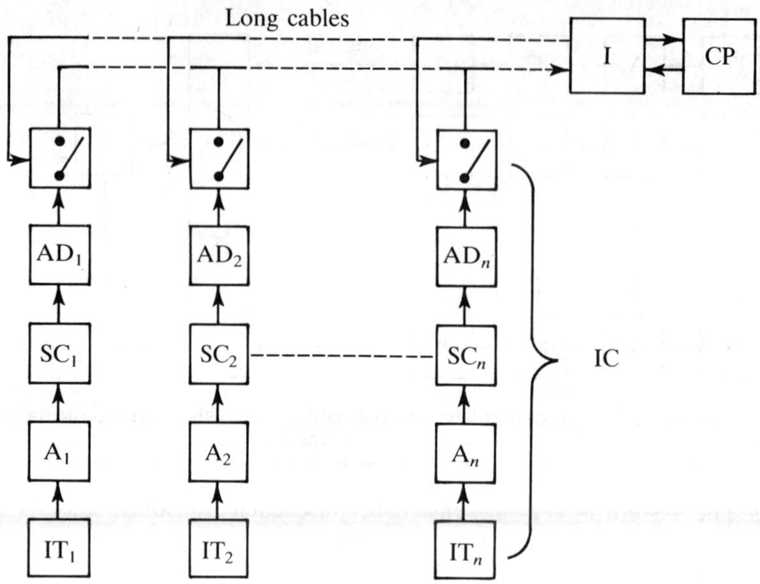

Figure 2.20 Distributed multiplexer data-acquisition system with only one coaxial interconnection cable

circumstances the information must be transported over extremely long distances. This is called telemetry.

2.4.5 Telemetry

In telemetric systems, either the information has to be transported over extremely long distances, or it is not possible to connect the data-processing part to the data-acquisition part. Examples of this can be found in medicine, in complicated machinery, or in a radioactive environment.

There are three possible methods of transmitting the information – radio channel, telephone cable or glass fiber cable. For satellite information transmission, a radio channel is the only possibility.

In medicine, a pressure telemetry system is a very sophisticated application. When the behavior of the bladder pressure versus filling volume must be monitored, a telemetry 'pill' system allows the registration of the pressure without inconveniencing the patient. The pressure sensor and the telemetry system are sealed into a miniature cylindrical ceramic container which is implanted in the patient. The information is transmitted applying an FM signal and received again by a receiver with a high signal-to-noise ratio. The pressure signal can be available in either analog or digital form. Fig. 2.21 shows a schematic block diagram.

Very large-scale integration (VLSI) techniques are of great value in these types of applications. We will consider these in more detail in Chapter 6.

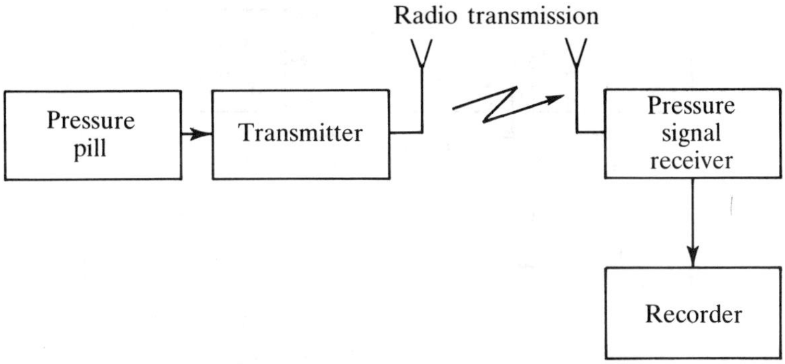

Figure 2.21 Block diagram of a telemetry pressure sensor configuration

2.5 AUTOMATED ELECTRONIC STRUCTURES

In general, automated electronic structures can be defined as being a combination of a number of electronic instruments controlled by a central unit. Not only can this central unit control the operation of the individual instruments but it can also determine the sequence of the measurements more or less automatically. Possible additional operations are sorting, recording and calculating. Automated measuring systems can be divided into three groups.

1 Programmer-operated systems

Programmer-operated systems are preset test and measuring set-ups with partly manual control or operation. The various test programs are preset manually, for instance, by means of electronic switches, and then initiated by a push-button or by automatic sequencing.

2 Controller-operated systems

These test and measurement systems have simple processing facilities, as for instance sorting and recording. No calculation facilities are present. The various test programs are stored on a magnetic tape or paper close to the central controller.

3 Computer-operated systems

These fully automated operated measuring systems have all processing facilities on board, such as sorting, recording and computing. The various measuring programs are stored in the computer memory.

Automated electronic measuring systems can be structured in a *star* structure or in *bus-line* system. An example of a star system is depicted in Fig. 2.22.

Each instrument is individually connected to the central control unit by its own bundle of control lines. In a star system, the programming information is given as a type of level and in a certain code. The code is applied as long as the function

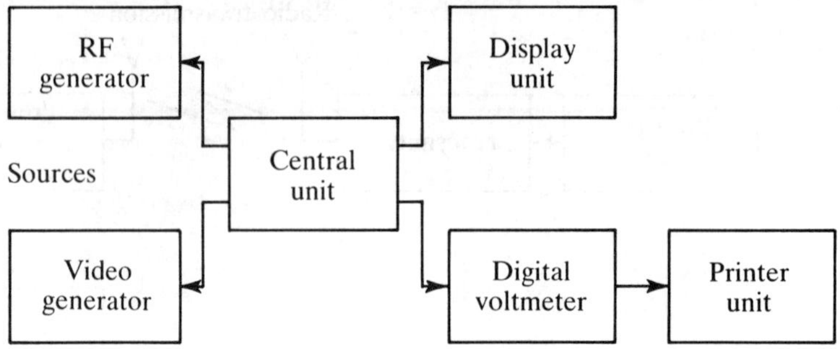

Figure 2.22 Star system

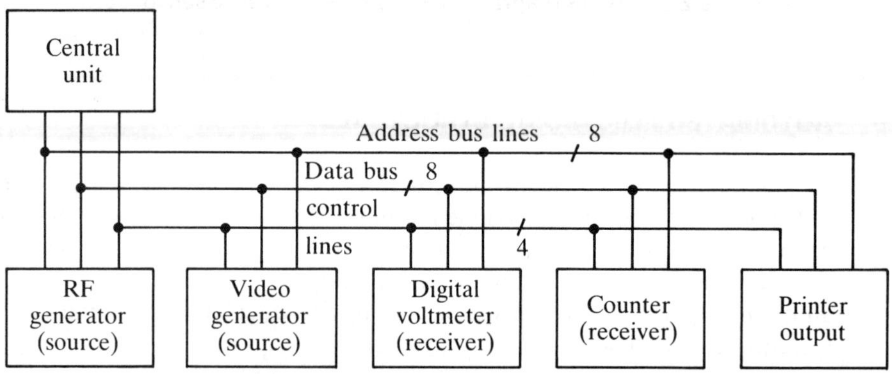

Figure 2.23 A bus-line oriented automated measurement system. The numbers represent the number of lines used

is required. Also a number of analog control lines can be implemented to set continuous functions. The input and output information can be given simultaneously via separate input and output lines.

In a bus-line oriented system all instruments are connected in parallel to the CPU via a common set of lines, called buses. In general three types of buses can be distinguished – the *address* (*line*) *bus*, which addresses the required instrument; the *control bus*, by which lines special functions can be initiated, as for instance the instructions 'input' or 'output', or read/write; and the *data bus*, via which the information is transported. Note that in smaller systems the address bus and data bus sometimes are combined into one. A bus-oriented system is depicted in Fig. 2.23. Assume this system is designed to measure automatically a part of a video system in a manufacturing process. Sources and receivers are indicated. The sources generate the required test signals, and the receivers measure the generated output signals.

For each test or measurement the appropriate instrument is instructed or 'activated' via input and output ports. This is executed by means of the appropriate input and output lines (I/O) and the address and control lines. In a bus-line system the programming information is given as sequences of binary numbers. The programming information is stored in the computer's memory. Every measurement program or routine is stored in memory in the program flow as a subroutine, which is called on by the program in the required sequence at the right time.

In the program, precautions have to be taken to ensure that only one instrument is activated at a time. Otherwise so-called bus-conflicts, such as a short-circuits, may occur.

Several aspects of automated systems will be discussed later on. In the following chapter electronic implementations of functions of measuring systems are presented.

2.6 PROBLEMS

2.1 What is meant by an active or intensive physical quantity? Give examples.

2.2 Discuss the measuring structure for a passive quantity.

2.3 A signal is given as a function of position. Which kinds of conversions have to be realized when the information is required to be recorded on magnetic tape?

2.4 Discuss the several reasons why it is an advantage to implement filters in a measuring chain.

2.5 Today in almost every measurement system an A/D converter can be found. What are the main reasons why these blocks are implemented and what is basically happening when an analog-to-digital conversion is performed? Can you compare it with an everyday tool?

2.6 Determine the impulse response $h(t)$ of the Laplace transformed system function $H = 1/(R + sL)$, where $R = 10\ \Omega$ and $L = 1$ mH.

2.7 The Laplace transformed system function is given as

$$Y(s) = \frac{7s + 23}{s^2 + 7s + 10}$$

Determine the impulse response $h(t)$.

2.8 Perform the same operation for

$$Y(s) = \frac{4s + 20}{s^2 + 4s + 13}$$

2.9 Find the impulse response $h(t)$, the step response $r(t)$, and the ramp response, $\mathrm{rmp}(t)$ for a series connection of a resistor and a capacitor, with $R = 1\ \mathrm{k\Omega}$, $C = 1\ \mu\mathrm{F}$ and $v_c(0) = 0$ V.

2.10 Express the Fourier series in equation (2.18) as a sum of exponentials.

2.11 Discuss the advantages of a bus-line oriented automated structure. Which different hardware-implemented functions are required?

Three

IMPLEMENTATION OF FUNCTIONS

3.1 INTRODUCTION

In Chapter 2 we presented a review of the types of measuring systems. This was based on the four different possible structures to be recognized in measuring systems. In discussing these structures, many different electronic functions were mentioned from a theoretical point of view. In this chapter, we will discuss the electronic implementations of functions. Most of these functions are frequently used in measuring systems. The following topics will be covered:

1 tools;
2 analog functions;
3 analog to digital conversion and vice versa;
4 digital functions;
5 miscellaneous.

3.2 TOOLS

In order to be able to present some concepts in a suitable context, we start with a presentation of a collection of 'tools', by which means functions can be described in a characteristic way. If these tools are already familiar, this section can be skipped. The concepts are presented in a very abbreviated form, without any mathematical proof and the appropriate literature should be consulted if so desired. A basic understanding of complex calculus is also assumed.

In Section 2.3 we have seen several possibilities for determining the characteristic behavior of a system. If we want to describe this behavior several tools are available.

To start with, we assume that the systems to be considered are linear and therefore fulfill the requirements as expressed in equation (2.6) and the conditions (a) to (c) following that equation. For convenience, the equation is repeated here.

$$a_0 S_i + a_1 \frac{dS_i}{dt} + \frac{d^2 S_1}{dt^2} + \cdots = b_0 S_o + b_1 \frac{dS_o}{dt} + b_2 \frac{d^2 S_o}{dt^2} + \cdots \tag{3.1}$$

If we assume that the input signal $S_1(t)$ is equal to $\exp(pt)$ where p is a complex quantity, defined as

$$p = \alpha + j\omega \tag{3.2}$$

then the differential equation (3.1) changes into an ordinary algebraic expression, because the derivative of an exponential expression is again the expression itself, apart from a possible constant.

Note that if $\alpha < 0$, then $\exp(pt)$ represents a decreasing periodic signal; if $\alpha > 0$ then $\exp(pt)$ represents an increasing periodic signal; and if $\alpha = 0$ the signal is a purely periodic one.

When we want to consider the ratio of the output quantity $S_o(p)$ and the input quantity $S_i(p)$ then this ratio represents an analytical function in the complex frequency domain of p. This ratio, usually denoted by $A(p)$, or $H(p)$, is called the *transfer function* of the system. This ratio can be written as

$$A(p) = H(p) = \frac{S_o(p)}{S_i(p)} \tag{3.3}$$

Depending on which quantities are involved, it represents an impedance, an admittance, or a dimensionless transfer function.

For a given circuit the required expression in p is obtained when the appropriate algebraic equations are applied for every node and/or mesh. The substitution for any impedance of an inductance, pL, and for any capacity, $1/pC$, offers the desired expression in p. It is often more convenient to write the equations directly in p, than writing the appropriate differential equations first.

Further, it can be shown that the transfer function $A(p)$ can always be denoted as the ratio of two polynomials, that is with a nominator and a denominator polynomial. If this is executed, $A(p)$ can be written as

$$A(p) = \frac{a_n}{b_m} \frac{(p - z_1)(p - z_2) \ldots (p - z_n)}{(p - p_1)(p - p_2) \ldots (p - p_m)} \tag{3.4}$$

When we want to have a plot for the magnitude of $A(p)$ versus frequency then the expression becomes

$$|A(p)| = \frac{a_n}{b_m} \frac{|p - z_1||p - z_2| \ldots |p - z_n|}{|p - p_1||p - p_2| \ldots |p - p_m|} \tag{3.5}$$

and a phase plot versus frequency can be derived from

$$\arg\{A(p)\} = [\phi_1 + \phi_2 + \ldots + \phi_n] - [\theta_1 + \theta_2 + \ldots + \theta_m] \tag{3.6}$$

where

$\phi_1 = \arg\{p - z_1\}$,
$\phi_2 = \arg\{p - z_2\}$,
etc.

and

$\theta_1 = \arg\{p - p_1\}$,
$\theta_2 = \arg\{p - p_2\}$,
etc.

For simplicity, we further assume that the complex frequency equals $p = j\omega$. Usually a log dB scale versus frequency is used in order to be able to depict several decades of frequency at once. Then it is easily seen from (3.5) that the following expression can be derived:

$$|A(j\omega)|_{\text{dB}} = 20\log|A(j\omega)| = \sum_{k=1}^{n} 20\log|j\omega - z_k| - \sum_{k=1}^{m} 20\log|j\omega - p_k| + C \tag{3.7}$$

where $C = 20\log(a_n/b_m)$.

Of special interest are the cases where $j\omega$ varies if ω varies from $-\infty < \omega < \infty$. This implies that in the complex p–domain the imaginary axis is fully covered. The plot of the transfer function in the p-domain then describes the behavior versus frequency of the system, for pure periodic sinusoidal signals. This graph is called a *polar plot* of $A(p)$ and completely describes the system's properties. Of course, in practical circumstances it makes sense only if $0 < \omega < \infty$. In the following example we shall demonstrate how a polar plot can be constructed.

* * *

Example 3.1

Consider the simple circuit depicted in Fig. 3.1, then the transfer function $A(p)$ between $V_o(p)$ and $V_i(p)$ yields

$$A(p) = \frac{V_o(p)}{V_i(p)} = \frac{1/pC}{R + 1/pC} \tag{3.8}$$

Multiplying this expression by pC gives

$$A(p) = \frac{1}{1 + pRC} = \frac{1}{1 + p\tau} \tag{3.9}$$

where the so-called time constant is $\tau = RC$ (units in seconds). If we substitute for $p = j\omega$ and thus assume in (3.2) $\alpha = 0$, then (3.10)

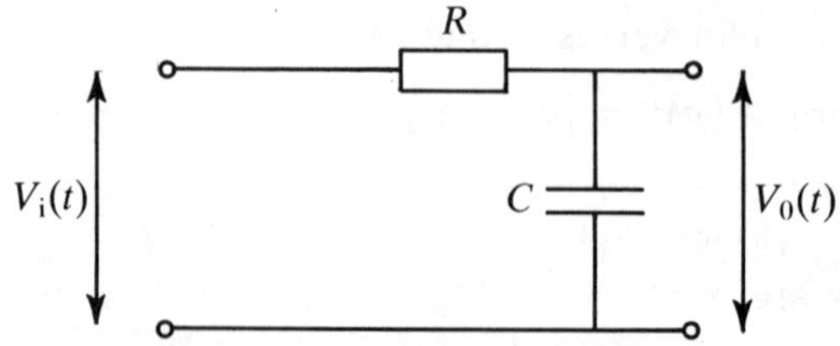

Figure 3.1 An RC circuit for Example 3.1

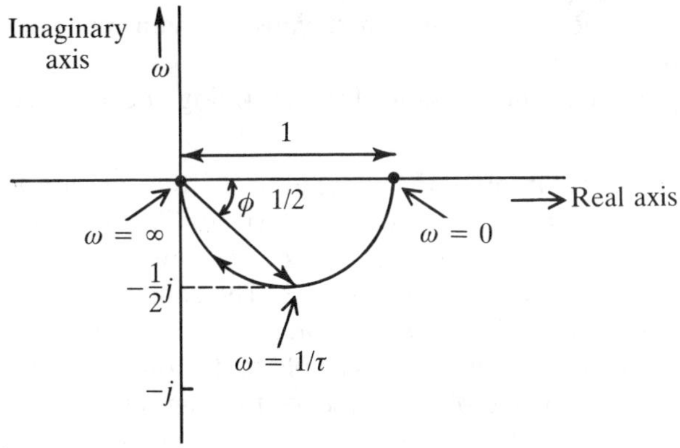

Figure 3.2 Polar plot for the circuit of Fig. 3.1

$$A(j\omega) = \frac{1}{1 + j\omega\tau} \tag{3.10}$$

In complex function theory it is shown that, if ω varies from minus infinity to plus infinity, and thus $p = j\omega$ is a straight line in the complex frequency domain, then the conformal plot of a function of the type of $A(p)$ proves to be a (semi)circle. The remaining information required to draw this plot is to find three characteristic points. The following characteristic frequencies provide these three points: $\omega = 0$, $\omega = 1/\tau$ and $\omega = +\infty$. Again from a physical point of view choosing $\omega = -\infty$ makes no sense at all.

If $\omega = 0$ we see immediately $A(j\omega) = 1$; if $\omega = 1/\tau$ then $A(j\omega)$ equals

$$A(j\omega) = \frac{1}{1 + j} = \tfrac{1}{2}(1 - j); \tag{3.11}$$

and if $\omega \to \infty$ then $A(j\omega)$ approaches zero via the negative imaginary axis. The result is drawn in Fig. 3.2.

* * *

Concerning this result we can make the following remarks:

1 From equation (3.11) we can derive immediately that the plot is in the right half-plane below the x-axis.
2 For $\omega - 0$ (thus DC voltage), the transfer function is simply 1, which can be seen immediately by inspection of the circuit.
3 For $\omega \to \infty$ the circuit behaves as a short-circuit, which will be the case for every circuit.
4 The frequency $\omega = 1/\tau$ is called the 3dB frequency, or the 0.707 frequency, or the $1/\sqrt{2}$ frequency, which is all the same. This will become clear when we draw the Bode plot (see below).
5 The polar plot completely describes the behavior of the circuit and shows the amplitude and phase of the transfer function simultaneously. Note that the

phase is defined as the shift in radians between input and output signal versus frequency.

6 Polar plots are only suitable for describing the response on sinusoidal input signals.

From the polar plot, the magnitude of the amplitude, $|A(p)|$, and phase, $\arg\{A(p)\}$, versus frequency can be derived. This can be executed by graphical methods, measuring at each frequency the amplitude and phase shift. This is illustrated for one frequency in Fig. 3.2. The combination of magnitude and phase plots versus ω are called a *Bode plot*.

We shall now use the same example to demonstrate the procedure for drawing a Bode plot. The Bode plot consists of two figures, hence we have to derive the amplitude and the phase of $A(j\omega)$ versus frequency ω. Again, in most cases it suffices to consider the same characteristic frequencies as in the previous example.

* * *

Example 3.2

In order to find the Bode plot we substitute the same frequencies in (3.10) and determine the amplitude and phase of $A(j\omega)$. For $\omega = 0$ we find $|A(j\omega)| = 1$, for $\omega \to \infty$ we find $|A(j\omega)| \to -j1/\infty$ and for $\omega = 1/\tau$ we find

$$|A(j\omega)| = \left|\frac{1}{1-j}\right| = \frac{|1|}{|1-j|} = \frac{1}{\sqrt{2}}$$

Generally in order to find the phase ϕ of $A(j\omega)$ it is convenient to describe (3.10) in the form $A + jB$ first. This yields

$$A(j\omega) = \frac{1 - j\omega\tau}{1 + \omega^2\tau^2} \tag{3.12}$$

then by definition $\phi = \arg\{A(j\omega)\}$ is

$$\arg\{A(j\omega)\} = \phi = \arctan\left\{\frac{\mathrm{Im}\{A(j\omega)\}}{\mathrm{Re}\{A(j\omega)\}}\right\} \tag{3.13}$$

or, in this example,

$$\phi(\omega) = \arctan\left\{\frac{-j\omega\tau}{1}\right\}$$

Substituting the three characteristic frequencies we find for $\omega = 0$, the phase shift equals 0 radians, for $\omega = 1/\tau$ the phase shift $\phi = -\pi/4$ and for $\omega \to \infty$ the phase shift is $-\pi/2$. The results are plotted in Fig. 3.3.

* * *

Concerning the Bode plot we make the following remarks:

1 The Bode plot provides information concerning the amplitude and the phase shift between input and output separately.

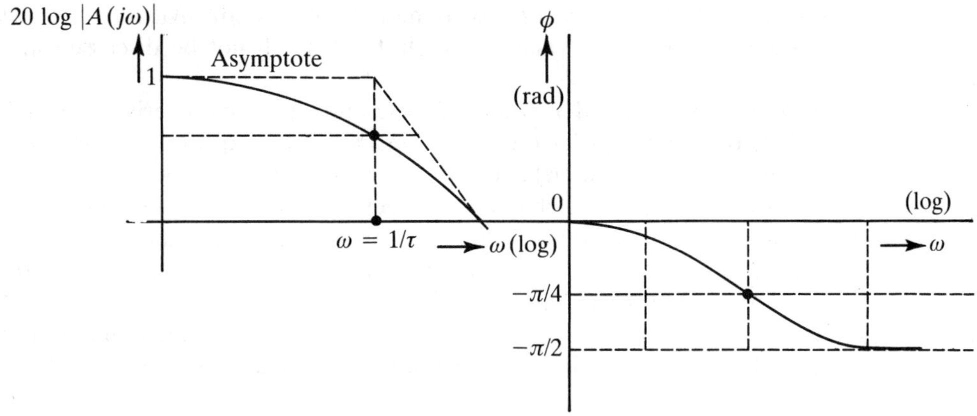

Figure 3.3 Bode plot for the circuit of Fig. 3.1

2 In order to be able to cover a number of decades, especially in higher-order systems, the amplitude of $A(j\omega)$ and the frequency ω, are usually depicted on logarithmic scales.

Assume this is executed for $\omega = 1/\tau$ in Example 3.2, then we find, applying the usual dB notation:

$$20\log|A(j\omega)|_{\omega=1/\tau} = 20\log|\{1/\sqrt{2}\}| = -20 \times ½ \times 0.3010 = -3 \text{ dB}$$

This is the already mentioned 3 dB frequency value and means that, at that frequency, the amplitude has become 3 dB less than the orginal amplitude with respect to the maximum value.

3 Often the Bode plot is indicated by its asymptotes only.
4 The minimum phase shift is zero radians and the maximum phase shift is $-\pi/2$ radians. This is characteristic for a circuit with one time constant only, and it is usually called a first-order circuit.
5 The slope with which the amplitude of $A(j\omega)$ here decreases is −6 dB/octave.
6 This circuit is also known as a first-order *lowpass filter*, because frequencies higher than $f = 1/2\pi\tau$ are attenuated with 6 dB/octave.
7 The Bode plot is only applicable for sinusoidal functions.

The results of both the polar plot and the Bode plot are summarized in Table 3.1.

Table 3.1 Summary of the results of polar and Bode plots for Examples 3.1 and 3.2

rad/sec	$A(j\omega)$	$\|A(j\omega)\|$	$20\log\|A(j\omega)\|$	$\arg\{A(j\omega)\}$
$\omega = 0$	1	1	0	0
$\omega = 1/\tau$	$½/(1-j)$	$1/\sqrt{2}$	−3	$-\pi/4$
$\omega \to \infty$	0	0	$-\infty$	$-\pi/2$

Finally, one other type of plot is met, known as the *Nichols chart*, which represents the amplitude versus phase shift. This will not be discussed in detail here.

A more sophisticated method of describing the characteristics of a linear system is known as the *pole-zero* plot. Consider again expression (3.4), which we already know describes the behavior of a linear system. It can be shown that, apart from a certain constant, the characteristic behavior of the transfer function can be described with a limited number of points, usually called poles and zeros. It appears that there exists a definite relationship between the position of the poles and zeros in the p-domain and the qualitative and quantitive properties of the system. More precisely, the values of the independent variable p for which the nominator equals zero are called the zeros. The values of p for which the denominator equals zero are called the poles.

Considering the complex frequency domain, it can be said that the zeros, $z_1, \ldots, z_n$, are the complex frequencies for which the transfer function $A(p)$ equals zero. The poles, $p_1, \ldots, p_m$, are the complex frequencies for which the transfer function becomes infinite. This is easily verified by equation (3.4).

In the complex domain, these poles and zeros can be plotted, providing a series of points. Usually the zeros are denoted with an 'o' and the poles with an '×'. With this type of plot, the transfer function is also described completely, apart from the constant a_n/b_m in equation (3.4). As will be seen, a pole–zero plot is more powerful than the polar or Bode plot. Note that the described method holds for linear systems only.

In an abbreviated form, we first present some properties of the pole–zero plots. Further, we assume that the systems are physically realizable.

1 Owing to the real coefficients $a_0, \ldots, a_n$ and $b_0, \ldots, b_m$, we always have to deal with real transfer functions for real values of p. This is only possible for real values of poles and zeros, or for those occurring as conjugated complex pairs. The function is then called a conjugated analytical function.

2 Splitting the expression for the transfer function (3.4) in partial fractions we get an expression as follows:

$$A(p) = \frac{V_o}{V_i} = \frac{a_n}{b_m}\left\{\frac{A_1}{p - p_1} + \frac{A_2}{p - p_2} + \ldots + \frac{A_m}{p - p_m}\right\} \tag{3.14}$$

or,

$$V_o = V_i\frac{a_n}{b_m}\left\{\frac{A_1}{p - p_1} + \frac{A_2}{p - p_2} + \ldots + \frac{A_m}{p - p_m}\right\} \tag{3.15}$$

From this expression it appears that V_o can be interpreted as the superposition of a number of voltages of the form

$$V_o(k) = V_i A_k/(p - p_k) \tag{3.16}$$

Because p can be interpreted as the Laplace operator for d/dt, (3.16) corresponds to the differential equation

$$\frac{dV_o(k)}{dt} - p_k V_o(k) = A_k V_i \tag{3.17}$$

If $V_1 = 0$ the solution yields

$$V_o(k) = C \exp(p_k t) \tag{3.18}$$

where C is a constant.

Now note that if p_k has a positive real part, $V_o(k)$ will increase in time, the system is unstable and behaves like an oscillator. From this it may be concluded that the system is only stable when the poles of the transfer function do have a real but negative part. The consequence is that for a stable system all poles have to be found in the left half-plane of the p–domain. This condition is not required for the zeros. In principle, a single pole may be found on the imaginary axis.

3 For a practical circuit, the number of poles is always greater than the number of zeros, because for $p = j\omega$ and $\omega \to \infty$ the modulus of the transfer function $|A(p)|$ is $< \infty$. For $m = n$ the transfer function yields the constant $|A(p)| = a_n/b_m$. For every practical circuit, due to inherently present parasitic capacities every circuit reaches zero when $\omega \to \infty$.

4 For passive two-ports it can be shown that the difference between the number of poles and zeros never exceeds one. Note that in passive two-ports, no active components such as transistors are involved. They consist of R, L and C components only. In these networks all the poles as well as the zeros have to be found in the left half-plane.

5 For a series connection of unilateral circuits the transfer function equals the sum of the pole-zero plots of the independent systems. If a pole coincides with a zero they annihilate each other.

$$A = A_1 A_2 A_3 \ldots \tag{3.19}$$

The transfer function of a parallel connection of circuits contains the poles of the separate circuits but in general not the zeros.

* * *

Example 3.3

Let us consider again the circuit of Fig. 3.1. From (3.9) it can be derived that only one pole is involved. More specifically we find $p_1 = -1/\tau$. This can be plotted in the pole–zero plane, as shown in Fig. 3.4.

From this figure the following conclusions can be drawn:

1 The system is described with one point.
2 When the frequency ω varies from 0 to $+\infty$, p shifts in the positive direction of the imaginary y-axis.
3 The angle θ indicates the phase shift between the input and output. Generally the contribution of a pole is negative and that of a zero is positive. It is easily seen that, in this example, for $\omega = 0$ the phase shift is zero, and for $\omega \to \infty$ the phase shift is $-\pi/2$.
4 In this example the distance between a point on the $j\omega$-axis and the pole

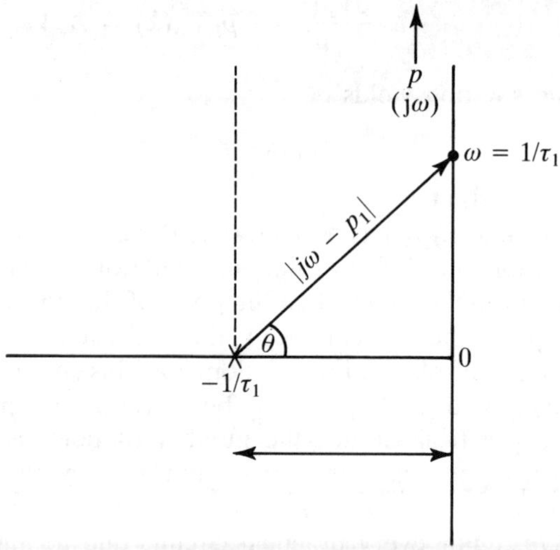

Figure 3.4 Pole-zero plot for the circuit of Fig. 3.1, with $\tau = \tau_1$

represents the amplitude of the transfer function at the indicated frequency. If more poles and zeros are involved then the result is the ratio of the products of the respective distances for every zero and every pole.

* * *

This pole-zero method is also applicable to signals other than pure sinusoidal ones. If for instance the step response must be found, then it suffices to multiply the transfer function $A(p)$ by $1/p$, because the Laplace transform of the step function is $1/p$.

To complete this review let us consider the case where *feedback* is involved. This is illustrated in Fig. 3.5. The principle of feedback can physically be defined as a fraction of the output (signal) that is coupled back into the input. We can distinguish a positive feedback and a negative feedback.

As will be discussed in the following chapter, negative feedback can largely improve the behavior of an amplifier. To determine the pole-zero plot of such a system we presume the poles and zeros of the separate systems are known. The transfer function of a system A_t, implying feedback, is

$$A_t = \frac{A}{1 - \beta A} \tag{3.20}$$

where

$$A = a\frac{(p - z_1)\ldots(p - z_n)}{(p - p_1)\ldots(p - p_m)} = a\frac{N_1}{D_1} \tag{3.21}$$

and

$$\beta = b\frac{(p - z_1')\ldots(p - z_n')}{(p - p_1')\ldots(p - p_m')} = b\frac{N_2}{D_2} \tag{3.22}$$

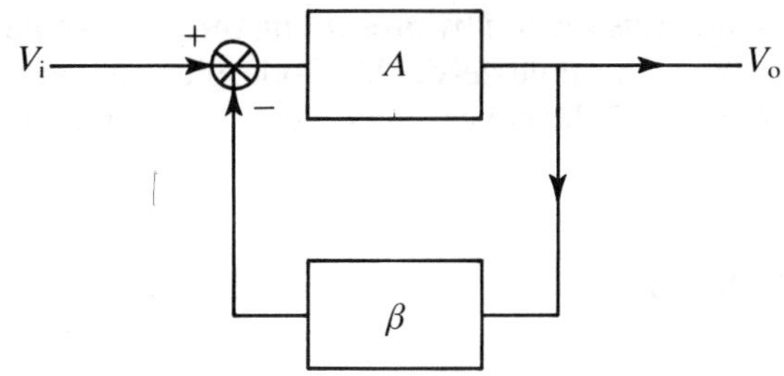

Figure 3.5 Block diagram of a feedback system

Substituting (3.21) and (3.22) in (3.20) yields

$$A_t = \frac{a\dfrac{N_1}{D_1}}{1 - ab\dfrac{N_1}{D_1}\dfrac{N_2}{D_2}} \tag{3.23}$$

or

$$A_t = \frac{aN_1 D_2}{D_1 D_2 - abN_1 N_2} \tag{3.24}$$

As can be seen from (3.24) the zeros of A_t are determined by the zeros of A and the poles of β. The poles of A_t can be derived from the equation $A\beta = 1$, where A_t becomes infinite, or

$$D_1 D_2 - abN_1N_2 = 0 \tag{3.25}$$

or

$$\frac{N_1 N_2}{D_1 D_2} = \frac{1}{ab} \tag{3.26}$$

Thus the poles of A_t are determined by equation (3.26). As stated before, the constants a and b are real, so we have to look for values where equation (3.26) proves to have real value. For these values, a graphical representation exists called *root loci* showing the changes in pole position as ab varies. Concerning these root loci, without any proof, we mention some properties:

1 Root loci start in the poles of A and β.
2 Root loci end in the zeros of A and β.
3 There exists a mirror symmetry with respect to the x-axis.
4 Generally a fraction of the real axis forms a part of the root loci.
5 For a number of g poles and h zeros existing within the finite region, $(g - h)$ zeros are located in the infinite.

Unfortunately it is not possible to measure poles and zeros. To determine whether a feedback system is stable the *Nyquist* criterion is used. The Nyquist criterion is stated as follows: *if the polar plot of $A\beta$ does not cover the point +1, the system is a stable one*. This is equivalent to the condition that all poles of the stable feedback system are located inside the left half-plane.

We conclude this section by presenting three convenient tables (Tables 3.2, 3.3 and 3.4) in which the pole-zero, the Bode plot and the polar plots for different systems are depicted. Moreover, some examples of implementations are given also.

3.3 ANALOG FUNCTIONS

Despite an increasing trend to digitize every electronic function where possible, we still are confronted with many analog functions in electronic measuring systems. As will be discussed in more detail in Chapter 6, this is mainly due to the analog nature of our environment. Think of a temperature, a pressure, a force, a voltage, radiation, magnetism, chemical reactions. They are all pure analog in their nature and behavior.

When we convert these physical quantities into an electrical signal by an appropriate transducer, usually the signal level is very weak (μV). Before a digitalization can be performed the signal has to be amplified to a suitable level.

To amplify a very weak signal, special precautions must be taken, concerning aspects such as offset and drift, noise levels and interference. All these aspects will be discussed in detail in the following chapters. Due to its general applicability one of the most popular building blocks in processing analog signals is the operational amplifier. We shall present a collection of important functions realized by operational amplifiers.

3.3.1 The operational amplifier

An operational amplifier is a basic building block, usually with a differential input and a single-ended output. By definition the amplifier has an infinitely high open-loop gain, zero output impedance, infinitely high input impedance, zero input current and perfect symmetry in impedance for both inputs. The infinite gain requires feedback and as a consequence the input voltage $V_i = V_o/G$ will become zero. The operational amplifier with the usually applied notations is depicted in Fig. 3.6.

The theoretical case where $V_i = 0$ and $i_i = 0$ is called a *nullor*. It will be understood that in practice this is not true, but it will be assumed in order to introduce a collection of basic functions more easily. Practically the gain, G, for instance, can attain values of 10^5–10^9.

To understand the function of the operational amplifier in its many applications, we will first look at its behavior when used with a simple feedback loop as is illustrated in Fig. 3.6. Compare this with Fig. 3.5 and regard the equivalence of the two circuits. The input marked with a minus sign is called the inverting input, as any signal applied to it will appear with opposite sign at the output. The other input is called the non-inverting input.

As a consequence of the above given definition the actual input voltage V_i between both amplifier inputs is zero, or with other words

$$V_p - V_n = 0 \tag{3.27}$$

Table 3.2 Simple systems with real poles and zeros. (Courtesy University of Delft)

	Transfer function	Poles and zeros	Behavior on a sine shaped input signal: Amplitude char.	Phase char.	Polar plot	Impulse response: p and n plot	Transfer char.	Example of implementation
1	1	$\omega \uparrow$, $\rightarrow c$	0; $\rightarrow \omega$ (log)	$\pi/2$; 0; $-\pi/2$; $\rightarrow$ (log)	1		1; 0; $\rightarrow t$	
2	$p\tau$		(log); 6dB/oct; $\rightarrow \omega$	$\pi/2$; 0; $-\pi/2$; $\rightarrow \omega$	$\omega \uparrow$; 0		0; $\rightarrow t$	ideal differentiator
3	$p\tau + 1$		6dB/oct; $\rightarrow \omega$	0; $\rightarrow \omega$	$\omega \uparrow$; 0; 1	$1/\tau$	$\rightarrow t$	
4	$\frac{1}{p\tau}$		-6dB/oct; $\rightarrow \omega$	0; $\rightarrow \omega$	∞; $\uparrow \omega$	(2)	$\rightarrow t$	ideal integrator
5	$\frac{1}{p\tau} + 1$		-6dB/oct; $1/\tau$; $\rightarrow \omega$	0; $\rightarrow \omega$	1; ∞; $\uparrow \omega$	$1/\tau$; (2)	$\rightarrow t$	
6	$\frac{1}{p\tau + 1}$		A; -6dB/oct; $\frac{1}{\tau} \cdot \frac{1}{RC}$; $\rightarrow \omega$	0; $\rightarrow \omega$	A; ∞; 0; ω		A; $\rightarrow t$	u_1; u_2

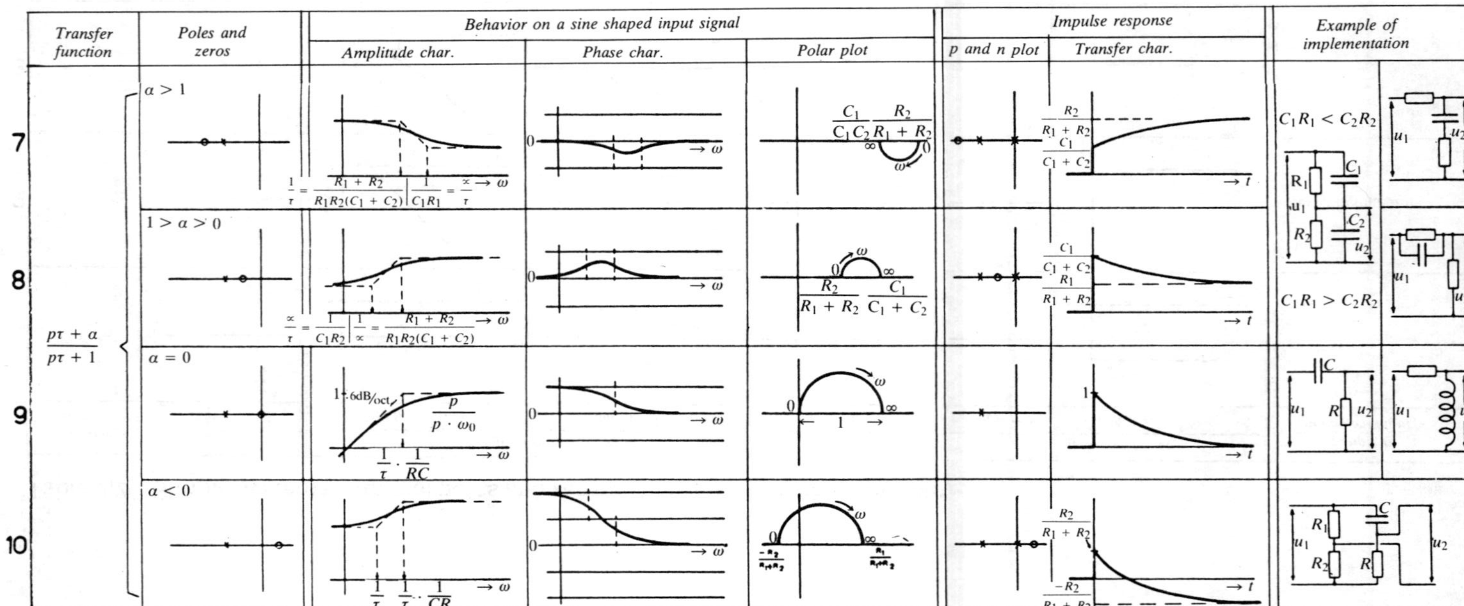
Transfer function
Poles and zeros
Behavior on a sine shaped input signal
Amplitude char.
Phase char.
Polar plot
Impulse response
p and n plot
Transfer char.
Example of implementation
7
8
9
10
$\frac{p\tau + \alpha}{p\tau + 1}$
$\alpha > 1$
$1 > \alpha > 0$
$\alpha = 0$
$\alpha < 0$
$\frac{1}{\tau} = \frac{R_1 + R_2}{R_1 R_2 (C_1 + C_2)}$
$\frac{1}{C_1 R_1} = \frac{\alpha}{\tau}$
$\frac{\alpha}{\tau} = \frac{1}{C_1 R_2}$
$\frac{1}{\alpha} = \frac{R_1 + R_2}{R_1 R_2 (C_1 + C_2)}$
1
6dB/oct
$\frac{p}{p \cdot \omega_0}$
$\frac{1}{\tau} \cdot \frac{1}{RC}$
$\frac{1}{\tau}$
$\frac{1}{\tau} \cdot \frac{1}{CR}$
$\rightarrow \omega$
$\frac{C_1}{C_1 C_2}$
$\frac{R_2}{R_1 + R_2}$
$\frac{C_1}{C_1 + C_2}$
$\frac{-R_2}{R_1 + R_2}$
$\frac{R_1}{R_1 + R_2}$
$\rightarrow t$
$C_1 R_1 < C_2 R_2$
$C_1 R_1 > C_2 R_2$
u_1
u_2

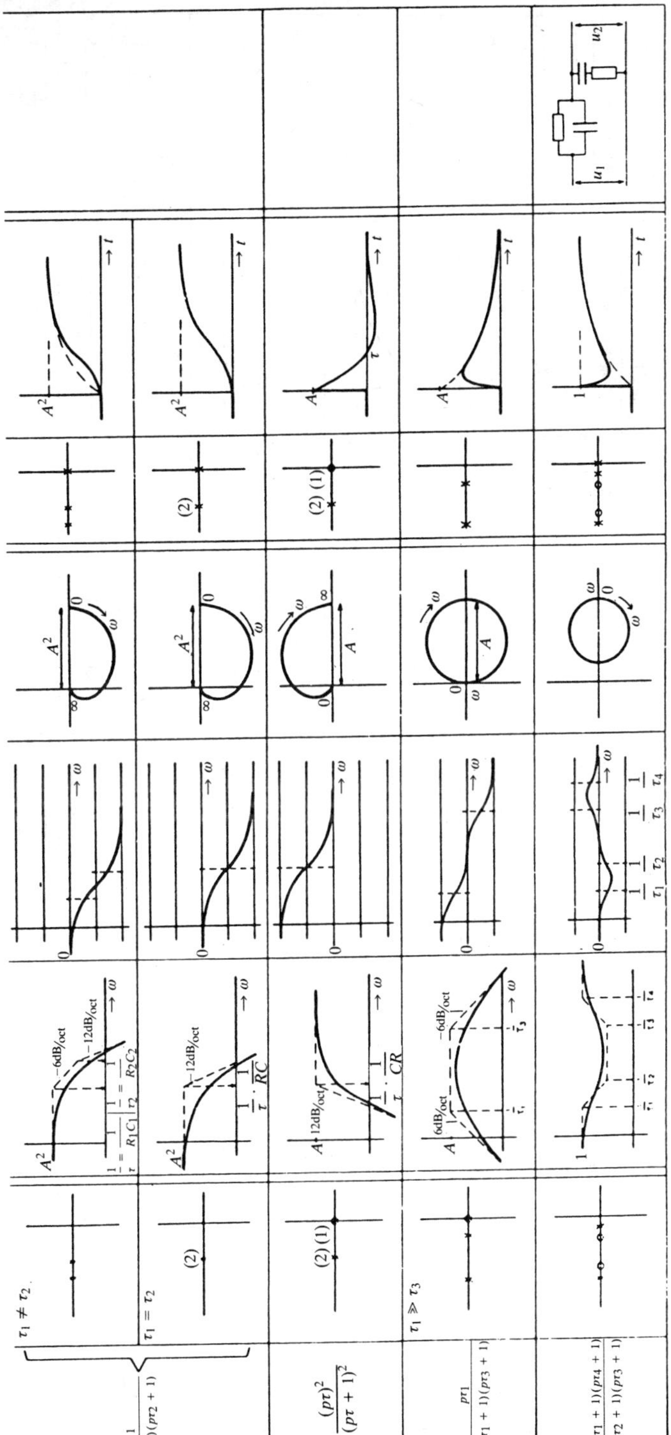
11
12
13
14
15
$\frac{1}{(p\tau_1 + 1)(p\tau_2 + 1)}$
$\tau_1 \neq \tau_2$
$\tau_1 = \tau_2$
$\frac{(p\tau)^2}{(p\tau + 1)^2}$
$\frac{p\tau_1}{(p\tau_1 + 1)(p\tau_3 + 1)}$
$\tau_1 \gg \tau_3$
$\frac{(p\tau_1 + 1)(p\tau_4 + 1)}{(p\tau_2 + 1)(p\tau_3 + 1)}$
(2)
(2) (1)
$\frac{1}{\tau} = \frac{1}{R_1C_1} \mid \frac{1}{\tau_2} = \frac{1}{R_2C_2}$
$\frac{1}{\tau} \cdot \frac{1}{RC}$
$\frac{1}{\tau} \cdot \frac{1}{CR}$
−6dB/oct
−12dB/oct
12dB/oct
6dB/oct
A^2
A
ω
t
u_1
u_2

Table 3.3 Simple system with real and complex poles and zeros. Simple correction circuits are given in 11–15

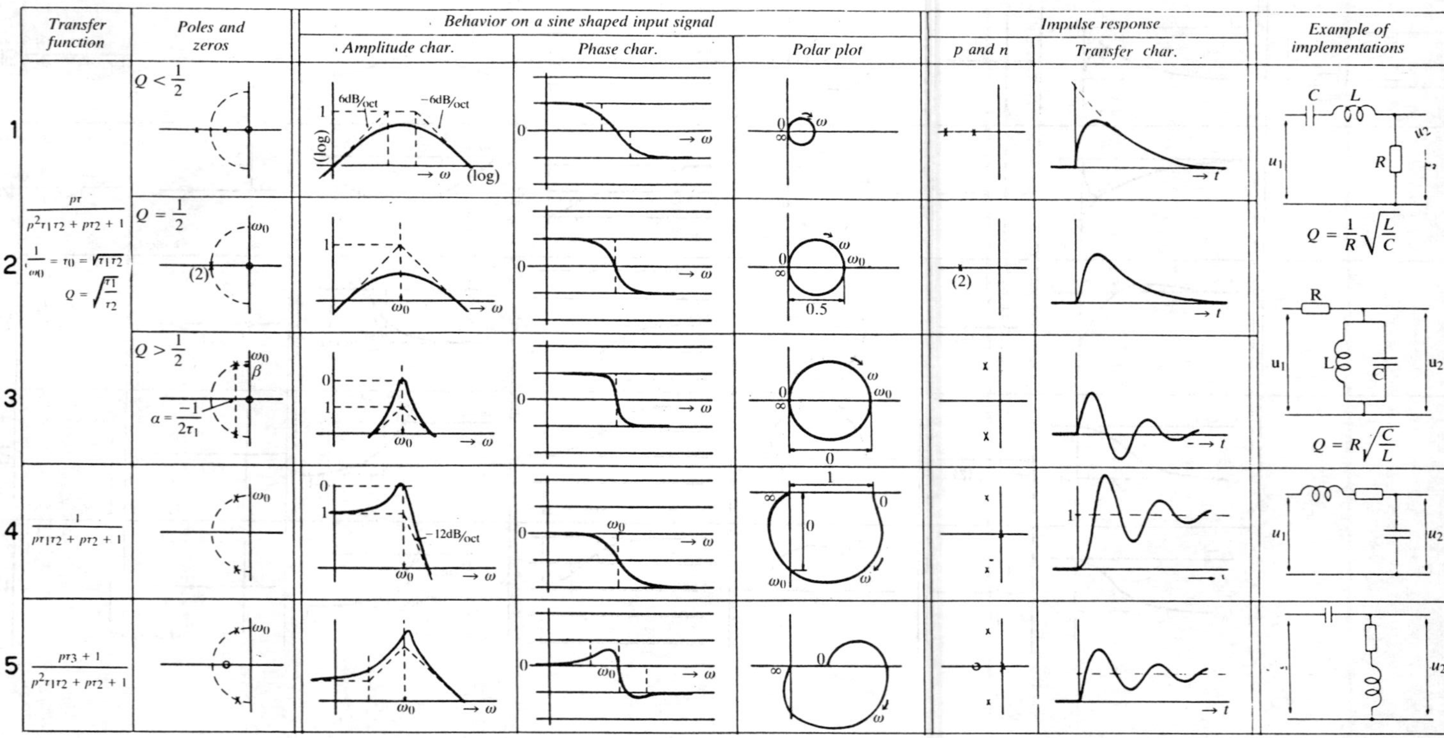

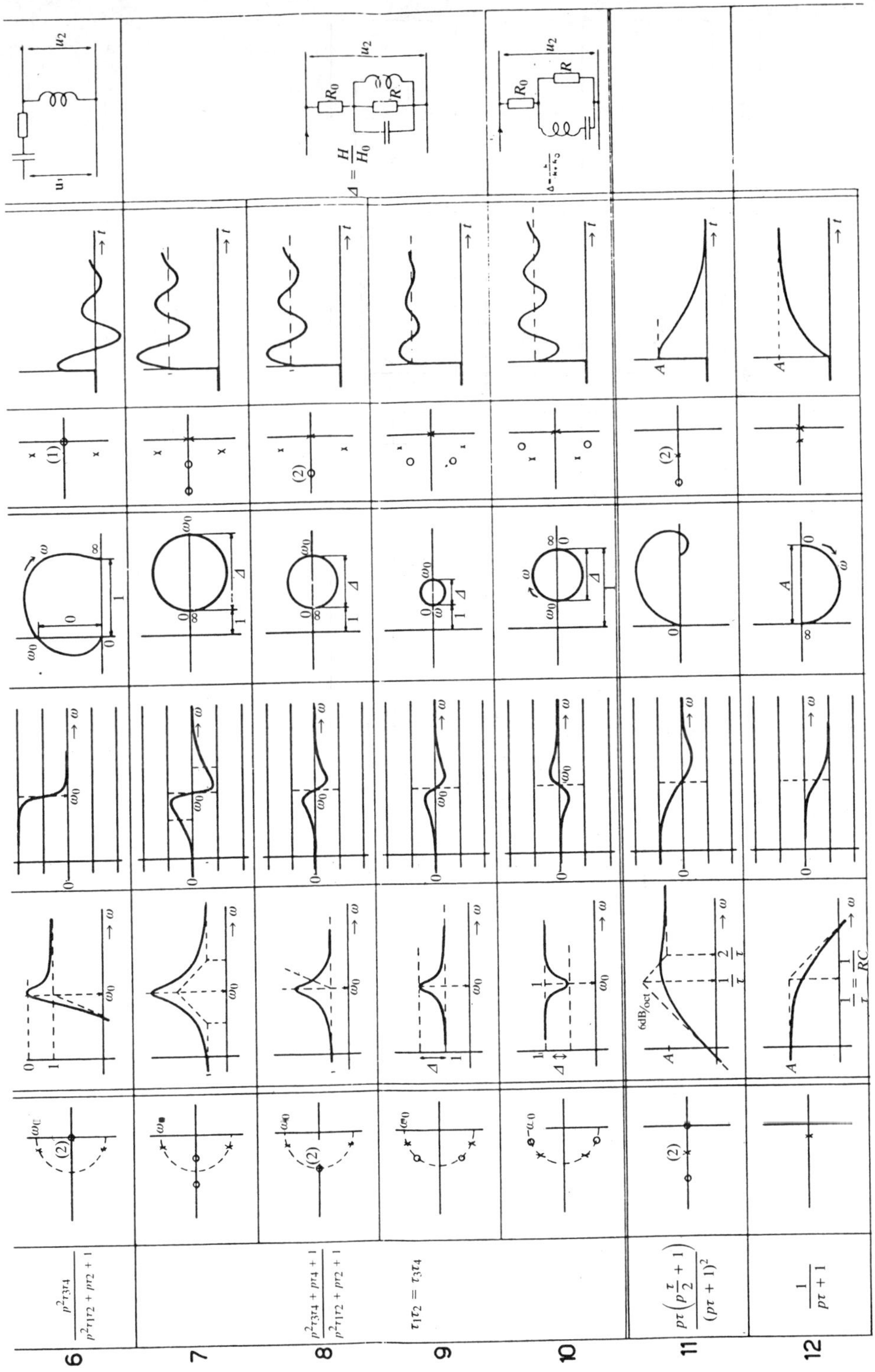
6
p²τ3τ4 / (p²τ1τ2 + pτ2 + 1)
7
8
(p²τ3τ4 + pτ4 + 1) / (p²τ1τ2 + pτ2 + 1)
Δ = H/H0
9
τ1τ2 = τ3τ4
10
11
pτ(pτ/2 + 1) / (pτ + 1)²
6dB/oct
12
1 / (pτ + 1)
1/τ = 1/RC

	Transfer function	Poles and zeros	Behavior on a sine shaped input signal: Amplitude char.	Behavior on a sine shaped input signal: Phase char.	Behavior on a sine shaped input signal: Polar plot	Impulse response: p and n	Impulse response: Transfer char.	Example of implementation
13	$\frac{p\tau_1 + 1}{p^2\tau_2\tau_3 + p\tau_3 + 1}$	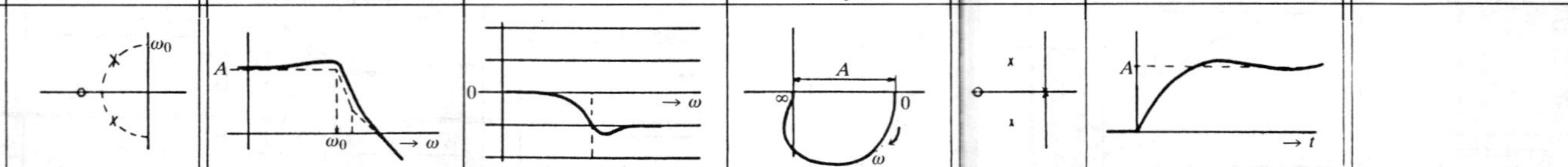						
14	$\frac{p\tau_1 + 1}{(p\tau_2 + 1)(p\tau_3 + 1)}$	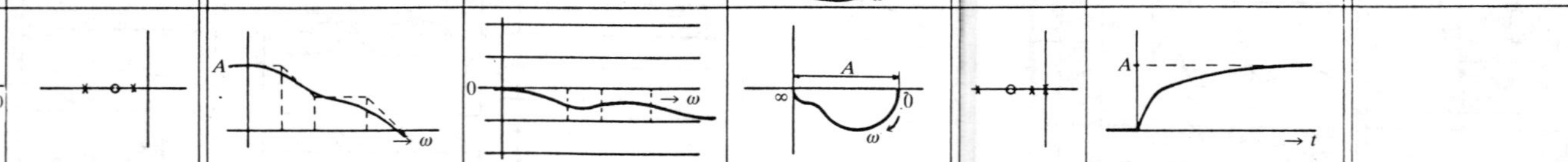						
15	$\frac{p(p\tau_1 + 1)}{(p\tau_2 + 1)(p\tau_3 + 1)}$	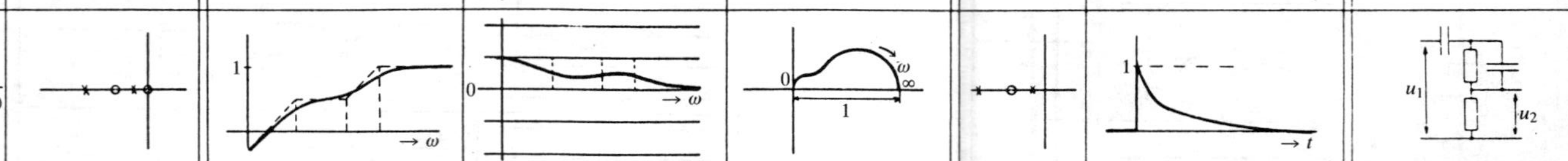						

Table 3.4 Examples with feedback. (Courtesy University of Delft)

	Without feedback		*With feedback*			
	Transfer function	*Poles and zeros*	$Eq \cdot \lvert A \cdot \beta \rvert = 1$	*'Root locus'*	*Oscillation*	*Examples of systems*
1	$\left(\frac{1}{p\tau + 1}\right)^2$	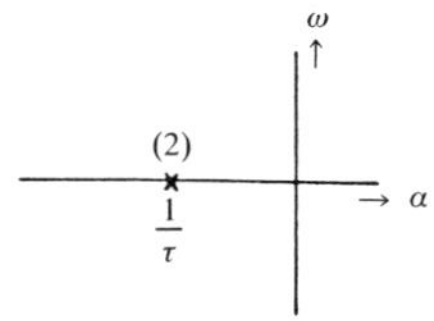	$\left(\frac{1}{p\tau + 1}\right)^2 = -\frac{1}{A}$	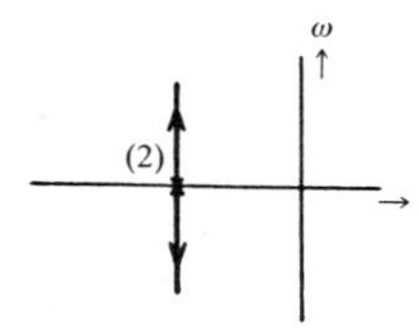	no	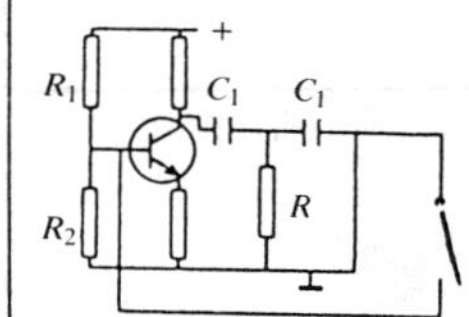
2	$\left(\frac{1}{p\tau + 1}\right)^3$	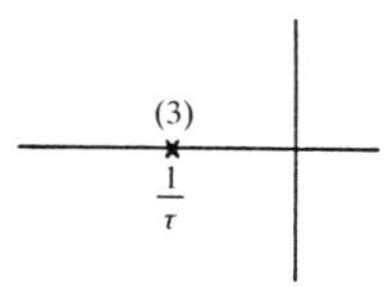	$\cdots = -\frac{1}{A}$	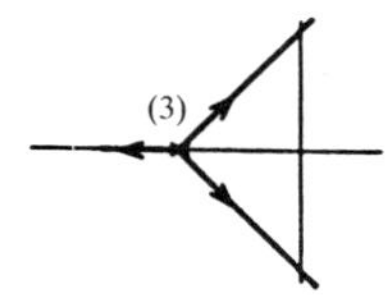	$\omega = \frac{\sqrt{3}}{\tau}$ $A = 8$	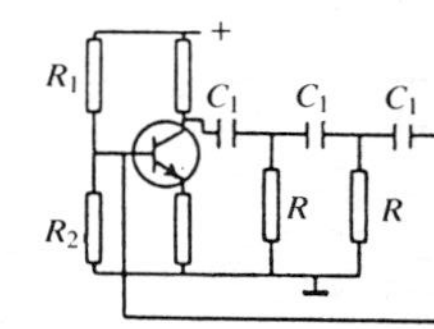
3	$\left(\frac{1}{p\tau + 1}\right)^4$	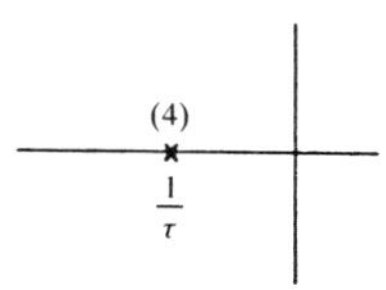	$\cdots = -\frac{1}{A}$	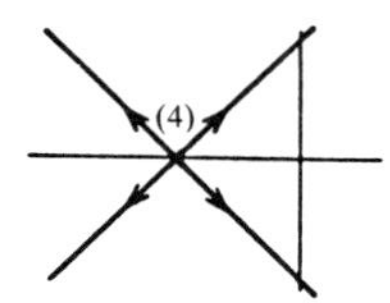	$\omega = \frac{1}{\tau}$ $A = 4$	

	Without feedback		With feedback			Examples of systems
	Transfer function	Poles and zeros	$Eq \cdot \lvert A \cdot \beta \rvert = 1$	'Root locus'	Oscillation	
4	$\left(\frac{p\tau}{p\tau + 1}\right)^2$	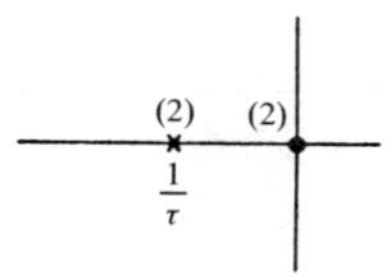	$\cdots = -\frac{1}{A}$	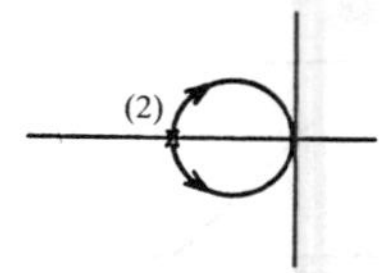	no	
5	$\left(\frac{p\tau}{p\tau + 1}\right)^3$	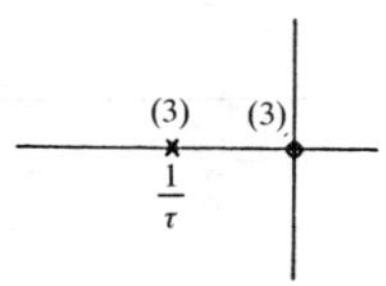	$\cdots = -\frac{1}{A}$	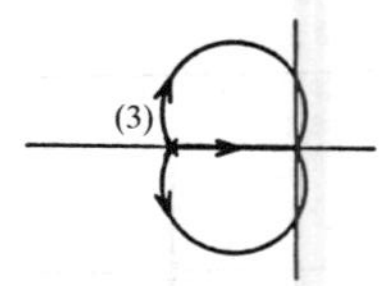	$\omega = \frac{1}{\tau\sqrt{3}}$ $A = 8$	
6	$\left(\frac{p\tau}{p\tau + 1}\right)^4$	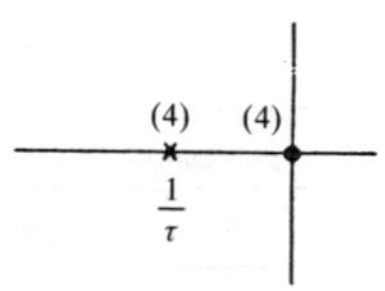	8 $\cdots = -\frac{1}{A}$	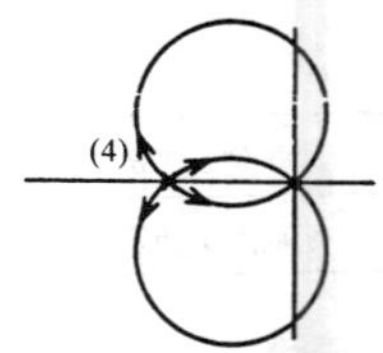	$\omega = \frac{1}{\tau}$ $A = 4$	
7	$\frac{p\tau}{p^2\tau_1\tau_2 + p\tau_2 + 1}$	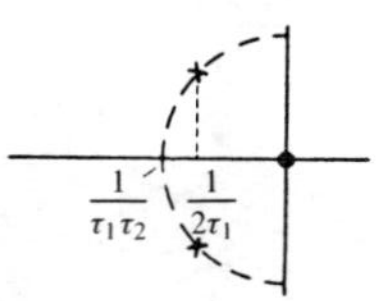	$\frac{p\tau}{p^2\tau_1\tau_2 + p\tau_2 + 1} = -\frac{1}{A\beta}$	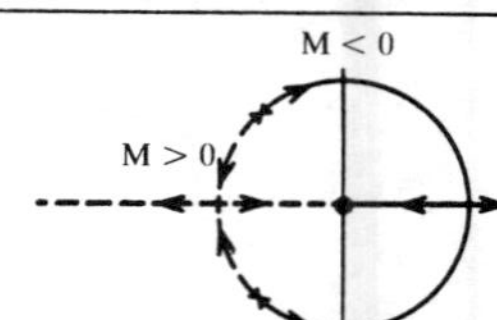	$\omega = \frac{1}{\tau_1\tau_2}$	

8	high frequency part $\dfrac{1}{(p\tau_{a3}+1)^2}\cdot\dfrac{1}{p^2\tau_{tr3}^2+p\dfrac{\tau_{tr3}}{Q}+1}$	$\tau_{a3}=\tau_{tr3}$ $Q=1.5$	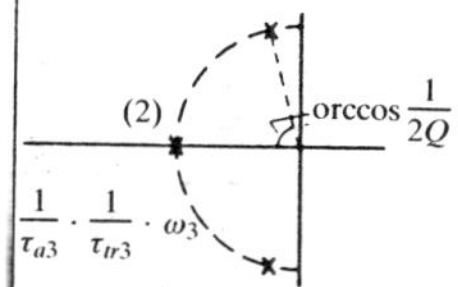	$\cdots = -\dfrac{1}{A\beta}$ β = real	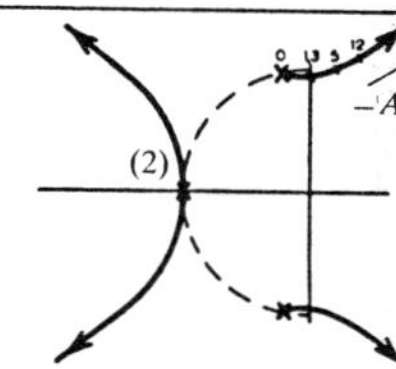	$\omega\approx\omega_3$ $A\beta\approx1.3$
9		$Q=0.7$	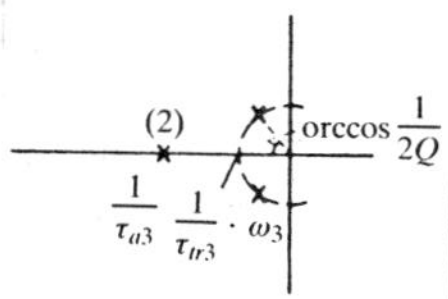	$\cdots = -\dfrac{1}{A\beta}$ β = real	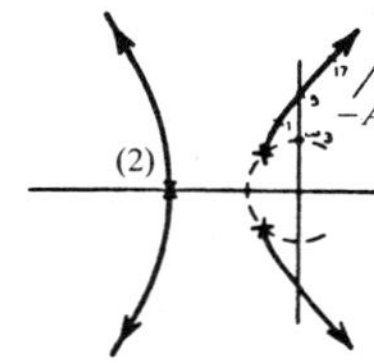	$\omega\approx2\omega_3$ $A\beta\approx5$
10		$\tau_{a3}=\frac{1}{5}\tau_{tr3}$		$\cdots = -\dfrac{1}{A\beta}$ $\dot{\beta}\approx\beta\dfrac{p\tau_{c1}+1}{p\tau_{c1}+1}$	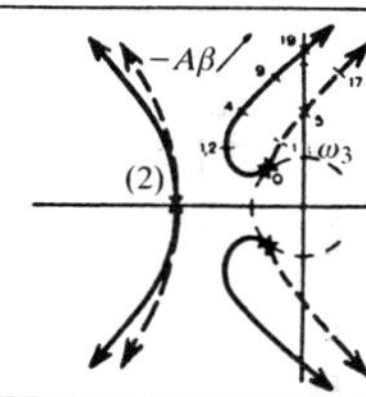	$\omega\approx3.3\omega_3$ $A\beta\approx19$

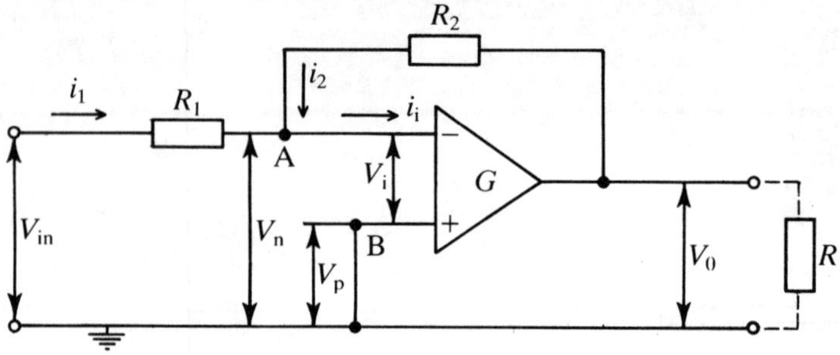

Figure 3.6 Operational amplifier with feedback loop

where V_p is the non-inverting input voltage at node B with respect to common and V_n is the inverting input voltage at node A with respect to the same common. Now consider node A and apply Kirchhoff's first law to it. Assuming an ideal amplifier, the current equation can be written

$$\frac{V_{in} - V_n}{R_1} + \frac{V_o - V_n}{R_2} = 0 \tag{3.28}$$

or

$$V_o = \frac{R_1 + R_2}{R_1} V_n - \frac{R_2}{R_1} V_{in} \tag{3.29}$$

and, as can be seen in Fig. 3.6, $V_p \equiv 0$. Thus from (3.27) $V_n = 0$ also. It is said that V_n is virtually earthed, which does not imply it is ground or earth, but that it has the voltage level of ground. Then applying equation (3.27) we can substitute in (3.29) $V_n = 0$, and this equation reduces to

$$V_o = -\frac{R_2}{R_1} V_{in} \tag{3.30}$$

In practical circumstances for $V_o = 10$ V and $G = 10^6$, V_i equals 10 μV. Again, with the same assumptions as above it can be seen that the input impedance of the amplifier itself (node A) has become $R_1 \,//\, R_2$ and with the load R_l we find $R_1 // R_2 // R_l$. To verify this, first note that the output impedance of the amplifier is zero and secondly, the ideal voltage source's impedance equals zero. Considering the input impedance of the circuit, or with other words the load impedance for the source V_{in}, this yields

$$R_{in} = R_1 = \frac{V_i}{i_1} \tag{3.31}$$

It should be kept in mind that, for minimum *offset error*, the condition of perfect symmetry of the impedance at the input nodes A and B should be fulfilled. *Offset* is defined as the voltage to be applied at the input to make the output voltage zero when no signal voltage is present at the input. This is related to the structure of the amplifier itself and will be discussed in more detail in Chapter 7. In the

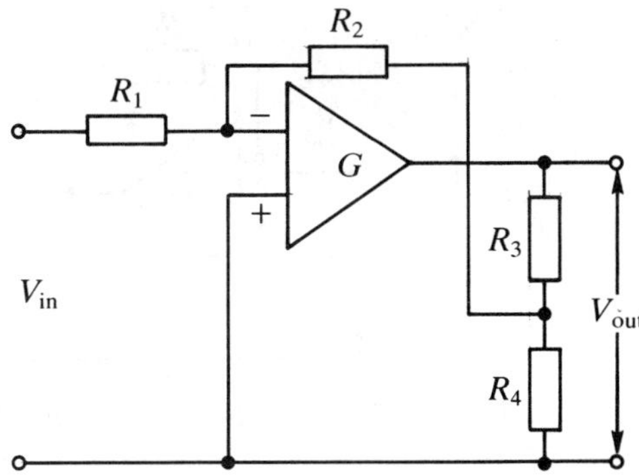

Figure 3.7 Operational amplifier with adjustable gain (courtesy Philips)

following sections we will discuss some frequently occurring important applications of the operational amplifier. Many variations exist, and they can not all be discussed in this book. They are easily found in textbooks dedicated to applications of the operational amplifier.

3.3.2 Preamplifier with adjustable gain

Figure 3.7 shows an amplifier configuration with adjustable gain. In this configuration, only a fraction of the output voltage V_o is used for feedback. This fraction, V_t equals

$$V_t = V_o \frac{R_4}{R_3 + R_4} \tag{3.32}$$

The output voltage becomes

$$V_o = -\frac{R_3 + R_4}{R_4}\frac{R_2}{R_1} \cdot V_i \tag{3.33}$$

As can be concluded, attenuation of the output voltage before applying feedback gives a higher gain. This has the special advantage that low-ohmic attenuators (some kilohms) can be used, providing a reduction in noise and hum. A high R_2, combined with the always present parasitic capacitors will result in a large time constant slowing down the speed performance of the amplifier. Speed performance is often expressed in *slew rate*, which is defined as the maximum possible swing of the output voltage per unit of time.

3.3.3 The non-inverting operational amplifier

The non-inverting amplifier is often used as a buffer amplifier to prevent any loading of a weak signal source. The configuration is depicted in Fig. 3.8. The voltage V_n at node A equals

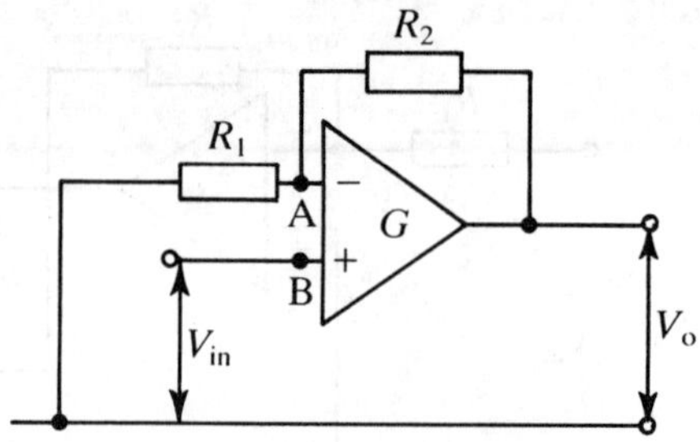

Figure 3.8 The noninverting amplifier

$$V_n = V_o \frac{R_1}{R_1 + R_2} \tag{3.34}$$

and with V_p and $V_{in} = V_n$ we find

$$V_{in} = V_o \frac{R_1}{R_1 + R_2} \tag{3.35}$$

or

$$V_o = \left(1 + \frac{R_2}{R_1}\right) V_{in} \tag{3.36}$$

3.3.4 The differential amplifier

As will be seen, this amplifier configuration is applied in cases where a difference voltage must be detected. Often the difference voltage of two signal sources is extremely small and superimposed on relatively high common-mode voltages. For instance, the difference between 5.1234 V and 5.1236 V is only 200 μV and 5.1234 V is considered as the common-mode voltage. The configuration is able to discriminate between the common-mode signal and the difference signal and is shown in Fig. 3.9.

The general expression for this configuration is found by superposition of the expressions for the inverting and non-inverting amplifier, or

$$V_o = \left(1 + \frac{R_2}{R_1}\right)\left(\frac{R_4}{R_3 + R_4}\right) V_{in2} - \frac{R_2}{R_1} V_{in1} \tag{3.37}$$

Note that $R_4/(R_3 + R_4)$ is the fraction of V_{in2} that comes available at node B. For minimum offset error the condition $R_1//R_2 = R_3//R_4$ must be fulfilled.

An exact differential expression for V_{in1} and V_{in2} is obtained by putting $R_1 = R_3$ and $R_2 = R_4$. The expression (3.37) becomes

$$V_o = \frac{R_2}{R_1}(V_{in2} - V_{in1}) \tag{3.38}$$

The above configuration is frequently applied to Wheatstone bridges. A drawback is the rather low input impedance, which is strongly improved in the so-called instrumentation amplifier. This configuration will be discussed in Chapter 6.

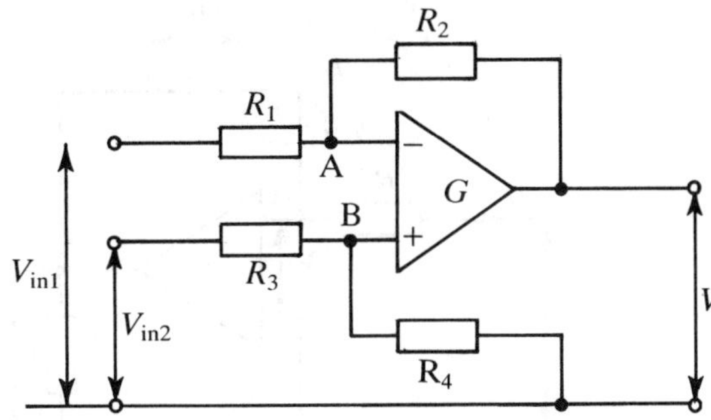

Figure 3.9 Differential amplifier configuration

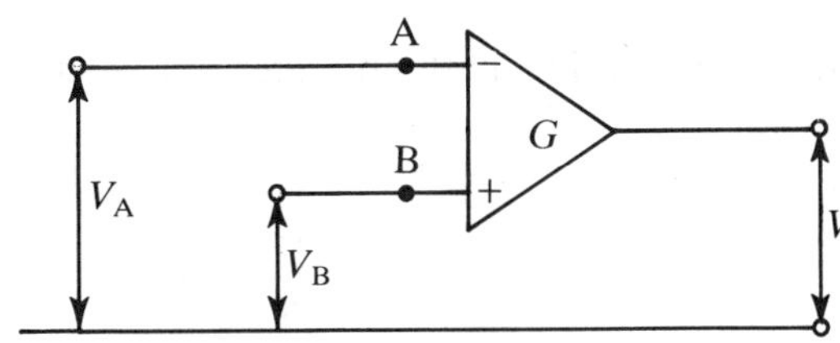

Figure 3.10 Comparator configuration

3.3.5 The comparator

A comparator or zero detector is another application of an operational amplifier (op amp). It is frequently used to compare two signals. This is illustrated in Fig. 3.10.

Because of the high gain of the amplifier a very small difference in input signal, of the order of 0.1 μV, will result in a large value of the amplifier's output voltage. Thus assume if V_A is only some tenths of a microvolt lower than V_B the output will be fully positive, and in the opposite case, when V_A is slightly higher than V_B, the output will be fully negative.

When V_B is connected to earth, the circuit compares V_A with zero, giving a sensitive zero detector which changes its output polarity when V_A passes zero. The error is only a small fraction of a microvolt.

The comparators, especially the high-speed types, are very sensitive to oscillations near the (zero) crossing-point. To prevent this, special precautions must be taken, for instance, applying a capacitive decoupling of the supply voltages and both input terminals.

3.3.6 The integrator

With a capacitor in the feedback loop, the op amp can be used as an integrator. The circuit is shown in Fig. 3.11. As long as the capacitor C in the feedback loop is not

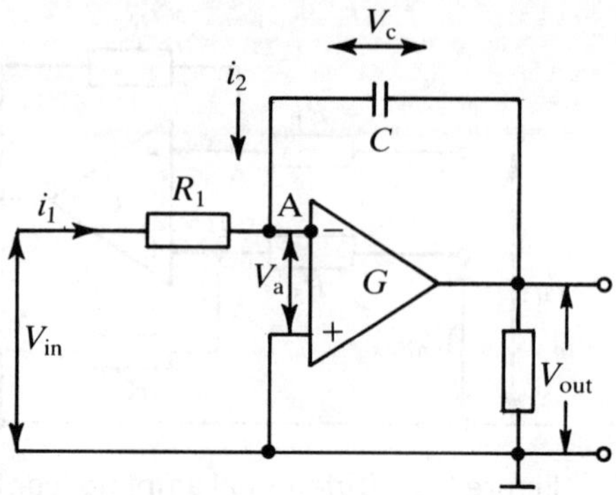

Figure 3.11 Operational amplifier used as an integrator (courtesy Philips)

fully charged, it acts as a DC path and does not change the characteristics of an op amp with a resistive feedback loop.

Applying again the current equation for the virtually earthed node A, we find

$$\frac{V_{in}}{R_1} + \frac{dQ_C}{dt} = 0 \tag{3.39}$$

and with $Q_C = CV_o$ we find

$$\frac{V_{in}}{R_1} + \frac{d(CV_o)}{dt} = 0 \tag{3.40}$$

or

$$V_a = -\frac{1}{R_1 C}\int V_{in}\,dt \tag{3.41}$$

The moment the supply voltage is achieved across the capacitor the integration will stop.

3.3.7 A constant current source

With operational amplifiers excellent current *sources* and *sinks* can be realized. A simple example of a current sink is given in Fig. 3.12 where the resistance $R_2 = R_1$ is considered as the load resistance. The expression for the generated current is

$$I_o = I_{ref} = \frac{V_{ref}}{R_1} \tag{3.42}$$

Very low currents can be generated, but a drawback is that the load has to be floating. There are many variations on this theme. As shown, a current sink is involved, but when the functions of the inputs of the op amp are reversed a current *source* is obtained.

The accuracy is mainly determined by the voltage reference V_{ref} and the resistor R_1.

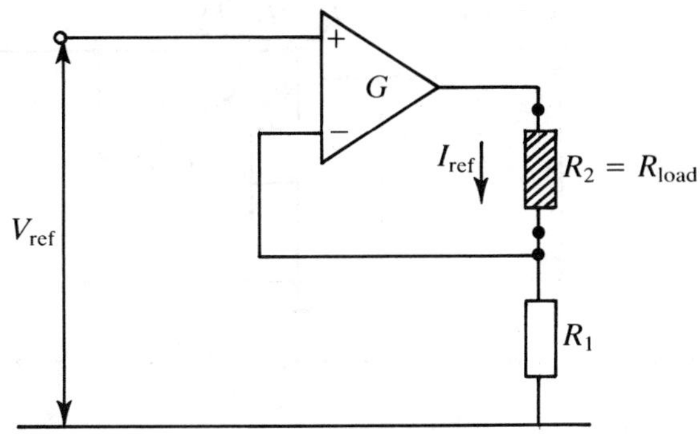

Figure 3.12 Operational amplifier used as a constant current sink (courtesy Philips)

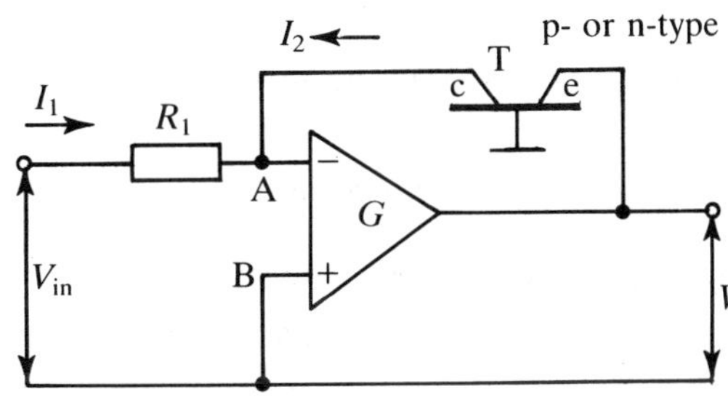

Figure 3.13 Circuit of a logarithmic amplifier. The bipolar transistor may be a p- or an n-type

3.3.8 The logarithmic amplifier

Occasionally a physical quantity has to be measured over a large range of values. This range is often called the *dynamic* and is defined as the ratio between the largest and smallest signal value to be measured. In this case, a logarithmic amplifier can be used to compress the signal within smaller boundaries. The principle of the logarithmic amplifier is illustrated in Fig. 3.13.

In the feedback loop, a bipolar transistor is implemented. A bipolar transistor shows an excellent exponential relation between collector current and base-emitter voltage, V_{BE}. This relation is expressed in the following equation and, with a good transistor, can exist over 7 decades (10^7) with an acceptable accuracy.

$$I_C = I_{SS} \exp(qV_{BE}/kT) \tag{3.43}$$

where

I_{SS} is the saturation current of the base-emitter diode
q specific electric charge
V_{BE} base-emitter voltage
k Boltmann's constant
T absolute temperature

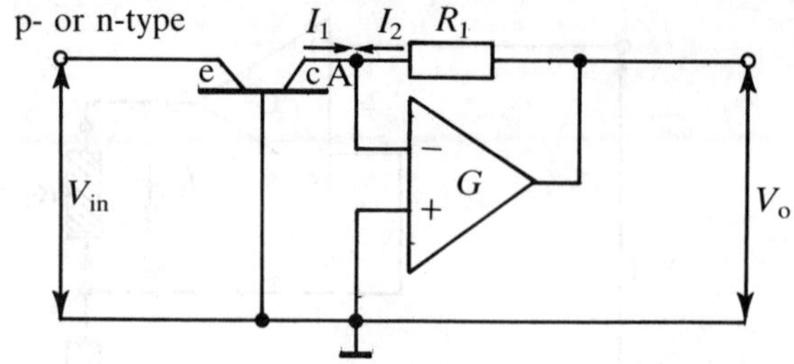

Figure 3.14 Exponential expansion of a signal. The transistor is either p- or n-type

If we apply the current equation to node A the result is

$$\frac{V_{in}}{R_1} + I_C = 0 \tag{3.44}$$

Substituting (3.43) in (3.44) and noting that $V_{BE} = V_o$ gives

$$\frac{V_{in}}{R_1} + I_{SS} \exp(qV_o/kT) = 0 \tag{3.45}$$

Rearranging (3.45) and taking logarithms we find

$$V_o = -\frac{kT}{q} \ln\left(\frac{V_{in}}{R_1 I_{SS}}\right) \tag{3.46}$$

So the output voltage is proportional to the logarithm of the input voltage. A p-type or n-type transistor can be applied, depending on the sign of the input signal V_i. Moreover, a parallel connection of a p-type and an n-type is also applicable. From expression (3.46) for this circuit the weak points can be figured easily. There is a large temperature dependency. This can be overcome by compensation techniques applying more sophisticated configurations.

If in Fig. 3.13 the transistor T and R_1 are interchanged we get an exponential amplifier with which an *exponential expansion* can be realized. The relation then yields

$$V_o = R_1 I_{SS} \exp(qV_{in}/kT) \tag{3.47}$$

This is illustrated in Fig. 3.14. Here also a large temperature dependency exists.

3.3.9 An analog multiplier

As we have already seen a logarithmic relation exists between the voltage and the current across a forward-biased pn-junction. This behavior can be used to convert a multiplier into an adder. The circuit is depicted in Fig. 3.15.

Assume four currents $I_1, \ldots, I_4$ are flowing through their respective diodes $D_1, \ldots, D_4$. The diodes may be replaced by bipolar transistors; then the respective V_{BE}'s represent the different diode voltages. At point A, a summing element is implemented for which

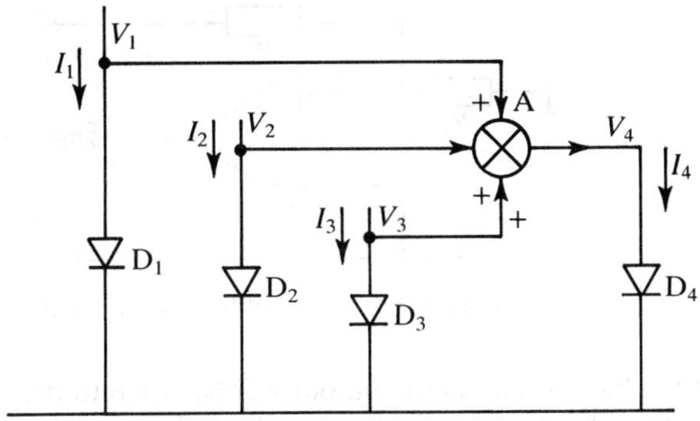

Figure 3.15 Principle of the analog multiplier

$$V_4 = V_1 + V_2 + V_3 \tag{3.48}$$

and substituting the following relation for the respective voltages,

$$V = \frac{kT}{q}\ln\left(\frac{I_1}{I_{SS}}\right) \tag{3.49}$$

we find

$$V_4 = (kT/q)\ln[I_1 I_2 I_3/(I_{SS1} I_{SS2} I_{SS3})] \tag{3.50}$$

or

$$V_4 = kT/q \ln[\alpha] \tag{3.51}$$

where $\alpha = I_1 I_2 I_3/(I_{SS1} I_{SS2} I_{SS3})$

then from (3.51) we find

$$\alpha = \exp(qV_4/kT) = I_4/I_{SS4} \tag{3.52}$$

and, ultimately, we find for the product of currents

$$\begin{aligned} I_4 &= I_{SS4}\,\alpha \\ &= I_{SS4} I_1 I_2 I_3/(I_{SS1} I_{SS2} I_{SS3}) \end{aligned} \tag{3.53}$$

Note that no temperature dependency is involved here. Several different variations on this principle exist, but they are all based on current multiplication.

3.3.10 A peak detector

A peak detector can be used in AC measurements to determine the root mean square (RMS) value of a sine wave. For this purpose, to obtain the RMS value, the peak value must be multiplied by 0.707. The basic circuit of a peak detector is shown in Fig. 3.16.

Since the diode is implemented in the feedback loop, the forward voltage drop is divided by the open loop gain. The diode will conduct whenever the input is

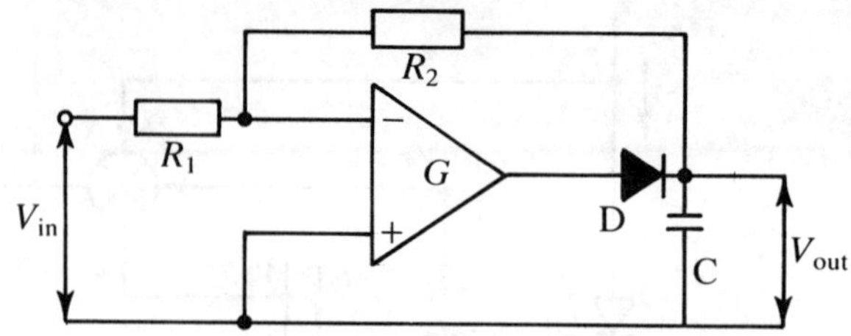

Figure 3.16 Circuit for a peak detector

greater than the output, so the output will be equal to the peak value of the input voltage. Once the diode is conducting, the following relation exists between input and output voltage, where the capacitor will store this maximum value of V_{in}.

$$V_{out} = -\frac{R_2}{R_1} V_{in} \tag{3.54}$$

If for instance the ratio of the resistors equals 0.707 the input voltage is multiplied by this value and the RMS value is obtained immediately.

3.4 ANALOG-TO-DIGITAL AND DIGITAL-TO-ANALOG CONVERSION

A/D and D/A conversion (sometimes abbreviated to ADC and DAC) are frequently used in all kinds of digitized measurement equipment, but can also be found in many consumer electronic products due to the ease of processing signals in a digital form. So at the beginning of a system we can distinguish an A/D conversion as soon as the signal is amplified to the desired level at which conversion can be executed. At the end of a system, depending on the application, a D/A conversion can be found when analog output is required.

In this section we will discuss both conversions, but first we will present a short introduction to coding, because that is the essence of the conversion process. In other words, a given signal has to be transposed into another domain by an unambiguous representation and vice versa.

3.4.1 Coding

Coding can be defined as a kind of conversion of data from one representation into another by a certain function. Often a reference source code is predefined. For instance, a constant voltage source can provide such a reference code. In the following we present a short introduction to coding of numbers.

Any real number can be represented by the expression

$$N = a_n g^n + a_{n-1} g^{n-1} + \cdots + a_1 g^{+1} + a_0 g^0 + a_{-1} g^{-1} + a_{-2} g^{-2} + \cdots + a_{-m} g^{-m}$$

or in an abbreviated form,

$$N = \sum_{i=-m}^{n} a_i g^i \tag{3.55}$$

where

g is base or radix
a is number or symbol.

If $m = 0$, the number is an integer.
If $n = 0$, the number is a fraction.

With $g = 10$ we have the decimal system, if $g = 2$ we have the binary system, if $g = 16$ we have the hexadecimal system, etc. The base number determines the amount of possible symbols.

A decimal number may also be represented by the expression

$$N_{10} = A \times 10^B \tag{3.56}$$

where

A is any number between zero and one, or $0 \leq A < 1$
B is any positive or negative integer $\ldots, -2, -1, 0, 1, 2, \ldots$

In its turn A is defined as

$$A = a_1 10^{-1} + a_2 10^{-2} + a_3 10^{-3} + \cdots + a_n 10^{-n} \tag{3.57}$$

where a_i is any positive integer between 0 and 9.

* * *

Example 3.4

Consider the number 98102. We can write this number as 0.98102×10^5, so $A = 0.98102$ and $B = 5$. We can write A as

$$A = 9 \times 10^{-1} + 8 \times 10^{-2} + 1 \times 10^{-3} + 0 \times 10^{-4} + 2 \times 10^{-5}$$

In the same way we can write any number to base 2 as

$$N = [a_1 \times 2^{-1} + a_2 \times 2^{-2} + a_3 \times 2^{-3} + \cdots + a_n \times 2^{-n}]2^B \tag{3.58}$$

where here, of course a_i can have the values 0 and 1 only. As will be seen, to indicate the sign of a number an extra bit has to be added.

* * *

Example 3.5

The binary representation of 5 can be written as 0101. When we write this number in the form of (3.58), this gives

$$N = [0 \times 2^{-1} + 1 \times 2^{-2} + 0 \times 2^{-3} + 1 \times 2^{-4}]2^4$$

To verify this, the expression can be rearranged back to the decimal representation.

To avoid the multiplying factor in (3.58), a binary number can be represented as

$$N = a_n 2^{n-1} + \cdots + a_3 2^3 + a_2 2^2 + a_1 2^1 + a_0 2^0 \qquad (3.59)$$

where again a_i can have the values 0 and 1 only.

* * *

When an *A/D conversion* is required the first representation is preferred, because the binary number is always a fraction of some maximum reference value. Let us illustrate that again with an example.

* * *

Example 3.6

Assume the fraction 0.625 = 5/8 must be represented in a binary form. Then, applying the notation of (3.58) this is simply 101, because

$$101 = [1 \times 2^{-1} + 0 \times 2^{-2} + 1 \times 2^{-3}]2^0$$

* * *

If we want to apply the two's complement to assign negative numbers we just complement each bit and add 1 to it.

* * *

Example 3.7

Assume we want to represent the decimal number −0.625 in the binary form (3.58), then the procedure is:

+5/8	0101	
	1010	complementing
	1	adding 1
	——	+
−5/8	1011	

The first bit represents the sign bit. When we add 5/8 and −5/8 the result is zero with a loss of the carry bit, or

5/8	0101	
−5/8	1011	
——	——	+
0	10000	

* * *

In ADCs and DACs the first bit is called the most significant bit (MSB) and has a weight of half the full scale of the converter. The second bit has a weight of 1/4

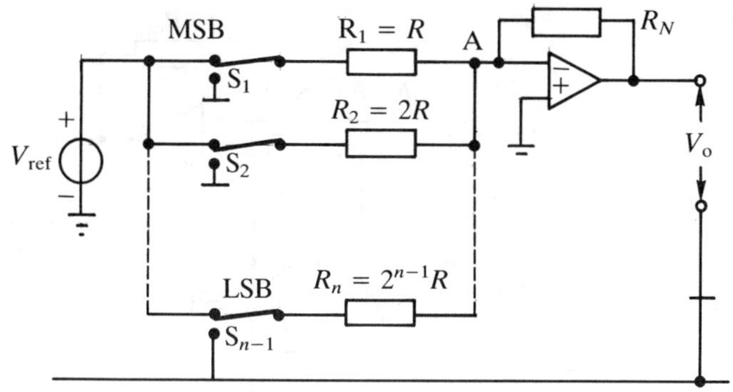

Figure 3.17 Circuit of a digital-to-analog converter

of full scale, etc. For instance, with 8 bits we can count up to 256, the MSB represents the number 127, so about half full scale. The last bit is called the least significant bit (LSB) and has the weight $1/2^n$ of the full scale value.

The *resolution* of an ADC is defined as the number of bits assigned by the size of the LSB. For a DAC it is the smallest possible incremental change in the output voltage (see also Chapter 4).

An alternative definition for the resolution of an ADC is the amount for the input voltage change required to increment the output with one code change of the LSB. Resolution can be expressed in percent of full scale or in binary bits.

* * *

Example 3.8

Consider a 12-bit ADC. The resolution is able to resolve 1 part in 2^{12}, i.e., 1 part in 4096. This equals 0.02441% of full scale. So a converter with 10 V full scale can resolve an input change of 10/4096 = 2.441 mV input change.

In the same way this holds for a DAC; an increment of one LSB at the input produces a voltage change of 0.02441% of full scale at the output.

Note that resolution is not related to accuracy at all.

The resolution of a converter can be related to the dynamic range in dB. For example, a factor of 2 equals about 6 dB. Therefore the dynamic range can easily be assigned in dB by the number of bits multiplied by 6. Thus a 12-bit converter has a dynamic range of $12 \times 6 = 72$ dB.

* * *

3.4.2 Digital-to-analog converter

As has been said already, in some applications it can be necessary to change binary data into an analog signal. The simplest form of a DAC is illustrated in Fig. 3.17.

With respect to node A this circuit behaves as a summing amplifier with weighted resistors in a *binary* fashion, so $R_1{:}R_2{:}R_3{:}\ldots = 1{:}2{:}4{:}\ldots$. They are all

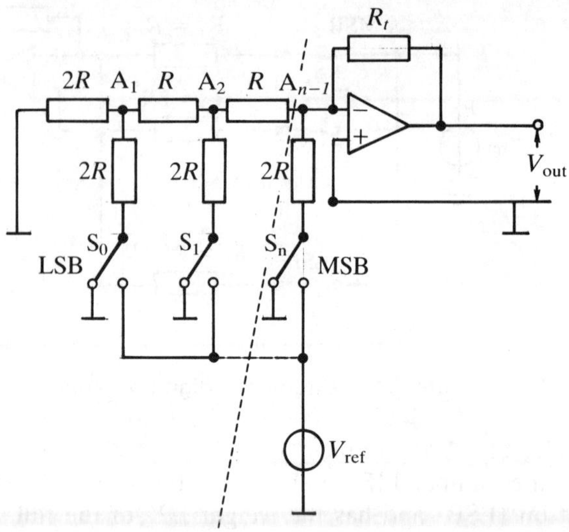

Figure 3.18 The ADC with an R-2R ladder network

connected via a logic electronic switch S_i, to a reference voltage V_{ref}, or to ground. A logic 1 applied to the respective switch connects the corresponding resistor to the reference voltage and increases the output voltage by the binary weighted increment involved. This process is expressed as

$$V_o = R_N\left[\frac{1}{R}S_1 + \frac{1}{2R}S_2 + \cdots + \frac{1}{2^{n-1}R}S_{n-1}\right]V_{ref} \tag{3.60}$$

The switches $S_1, \ldots, S_{n-1}$ represent the binary voltage levels. This circuit has several disadvantages. The value of each input resistor is twice that of the preceding one, so the absolute values become quite large. Moreover, matching, tolerances and temperature stability of the resistors influence the accuracy to a large extent. Finally, the input impedance of this circuit is low and changes continuously, producing offset errors.

A much better solution, avoiding several of these problems, is to make use of the R–$2R$ ladder network configuration. This circuit is depicted in Fig. 3.18. At each node $A_1, \ldots, A_{n-1}$ 'looking to the left' the same impedance R is seen and consequently the current flowing from right to left, generated by the reference voltage V_{ref}, is divided in two at each junction.

The input impedance of the operational amplifier has become a constant, independent of the binary word represented by the logical switch positions S_i, which are equivalent to a logic 0 or 1. For the conversion the expression appears to be

$$V_o = V_{ref}\frac{R_t}{R}\sum_{i=0}^{n} S_i 2^{i-n-1} \tag{3.61}$$

The above two converter types are pure binary. It is possible to adapt both circuits to other code conversions by implementing other resistor ratios or by other kinds of resistor networks.

3.4.3 Analog-to-digital converter

The function of an analog-to-digital converter is to convert the analog input signal into a discrete number of normally binary steps that can easily be processed, or presented in decimal form on a numerical display. The forms into which the analog signal is converted may be of different nature, e.g., a voltage, a time or a charge, and depend on the type of ADC involved. As will be seen, in all cases each step should correspond to an accurately known voltage related to a very accurate stable reference voltage to which a comparison should be made. If the step is a DC voltage this is quite a simple matter, but when the step is of another nature, such as time or charge, for instance, the comparison can be made by converting the reference voltage into the same steps.

For analog-to-digital converters, different principles of operation are available. The most widely used systems are the following:

1 Direct compensation
In this principle of operation, the input signal is compared with an internally generated voltage, which is increased in steps starting from zero. The number of steps needed to reach full compensation is counted and represents the desired output value.

2 Voltage-to-time conversion
Both input voltage and reference voltage are converted into a time and the ratio of the two times is counted and represents the output value.

3 Voltage-to-frequency conversion
The input voltage is converted into a number of pulses and the number of pulses in a well-defined constant time is counted and represents the output value.

3.4.3.1 The dual-slope integration ADC

In this section we will discuss an application of the voltage-to-time conversion principle only. This principle offers simple, accurate and very linear ADCs.

The dual-slope integration ADC is one of the most well known and widely used types. The principle of operation is based on integration of the input voltage for a constant time followed by measurement of the time required to discharge the integrating capacitor with a constant current. The principle is depicted in Fig. 3.19. At the start of a measuring period the logic circuit closes the switch S, and at the same time starts the timer, counting a fixed number of clock pulses corresponding to a fixed time T_1, e.g., during $N = 200\,000$ clock pulses. During the time T_1, V_{in} is connected to the input of the amplifier, the sampling period, and V_{in} will be integrated in the capacitor C.

It will be clear that, at the end of this period, V_C is proportional to V_{in}. At the input of the same period T_1, the logic disconnects the input V_{in} by the switch S, and connects the input now to the reference voltage V_{ref}, which should have the opposite polarity to the input V_{in}. The counter is reset to zero. At that moment, C starts to discharge with a constant current, because V_{ref} is a constant. The

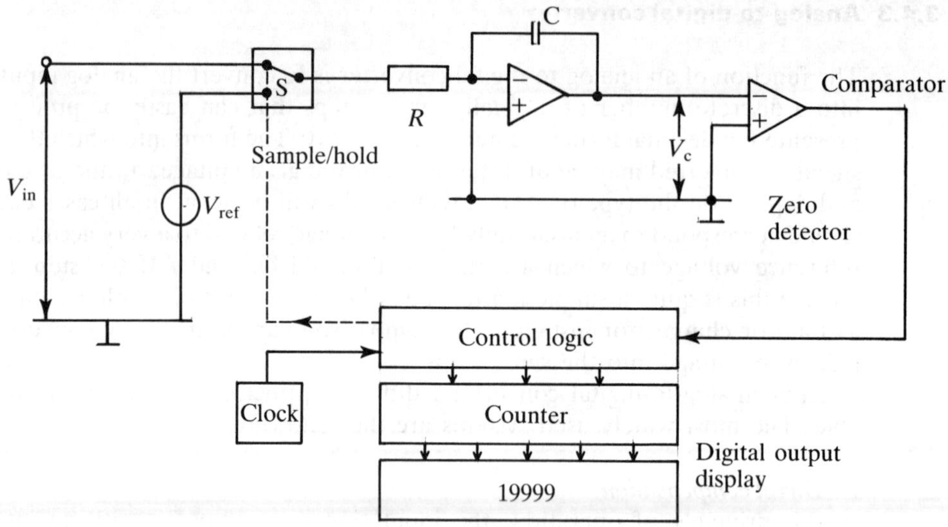

Figure 3.19 Principle of a dual-slope integration ADC

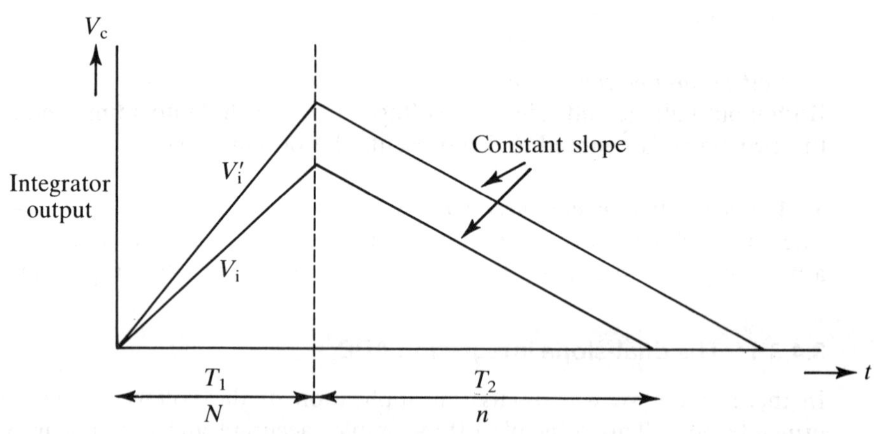

Figure 3.20 Voltage versus time graph of the dual-slope ADC for two input values

integrator integrates the reference voltage down to the comparator's threshold value at which time the counting is stopped. Assume the number of counts is n. This is depicted for two different input values in Fig. 3.20.

This process can be described by an equation with respect to charge on the capacitor C, or

$$\int_0^{T_1} \frac{V_{in}}{R} dt - \int_{T_1}^{T_2} \frac{V_{ref}}{R} dt = 0 \tag{3.62}$$

In the time interval T_1, V_{in} is supposed to be constant, then (3.62) can be written as

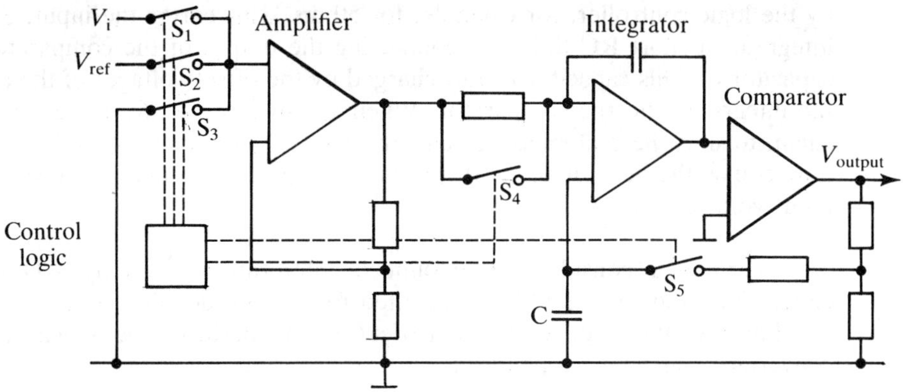

Figure 3.21 Simplified autozeroing circuit used with a dual-slope ADC

$$V_{in} T_1 = V_{ref} T_2 \tag{3.63}$$

or

$$V_{in} = \frac{T_2}{T_1} V_{ref} \tag{3.64}$$

If this equation is expressed in the number of counts, we get

$$V_{in} = \frac{n}{N} V_{ref} \tag{3.65}$$

So V_{in} can be read immediately from the counter. The advantages of this type of AD conversion can be summarized as follows.

The several errors of the integrator are present in the whole timing period T_1 and T_2 and will therefore cancel. The accuracy of the conversion will depend only on the accuracy of the reference voltage and not on the clock frequency. The only requirement for the clock is to be stable within the conversion period. The accuracy of the RC time is of less importance. For this type of AD conversion extremely high accuracy and linearity can be obtained.

3.4.3.2 Autozeroing and AD conversion

Finally, we will introduce here the concept of *autozeroing*, which is another very versatile feature in electronic measurement systems.

Each user of an instrument expects the instrument to indicate zero when the input is short-circuited. For instance, in a digital voltmeter with a maximum reading of 1999, a zero error of 0.05% of full-scale deflection (FSD) is sufficient to give an error reading of 0001. For optimum accuracy especially at low values, zeroing is required and automatic zeroing would be very convenient. This is implemented in most instruments nowadays. As an example, this feature, combined with a dual-slope ADC, is depicted in Fig. 3.21.

Before the real measurement is executed, the switches S_3, S_4 and S_5 are closed by the logic controller, for example, for 50 ms. This earths the input, giving the integrator a short RC time and connecting the output of the comparator to the capacitor C. This capacitor is now charged by the offset voltages of the amplifier, the integrator and the comparator. When the switches S_3, S_4 and S_5 are opened again, to start the real measurement, the total offset voltage of the circuit, which now equals the zero error, is stored in the capacitor C and as a result the real input voltage is measured correctly.

There is a considerable amount of literature concerning AD and DA conversions and this should be consulted for a more detailed study.

For the interested reader a review of the performance specifications for converters is given in Appendix 1.

3.5 DIGITAL FUNCTIONS

Digital functions have had a large impact on measuring systems and have changed the world of measuring systems' performance both inside and outside the instrument. It is beyond the scope of this textbook to present a detailed treatment of digital electronics. In general, it can be said that any analog-implemented function has a digital counterpart. And considered on a microscopic scale, all digital functions are analog.

In this section, we shall discuss a few examples of combinational circuits. These examples occur frequently in electronic measurement systems. A basic knowledge of Boolean algebra and logic elements is assumed. The following items will be discussed:

A decoder
An encoder
A multiplexer

3.5.1 A decoder

A decoder is defined as a logic building block that accepts a binary input and converts it into a certain output number one at a time. A simple example of a 1-out-of-4 decoder is shown in Fig. 3.22. The Boolean functions for the four output variables are

$$\begin{aligned} S_0 &= \overline{A + B} \\ S_1 &= \overline{A + \bar{B}} \\ S_2 &= \overline{\bar{A} + B} \\ S_3 &= \overline{\bar{A} + \bar{B}} \end{aligned} \tag{3.66}$$

A decoder is very useful for addressing purposes in memories, devices, equipment, etc. The truth table is shown in the same figure.

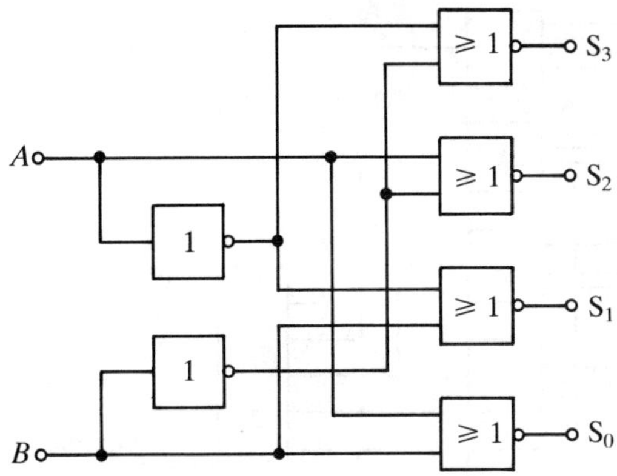

Input		Output			
A	B	S_0	S_1	S_2	S_3
0	0	1	0	0	0
1	0	0	1	0	0
0	1	0	0	1	0
1	1	0	0	0	1

Figure 3.22 Logic diagram and truth table of a 1-out-of-4 decoder

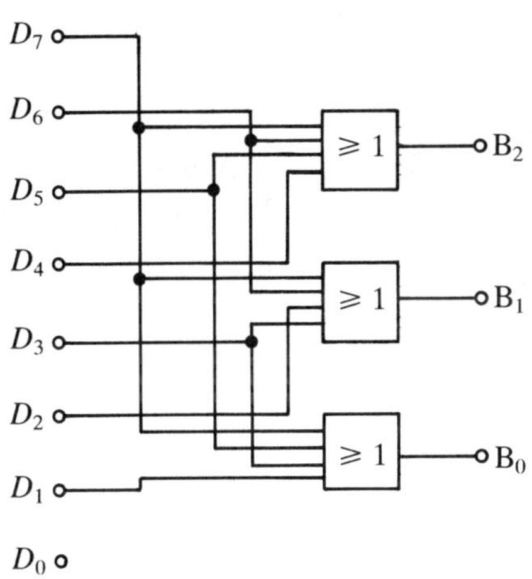

Input								Output		
Octal digits								Binary digits		
D_7	D_6	D_5	D_4	D_3	D_2	D_1	D_0	B_2	B_1	B_0
0	0	0	0	0	0	0	1	0	0	0
0	0	0	0	0	0	1	0	0	0	1
0	0	0	0	0	1	0	0	0	1	0
0	0	0	0	1	0	0	0	0	1	1
0	0	0	1	0	0	0	0	1	0	0
0	0	1	0	0	0	0	0	1	0	1
0	1	0	0	0	0	0	0	1	1	0
1	0	0	0	0	0	0	0	1	1	1

Figure 3.23 Logic diagram and truth table of an octal-to-binary encoder

3.5.2 An encoder

An encoder is a logic circuit that accepts any number of inputs and converts them into a binary code. As can be seen from this definition, it performs the reverse function to the decoder. An example of an encoder is the octal-to-binary encoder shown in Fig. 3.23. The truth table is depicted in the same figure. Note that the names for a decoder and an encoder are often interchanged.

Encoders are used to realize conversions between different codes, e.g., from one BCD (binary code digital) to another BCD code.

The following Boolean functions are realized:

$$\begin{aligned} B_0 &= D_1 + D_3 + D_5 + D_7 \\ B_1 &= D_2 + D_3 + D_6 + D_7 \\ B_2 &= D_4 + D_5 + D_6 + D_7 \end{aligned} \tag{3.67}$$

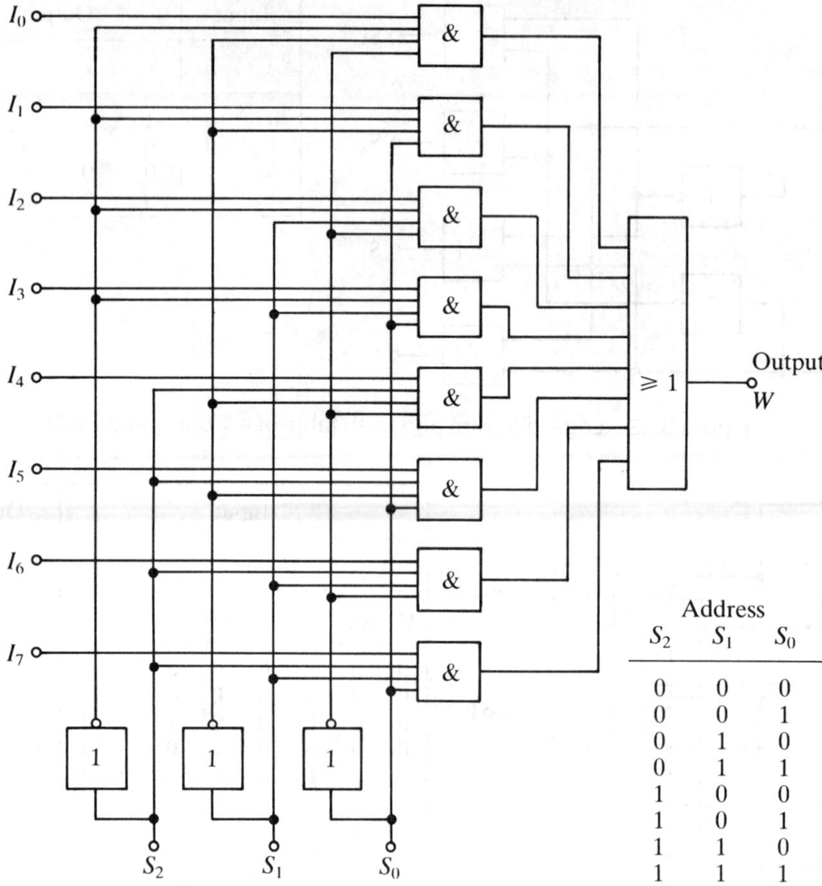

Address			Selected
S_2	S_1	S_0	line
0	0	0	0
0	0	1	1
0	1	0	2
0	1	1	3
1	0	0	4
1	0	1	5
1	1	0	6
1	1	1	7

Figure 3.24 An eight-input multiplexer (see Appendix 2 for standard symbols list)

3.5.3 A multiplexer

The third group of combinational circuits are the multiplexers and demultiplexers. The word 'multiplexing' is derived from telecommunication techniques, where a large number of voice channels are transmitted over a small number of lines. In the same fashion, the digital multiplexer can be used to transmit a large number of digital signals over a few lines. Sometimes the multiplexer is termed a 'data selector'. The reverse operation is called demultiplexing.

A multiplexer can be defined as a combinational circuit that selects data from 2^n input lines and transmits them through a single output line. This definition closely resembles the definition of the encoder, but the difference is the presence of a number of additional input lines called the address or selector lines. An example of an 8-input digital multiplexer is shown in Fig. 3.24.

The eight input lines are connected to eight 4-input AND gates which are controlled by the address inputs S_0, S_1 and S_2. As can be derived from the truth

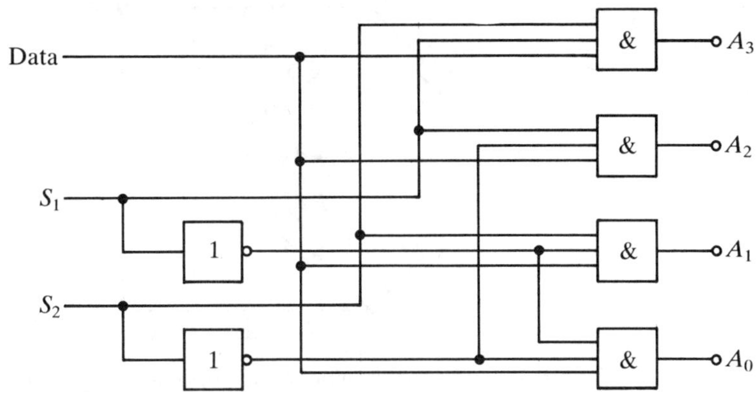

Figure 3.25 A four-input digital demultiplexer

table only one input at a time is connected to the output via the 8-input OR gate. Often the AND gates have a fifth input to provide an enable input, by which all inputs can be switched on and off from the output simultaneously.

The reverse operation is called demultiplexing and an example of a 4-output digital demultiplexer is given in Fig. 3.25. This circuit is self-explanatory.

3.6 MISCELLANEOUS

In this section, we will discuss the well-known Wheatstone bridge configuration often applied at the input of a measuring chain. Because silicon integration technology has made possible completely new implementations of these bridges with new features and characteristics, we shall present a discussion of an integrated bridge configuration. We will start with the conventional concept.

3.6.1 The conventional Wheatstone bridge configuration

In many cases, the Wheatstone bridge provides the best configuration for input transduction to obtain maximum sensitivity. Moreover as will be seen in Chapter 6, an auxiliary energy source is required for modulating transducers. This means that for such a type of transducer, a so-called common-mode signal is present. Applying a Wheatstone bridge can supress this common-mode signal to a large extent. The bridge composed of four resistive elements is illustrated in Fig. 3.26. Basically, the four branches can be any type of element sensitive to the physical quantity to be measured.

A balanced bridge configuration shows an inherent insensitivity to the bridge supply voltage and the impedance of the supply voltage and the detector. This can be derived from the equilibrium condition for a bridge configuration. In general, the output voltage V_o equals

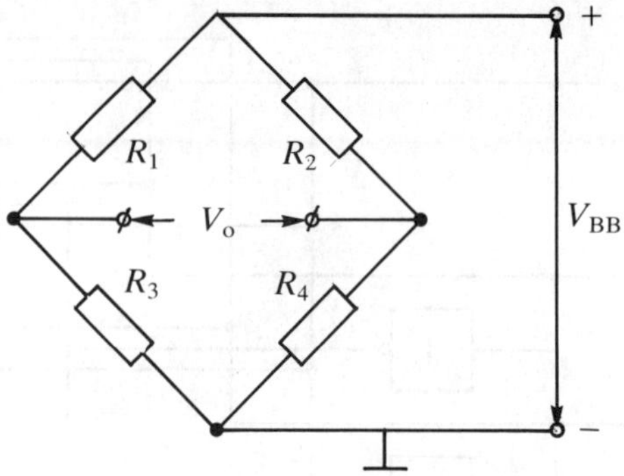

Figure 3.26 Resistive Wheatstone bridge configuration

$$V_o = V_{BB}\left[\frac{R_3}{R_1 + R_3} - \frac{R_4}{R_2 + R_4}\right] \tag{3.68}$$

where V_{BB} is the supply voltage, and the equilibrium condition gives

$$\frac{R_3}{R_1 + R_3} - \frac{R_4}{R_2 + R_4} = 0 \tag{3.69}$$

or

$$R_1 R_4 = R_2 R_3 \tag{3.70}$$

This expression is known as the equilibrium condition. In this fashion, a bridge configuration may be considered as a passive *nulling* element, because the measurand is that output value of a branch setting which satisfies (3.70). In that case, the influence of the supply voltage V_{BB} is put aside. Compare this with the nullor concept of an operational amplifier discussed in Section 3.3.1.

It will be obvious that when the deflection method is applied, the output signal is linearly dependent on the supply voltage, V_{BB}, and so the measurand is in accordance with equation (3.68).

For the Wheatstone bridge the following parameters are important:

1 The *differential sensitivity* is defined as the ratio of the change in output voltage due to a change in one of the sensor (input) elements, or for a change in R_1, and neglecting the sign we find

$$S_{diff} = \frac{dV_o}{dR_1} = V_{BB}\frac{R_3}{(R_1 + R_3)^2} \tag{3.71}$$

2 More important for a bridge configuration is the *relative sensitivity* defined as

$$S_B = \frac{dV_o}{dR_1/R_1} = \frac{dV_o}{dR_1}R_1 = V_{BB}\frac{R_1 R_3}{(R_1 + R_3)^2} \tag{3.72}$$

Substituting $F = R_1/R_3$ in equation (3.72) yields

$$S_B = V_{BB}\frac{F}{(1 + F)^2} \tag{3.73}$$

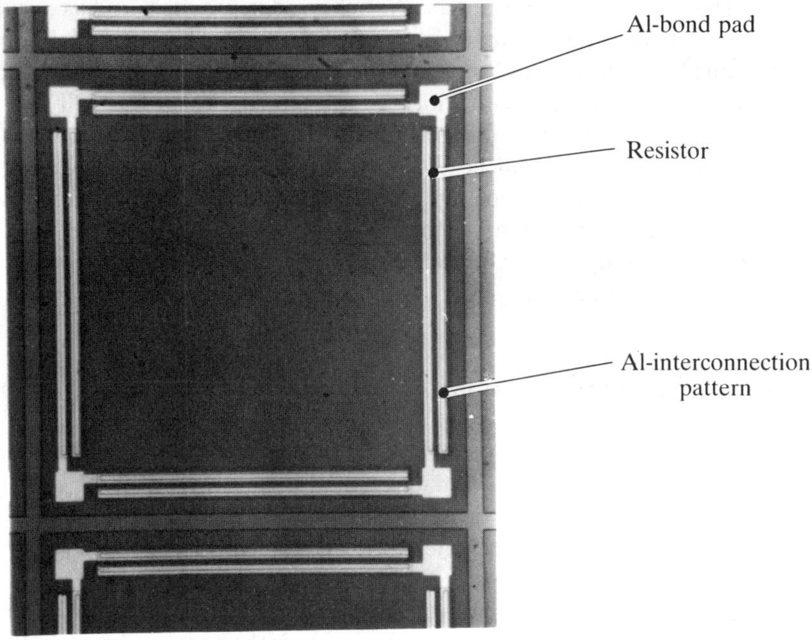

Figure 3.27 Integrated silicon Wheatstone bridge configuration

This expression has a maximum for $F = 1$, or for $R_1 = R_3$, which then results in

$$S_B = \frac{1}{4} V_{BB} \tag{3.74}$$

Thus maximum sensitivity is achieved as the transductor elements are pairwise equal, or

$$\begin{aligned} R_1 &= R_3 \\ R_2 &= R_4 \end{aligned} \tag{3.75}$$

By inspection of (3.73) it will be clear that a high supply voltage will also increase the sensitivity.

The concept of resolution (defined in Section 3.4) can also be specified here. It is defined in this context as the smallest change which can be processed by the system.

3.6.2 An integrated Wheatstone bridge

When a resistive bridge configuration is integrated in silicon, perfect matching and temperature tracking for all branch resistors may be expected. This is because the four resistors are integrated on one chip in one batch process. The temperature coefficients are equal and perfect temperature tracking in silicon is guaranteed. An example of such a bridge is shown in Fig. 3.27. The four resistors are integrated on a 4 mm^2 silicon chip and have equal resistances.

This ideal behavior is not always achieved in practice, and a considerable effort is being made in the manufacturing process of all types of integrated bridge transducers. Often a kind of trimming is executed to meet the specifications with respect to temperature behavior. Temperature effects are still one of the main reasons for inaccuracy in bridge transducers. These errors are normally known as drift phenomena and will be discussed in Chapter 7 in more detail.

Besides trimming techniques, compensation techniques are available. But a more recent development for bridge transducers proposes to eliminate the influence of temperature by accomplishing thermal feedback. This is discussed in the following section.

3.6.3 The principle of the constant voltage, constant current bridge configuration

Generally the bridge output voltage V_o is a function of more than one parameter. We have already seen the influence of the supply voltage V_{BB} in (3.68), but a change in one of the transducer elements R_i can also be caused by other physical parameters than the desired one alone. These aspects will be discussed in detail in Chapter 6.

Assume that the output voltage V_o is a function of four parameters, which might be related to each other; in other words they cannot be considered as strictly independent. In the following equation the suggested parameters are: the supply voltage, V_{BB}, the supply current, I_{BB}, one or more of the resistive transducer elements, R_i and the ambient temperature, T_{amb}:

$$V_o = f(V_{BB}, R_i, I_{BB}, T_{amb}) \tag{3.76}$$

Of course the preferred output voltage is obtained when the output voltage is a function of a single measurand affecting R_i only. When V_{BB}, I_{BB} and T_{amb} have become constants, then this condition is fulfilled and we get the following expression:

$$V_o = f(R_i) \tag{3.77}$$

In many cases it appears that the ambient temperature changes are the most disturbing factors affecting all other parameters. In that case, it suffices to maintain a constant environmental temperature (T_{amb}).

Now consider again the integrated bridge configuration of Fig. 3.27 and assume the four equal resistive elements R_i have a positive and linear temperature coefficient. It can then be argued that if the ratio of V_{BB} and I_{BB} is kept constant, the average level of the magnitude of the four resistor values is constant as well.

A constant ratio of the average value of all resistors requires that both the supply voltage and the supply current in the bridge are kept constant. This can be realized by integrating a second bridge and applying a thermal feedback in the chip. This double bridge configuration is depicted in Fig. 3.28. Both bridges have equal impedance due to the constant length-to-width ratio of all eight branch resistors. Due to a relatively high thermal conductivity of silicon (84 W/mK), both bridges are coupled thermally by the substrate and this guarantees a complete thermal connection.

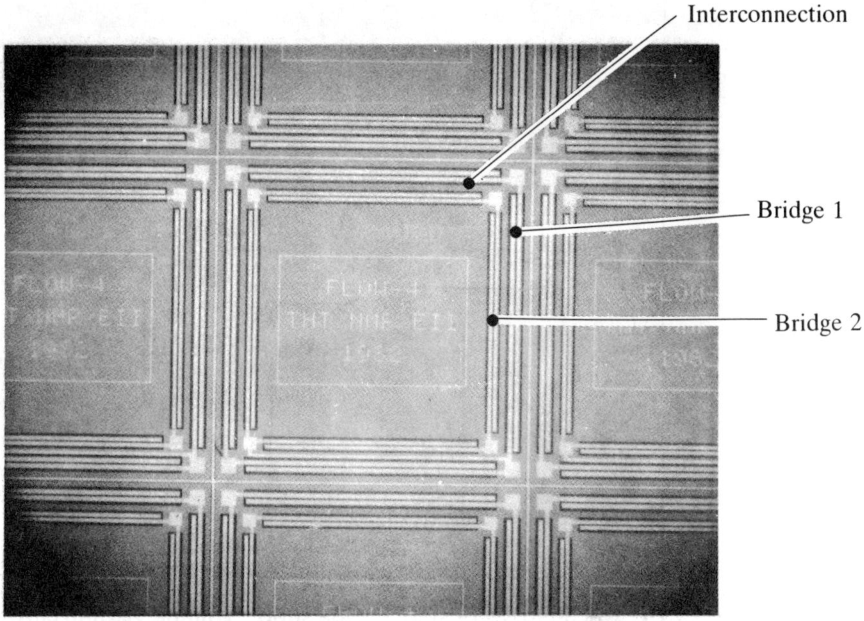

Figure 3.28 Integrated silicon double bridge configuration. Both bridges have equal resistances due to a constant length-to-width ratio in all eight branch resistors

The circuit needed to produce a constant V_{BB} and I_{BB} is depicted in Fig. 3.29. The first, outer bridge is connected to a current source delivering I_{BB}. The second, inner bridge is connected to a voltage-controlled current source.

Its voltage level is determined by

$$I_{BB2} = g_m G(V_{BB} - V_{ref}) \tag{3.78}$$

where g_m is the transconductance of the voltage-controlled current source. Each deviation of V_{BB} from the reference voltage V_{ref} is amplified via the 'comparator' configuration A and fed into the second bridge by the voltage-controlled current source. Hence each deviation results in a change in power dissipation in the second bridge, causing a temperature change in the chip and thus in a change in the first bridge. This change in temperature will continue until V_{BB} is exactly equal to V_{ref}. Consequently, the supply voltage and current of the outer bridge are constant under all measuring conditions when negative feedback is assumed.

Using this configuration the measurand developed in the outer bridge will be a function of the current division I_1 and I_2 alone, where the sum of both currents is constant.

This principle is called *thermal biasing*, because a fixed operating point for the sensor is obtained, concerning the supply voltage and current, the ratio of V_{BB} and I_{BB}, which equals the average sensor resistance R_s and the sensor temperature T_s.

It is not strictly necessary that both bridges are equal, or that the inner circuit is a bridge configuration, but the applied configuration guarantees maximum performance in response time and thermal coupling. It can be concluded that

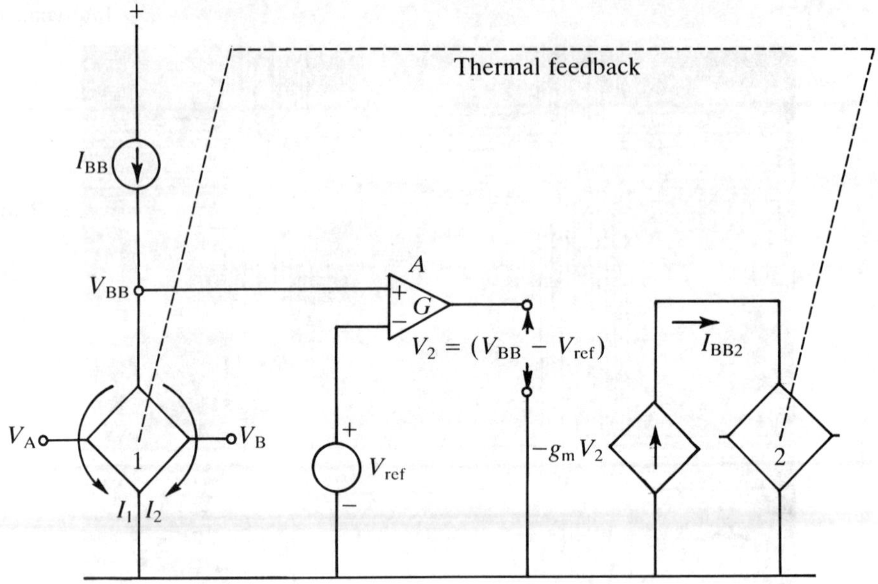

Figure 3.29 Circuit for a constant bridge supply voltage V_{BB} and current I_{BB}

this principle of operation is well suited for many different input transducers manufactured in silicon technology.

3.7 PROBLEMS

3.1 Determine the transfer function $H(j\omega) = V_o/V_{in}$ for the circuit depicted in Fig. 3.30.

3.2 In Fig. 3.31 an active filter is shown.
(a) Determine the transfer function $H(j\omega) = V_o/V_{in}$.
(b) For which ω is $H = 0$?
(c) For which ω does $H(j\omega)$ obtain a real value?

3.3 Determine the Bode plot and the Polar plot for the circuit depicted in Fig. 3.30.

3.4 Determine the pole-zero plot for the circuit depicted in Fig. 3.31.

3.5 Give a circuit for the non-inverting summation of three voltages V_1, V_2 and V_3 and give an expression for the output voltage. The gain factor should equal 12. Design the circuit for minimum offset error.

3.6 Design a circuit to determine the root mean square value of a full sine wave function. If an arbitrary AC signal is given, which solution is preferred?

3.7 A precision current source of 50 mA is required. The available operational amplifier can deliver 15 mA. Design a circuit and derive an expression for the current. Discuss the factors which determine the accuracy.

3.8 Design a voltage regulator for 2 V and 1 A. The supply voltage is 3.5–15 V. Try to avoid the implementation of a zener diode.

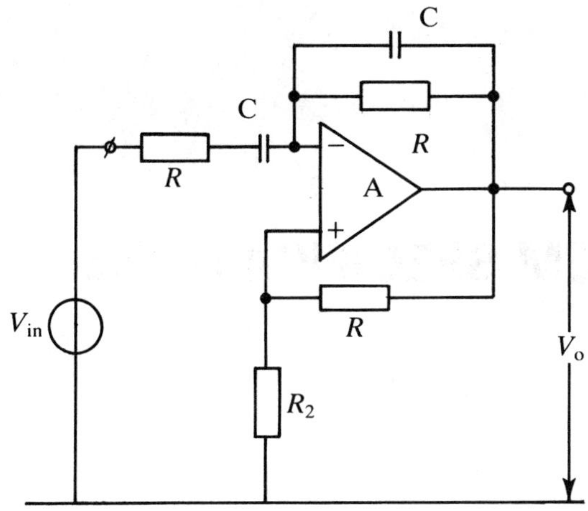

Figure 3.30

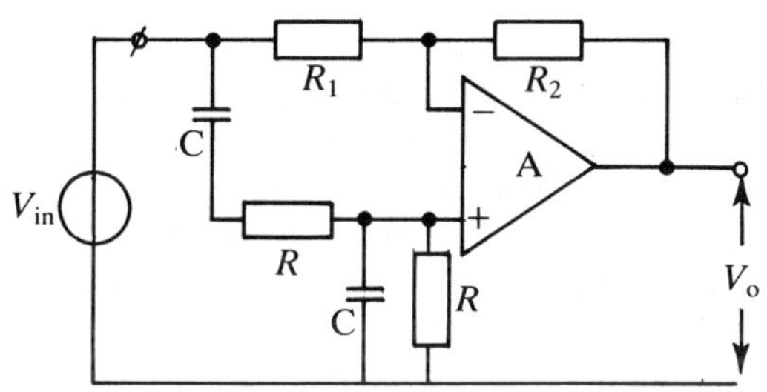

Figure 3.31

3.9 Design a digital multiple input system which can measure 32 different temperatures with an inaccuracy of <1%. Every 24 hours 2048 measurements have to be executed. The measurements are temporarily stored in a memory and have to be read the next day.

What is the required memory capacity and how many memory address lines are required? If a digital multiplexer is applied how many mux control lines are then required?

Four

SYSTEM SPECIFICATIONS

4.1 INTRODUCTION

In the previous chapters we have discussed several aspects of measurement set-ups. Recent developments show an increasing impact of, and dependency on, electronic measurement systems. This occurs both in the industrial environment, as well as in the laboratory. For instance, maintaining a high standard of quality during a manufacturing process makes it obligatory to perform extensive measurements.

In general the overall specifications of a measurement system are highly application- and marketing-dependent. In all cases we want to know if the measurement system complies with the claimed specifications. If the system either fails to meet or goes beyond its *predefined* specifications, it is said that the system fails.

Several system specifications are in common use. Very often the technical specifications of a system are badly specified and confusing. Here we attempt to clarify the matter.

A global description of a system involves the following aspects:

- technical specifications
- quality and reliability
- ergonomic aspects
- environmental constraints
- price–performance ratio
- maintenance requirements

Most of these aspects will be discussed more fully in Chapter 5, concerning reliability, where their interrelations will be shown as well. Some aspects of ergonomics are discussed in Chapter 11. In this chapter, we shall discuss the most important specifications to which every technical system can be subjected. To avoid or reduce certain errors, a discussion of some improving measures will also be presented.

4.2 TECHNICAL SPECIFICATIONS

System specifications characterize the system's behavior under predefined conditions and are usually provided by the manufacturer. Within certain well-defined limits deviations are allowed. These accepted deviations are called *tolerances*. Specifications can cover large and different areas of conditions and in order to be able to obtain the desired information, a detailed knowledge of system behavior is required.

4.2.1 Application area

This concerns the physical quantity (or quantities) to be measured or the process to be monitored and/or measured. Other aspects include the required supply voltages such as AC or DC, the environmental conditions, and rangings. This is illustrated in the following example.

* * *

Example 4.1

A digital temperature measurement system is specified as follows:

- Temperature range −50°C to +150°C
- Operating temperature −10°C to +40°C
- Storage temperature −40°C to +75°C
- Power 100 V AC ± 10%, 120 V AC ± 10%, 240 V AC ± 4%, selected by internal switches, 50–400 Hz, 5 W max.
- Display 7.5 mm high, 3½-digit, 7-segment LCD, with decimal point location and °C indication.
- Analog output: 0–10 V for complete range and load >1 kΩ,
0–20 mA or 4–20 mA, load < 600 Ω.
- Relative humidity <90%

All other specifications are only applicable within this range.

* * *

4.2.2 Resolution

We have already discussed resolution in the previous chapter, but for completeness the definition is repeated here in a slightly different way. Resolution is defined as the smallest change at the input which can be processed by the system. Resolution is defined in different fashions, e.g.:

1 the smallest change of the input value, $\delta x_{\min}$

2 the smallest change of the input value related to the maximum input value, $\delta x_{min}/x_{max}$
3 the reverse of the above definition, or $[\delta x_{min}/x_{max}]^{-1}$

The last two definitions are sometimes called relative resolutions. There is no conformity in the application of these definitions.

* * *

Example 4.2

Firstly consider a four-digit decimal display with a fixed decimal point, one position from the right. The specified resolution is 0.1 unit. The maximum value to be displayed is 999.9, consequently the relative resolution is about 10^{-4} or 10^4.

Secondly we have to determine the relative resolution of a digital voltmeter (DVM) in voltage mode, in current mode and in resistance mode. From the specifications delivered by the manufacturer we can read the respective resolutions in the different modes. Assume they are specified as follows:

10 mV	at	±200 V
0.1 μA	at	2 mA
0.1 Ω	at	2 kΩ

Then the respective relative resolutions are:

in voltage mode:	$200/10 \times 10^{-3} = 2 \times 10^4$
in current mode:	$2/0.1 \times 10^{-3} = 2 \times 10^4$
in resistance mode:	$2 \times 10^3/0.1 = 2 \times 10^4$

Note that in each mode the same relative resolutions are obtained.

* * *

4.2.3 Accuracy and inaccuracy

The *accuracy* of a system is a figure of merit which describes the probability that the measurand is correct. A lot of confusion exists in this statement, because often the specified accuracy of a system is related to its *inaccuracy*. If it is claimed the accuracy is 1% we certainly do not mean the accuracy is 1%, but the inaccuracy is 1% and hence the accuracy is 99%. Hence the accuracy equals (1 – inaccuracy) and so they are the complement of each other.

The inaccuracy of a measuring system is strongly dependent on the way the information is displayed, because measuring ultimately means displaying a quantity on a certain symbolic scale. It will be obvious that inaccuracies are involved in every system. Causes of inaccuracies include calibration faults, tolerances of components, interference, mismatching of stages, applying the wrong method, environmental conditions, etc. We can distinguish two types of error: absolute and fractional.

1 Absolute error

(a) The absolute error can be defined as the difference between the measurand, m, and the true value, tv, or

$$\text{absolute error} = \text{m} - \text{tv} \tag{4.1}$$

(b) The absolute value of the error is defined as the absolute error, neglecting the sign, or

$$\text{absolute error value} = |\text{m} - \text{tv}| \tag{4.2}$$

(c) The fractional absolute error is defined as the ratio of the absolute error and the value of full scale deflection, fsd, often expressed as a percentage. In mathematical form

$$\text{fractional absolute error} = \frac{\text{m} - \text{tv}}{\text{fsd}} \times 100\% \tag{4.3}$$

* * *

Example 4.3

Suppose the measured value is 19.8 V, the true value is 19.9 V and full scale deflection is 100 V, then

$$\text{absolute error} = 19.8 - 19.9 = -0.1 \text{ V}$$

$$\text{absolute value of error} = |-0.1| = 0.1 \text{ V}$$

$$\text{fractional absolute error} = (0.1/100) \times 100\% = 0.1\%.$$

* * *

2 Fractional error

The fractional error is defined as the ratio of the absolute error value and the measurand, often denoted in percentage value, hence

$$\text{fractional error} = \frac{|\text{m} - \text{tv}|}{\text{m}} \times 100\% \tag{4.4}$$

* * *

Example 4.4

Applying the same values as in Example 4.3 we find

$$\text{fractional error} = |[(19.8 - 19.9)/19.8]| \times 100\% = 0.5\%$$

* * *

The manufacturer has to provide the inaccuracy of his measuring system; in its turn every measurement must be accompanied by its inaccuracy, otherwise its value is meaningless.

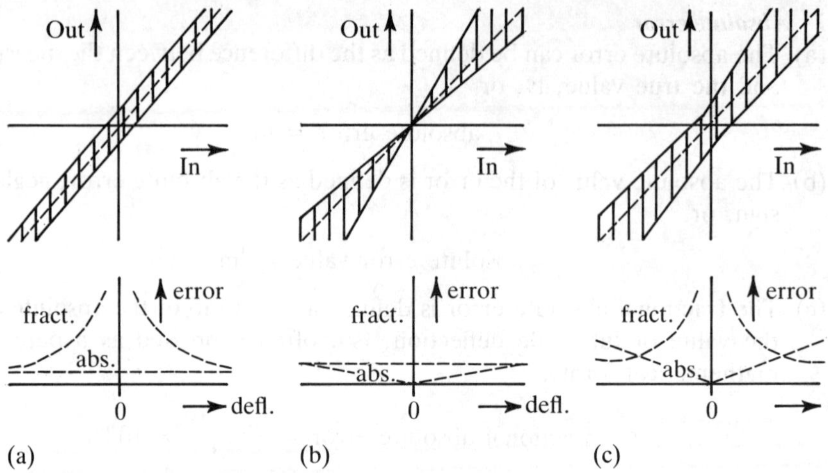

Figure 4.1 The differences between additive and multiplicative errors: (a) additive error with constant absolute error; (b) multiplicative error with constant fractional error; (c) total inaccuracy

Constant independent errors can be considered as additive errors, while errors depending on the measured value can be considered as multiplying errors. Multiplying errors often deliver a constant fractional error and are strongly related to the inaccuracy of the display method applied. This is illustrated in Fig. 4.1, where three types of errors are depicted with their functional behavior and their respective absolute and fractional behavior as a function of the deflection.

4.2.4 Sensitivity

Concerning the Wheatstone bridge in Section 3.6 we have already met a definition of sensitivity. Here we shall define sensitivity in a broader scope. For every measuring system the sensitivity can be defined as the ratio of a change y at the output, caused by a change x at the input, or

$$S = y/x \tag{4.5}$$

Sensitivity can be expressed in V/A, cm/V, or volts/div, for example, for a current-to-voltage converter, a recorder, or an oscilloscope respectively. In most cases, a measuring system appears to be sensitive to other parameters than the desired one. These sensitivities are called *parasitic* sensitivities. They are sometimes specified with respect to the sensitivity of the quantity to be measured.

* * *

Example 4.5

Consider a piezoelectric positioner with a voltage sensitivity of 1 mV/μm and a temperature sensitivity of −10 μV/K. From this it can be derived that a change of

10 μV corresponds to a change in position of 10 nm. The (parasitic) temperature sensitivity then gives −10 nm/K.

* * *

Example 4.6

Consider a simple analog temperature measurement system with a sensitivity of 100 mV/K. In the system an amplifier is involved which appears to have a temperature sensitivity resulting in an output change of 0.1 mV/K. The parasitic temperature sensitivity can be expressed in 1/1000 K/K or in 0.1 mK/K. In other words, a temperature change in another part of the system than the temperature sensor itself can affect the output.

* * *

Ultimately by electronic means, the sensitivity can be made extremely high, theoretically without any limit. But going beyond certain limits no longer makes sense, because the real physical limit in sensitivity is *noise*. Noise is just a collective noun for a number of mechanisms which are inherently present in every component. The subject of noise is dealt with in Chapters 9 and 10.

4.2.5 Linearity

As can be seen from (4.5) if strict linear behavior of the transfer characteristic is involved, the sensitivity is independent of the x value. If the system is non-linear then the sensitivity is a function of x and y. It will be clear that in practical circumstances a linear behavior is preferred, e.g., because the sensitivity can be specified with one number. This is illustrated in Fig. 4.2.

The *non-linearity* of a system can be defined as the maximum deviation of the

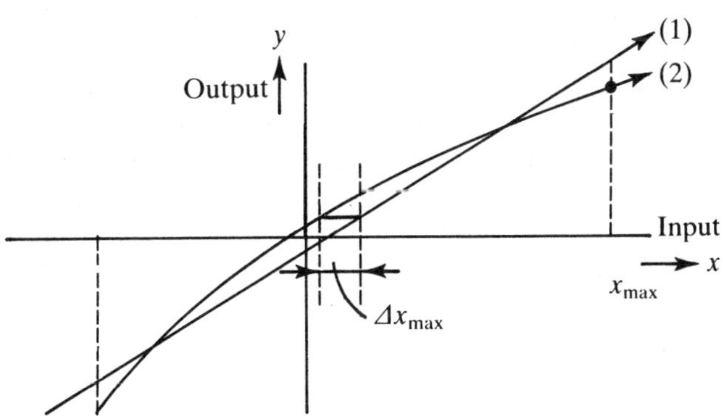

Figure 4.2 Plot of output versus input, $y = f(x)$. (2) is the real transfer characteristic, (1) an approximated transfer characteristic

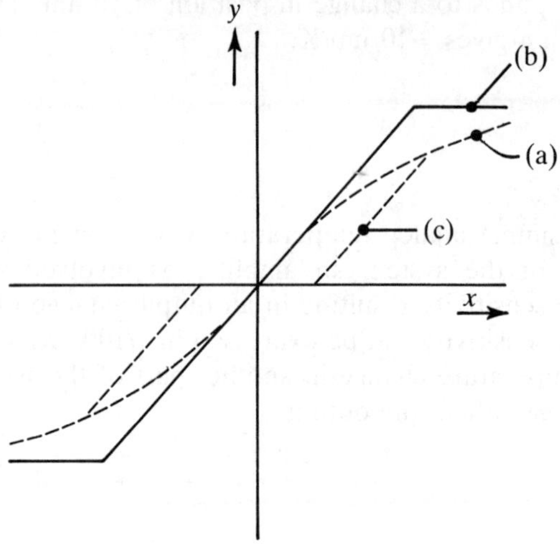

Figure 4.3 Examples of static nonlinearities: (a) saturation; (b) jamming; (c) a dead zone

specified approximated linearity. In Fig. 4.2 the non-linearity is denoted as Δx_{max}, or expressed relatively,

$$\text{non-linearity} = \frac{\Delta x_{\mathrm{max}}}{x_{\mathrm{max}}} \tag{4.6}$$

This value can be expressed as a percentage.

In Fig. 4.3 other examples of frequently occurring static non-linearity mechanisms are depicted, viz.

(a) saturation, e.g., in a transistor characteristic;
(b) jamming, e.g., the maximum supply level is achieved;
(c) dead zone, e.g., when hysteresis is involved.

These non-linearities are called static because they can occur when slowly varying signals are involved. When rapid changes are involved another type of non-linearity comes into scope, called slew-rate.

Slew-rate is defined as the time rate of change of the output voltage for a voltage step applied at the input. For instance, the slew-rate of an amplifier is specified as 100 V/μs. For sine wave-shaped signals an alternative definition is used; it is the maximum frequency at which the peak output voltage swing can be obtained without distortion. The effect of slew-rate is shown in Fig. 4.4.

As illustrated, changes in the input are time-limited at the output; in other words, the slope is always less than $\pi/2$.

4.2.6 Offset and drift

Offset can be defined as the deviation of the output signal from zero when the input is zero. A difference is made between input offset and output offset. Drift can be

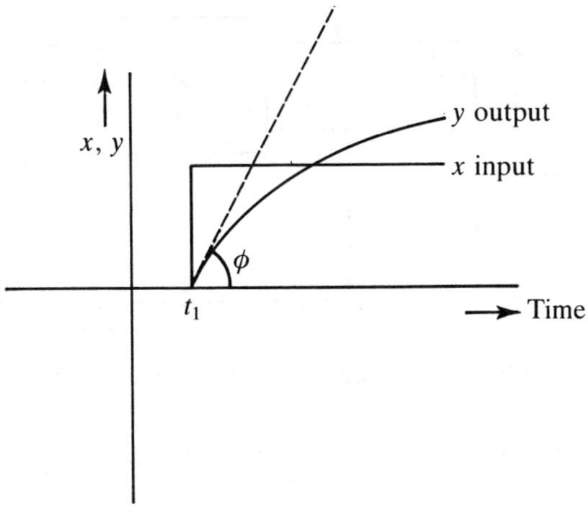

Figure 4.4 The effect of slew rate of an amplifier

defined as a slow variation of a zero-setting. We will not go into detail here, because a comprehensive discussion of offset and drift is presented in Chapter 7. Here we provide an example of the specifications of an operational amplifier.

* * *

Example 4.7

Every databook of operational amplifier specifications contains information concerning the offset and drift of a specific amplifier. For instance, for instrumentation amplifiers the following specifications are known:

	typical	*max*
input offset voltage	±40.0 μV	100.0 μV
drift	± 0.4 μV/K	1.8 μV/K
long-term stability	0.4 μV/month	2.0 μV/month
offset current	10.0 nA	75.0 nA
slew-rate	1.7 V/μsec	1.9 V/μsec

* * *

4.2.7 Rejection factor

In Section 3.3.1 we introduced the concept of the differential amplifier in a theoretically ideal form with infinitely high open-loop gain, infinitely high input impedance, zero input current and zero output impedance. As a building block it is usually called an operational amplifier. Of course, a real operational amplifier does not have these ideal properties and to characterize the real behavior some characteristic figures of merit are used.

We have seen also that one of the most powerful characteristics of a

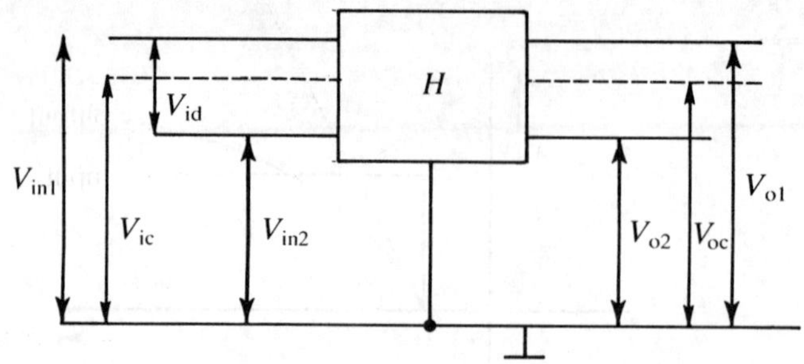

Figure 4.5 Operational amplifier with a double-ended output as a black box

differential amplifier is its capability to amplify a very small voltage difference between two relatively large signals, without amplifying the two large signals. The advantage of such a building block will be quite clear, especially in circumstances where a large common-mode signal is present at both inputs, no matter whether it is a signal level or interference.

Let us now consider a differential amplifier with a differential input and output, as a black box. This is illustrated in Fig. 4.5 and is the most general case.

For the input voltages the following definitions will be applied:

$$V_{id} = V_{in1} - V_{in2} \tag{4.7}$$

which is called the differential-mode input voltage. The common-mode input voltage is defined as

$$V_{ic} = \tfrac{1}{2}(V_{in1} + V_{in2}) \tag{4.8}$$

When the amplifier does not have perfect symmetry, for whatever reason, we can state that a common-mode voltage at the input can generate a differential-mode voltage and a common-mode voltage at the output. In the same way a differential-mode signal at the input can generate a differential-mode and a common-mode signal at the output, or

$$V_{od} = A_{dd} V_{id} + A_{cd} V_{ic} \tag{4.9a}$$

$$V_{oc} = A_{dc} V_{id} + A_{cc} V_{ic} \tag{4.9b}$$

where:

$$V_{od} = V_{o1} - V_{o2} \tag{4.10a}$$

$$V_{oc} = \tfrac{1}{2}(V_{o1} + V_{o2}) \tag{4.10b}$$

and by definition

A_{dd} is the differential to differential gain
A_{dc} is the differential to common-mode gain
A_{cd} is the common-mode to differential gain
A_{cc} is the common-mode to common-mode gain

and by definition

$$\left.\begin{aligned} A_{dd} &= V_{od}/V_{id} \text{ at } V_{ic} = 0 \\ A_{dc} &= V_{oc}/V_{id} \text{ at } V_{ic} = 0 \\ A_{cd} &= V_{od}/V_{ic} \text{ at } V_{id} = 0 \\ A_{cc} &= V_{oc}/V_{ic} \text{ at } V_{id} = 0 \end{aligned}\right\} \quad (4.11)$$

From (4.9) we see that generally the differential-mode output voltage and the common-mode output voltages are the superposition of two contributions. The ideal situation exists when A_{cd} equals zero. Note that the gain factors A_{dc} and A_{cc} are of less importance, although both should be as small as possible. With help of (4.9) we can now define the following two figures of merit for an operational amplifier:

1 The *discrimination factor, F*, which is defined as the ratio of the differential-to-differential gain and the common–mode-to-common–mode gain, or

$$F \stackrel{\text{def}}{=} \frac{A_{dd}}{A_{cc}} \quad (4.12)$$

The discrimination factor, F, is of importance only when a double-ended output is involved, as is illustrated in Fig. 4.5. The factor F makes no sense when no common-mode signal at the output is present. But a common-mode voltage at the input can always generate a differential mode signal at a single-ended output. The most commonly applied operational amplifier configurations have a single-ended output.

2 The *rejection factor*, or more precisely the *common-mode rejection factor* (*CMRR*), is defined as the ratio of the differential to differential gain and the common-mode to differential gain, or

$$H \stackrel{\text{def}}{=} \frac{A_{dd}}{A_{cd}} \quad (4.13)$$

or in dB

$$20\log|H| = 20\log\left|\frac{A_{dd}}{A_{cd}}\right| \quad (4.14)$$

The CMRR is normally expressed in dB and is the most crucial factor of a differential amplifier. It describes the ability of the amplifier to eliminate a common-mode signal connected at the input. This is strongly related to the *symmetry* of the amplifier itself and its input configuration. It will be clear that the CMRR should have a large value, e.g., 120 dB is a good value, and 70 dB is rather poor. Sometimes the reciprocal value of H is used.

Finally, it will be clear that the differential to differential gain, A_{dd} must be high, in contrast to A_{cc}, which must have an extremely low value. Thus a high value for F is preferred. It should be noted that F always can be improved, while the rejection factor H, as a system-dependent quantity, is a fixed number. The CMRR H is always provided by the manufacturer, whereas F is not.

When the CMRR H is considered at the condition of the same output voltages both for A_{dd} and A_{cd}, then the CMRR can be expressed directly in voltages. By substitution of (4.11) we then find for H with $A_{dd} = V_{od}/V_{id}$ and $A_{cd} = V_{od}/V_{ic}$,

$$H = \left.\frac{V_{ic}}{V_{id}}\right|_{V_{out}=\text{const.}} \tag{4.15}$$

As will be shown in Chapter 8, it is important that the conditions under which the CMRR is measured are specified. In that chapter we shall repeat some important aspects of the differential amplifier. An example of a series system will illustrate the above given concepts in some more detail.

* * *

Example 4.8

In Fig. 4.6 a series system is drawn for which the CMRR must be determined. Assume the first system has a CMRR of H_1, a discrimination factor F_1, and a gain factor A_{dd1}. The second system has a single-ended output and has a CMRR H_2 and a gain A_{dd2}. Due to the single-ended output there is no F_2 involved.

Applying (4.9a) to the differential output voltage of amplifier 1 we can write

$$V_{od} = A_{dd1} V_{id} + A_{cd} V_{ic} \tag{4.16}$$

or

$$V_{od} = A_{dd1}(V_{id} + V_{ic}/H_1) \tag{4.17}$$

because $1/H_1 = A_{dd1}/A_{cd1}$.

For V_{oc} we find with (4.9b)

$$V_{oc} = A_{dc1} V_{id} + A_{cc1} V_{ic}$$

or

$$V_{oc} \approx A_{cc1} V_{ic} \tag{4.18}$$

because $A_{dc1} V_{id} \ll A_{cc1} V_{ic}$

For the second stage we find

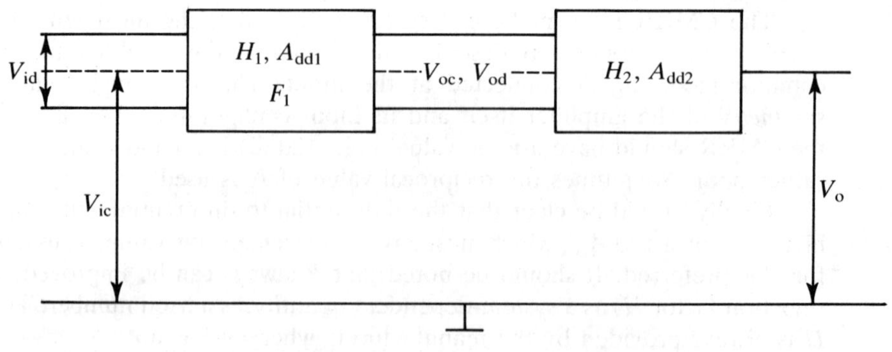

Figure 4.6 Series connection of two systems

$$V_o = A_{dd2}V_{od} + A_{cd2}V_{oc} \tag{4.19}$$

Substitution of (4.17) and (4.18) in (4.19) and rearranging, we find for the output voltage V_o of the second stage

$$\frac{1}{H_t} = \frac{1}{H_1 + F_1 H_2} \tag{4.20}$$

* * *

Example 4.9

A system with a specified CMRR of 10^4 is used to compare two voltages of about 10 V. Then the lowest voltage difference which can be detected with this system is determined by the rejection factor. This implies that a common-mode input of 10 V results in an output of ±1 mV. In other words, the accuracy is guaranteed to $10/10^4 = 1$ mV.

If the same system is applied to measure a voltage difference of 10 mV with a common-mode signal of again 10 V, then the inaccuracy is again 1 mV, or the inaccuracy as a percentage can be derived from

$$V_{in1} - V_{in2} \pm V_{CM}/H = 10 \pm 1 \text{ mV}$$

and this equals a rather poor accuracy of 90%, (inaccuracy of 10%).

* * *

4.2.8 Dynamic range

The (dynamic) range of a system is defined as the ratio of the maximum and minimum signal level which can be handled by the system within a specified accuracy.

The maximum signal level will be constrained by the operational limits in which the transfer of the signal is defined. For instance, the supply voltage will form a natural limit. The minimum signal level to be handled is limited by the so-called boundary sensitivity, again within specified accuracy. Natural limits here are for instance thermal voltages, noise and hysteresis. Usually the dynamic range is expressed in a single number or in dB, for instance 10^3 equals 60 dB.

Some aspects of the dynamic range can be clarified best by discussing an example in which the measurement of a DC voltage is performed.

* * *

Example 4.10

For the DC voltage measurement system shown in Fig. 4.7, with a required fractional inaccuracy of 4%, the dynamic range must be determined. The system consists of four stages, namely:

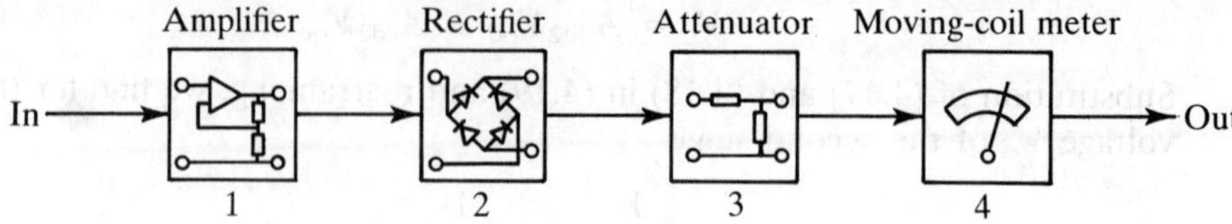

Figure 4.7 Measurement set-up for determining the dynamic range of DC voltages (courtesy Delft University Press)

1 An amplifier stage with a gain setting between 1 and 1000 times. The equivalent noise voltage is 10 μV, and the maximum voltage range for both input and output is 10 V.
2 A bridge rectifier consisting of four diodes with a dead zone of 0.1 V. The maximum allowed reverse voltage is 100 V.
3 An attenuator, composed of resistors to match the moving-coil meter to the system.
4 A moving-coil meter with an error class of 0.5, which shows full-scale deflection at 100 mV and 100 μA.

The claimed fractional inaccuracy requires an inaccuracy of less than 1% for every stage involved. We will consider every stage in detail. For the first stage we can derive from the data that a required inaccuracy of 1% equals an inaccuracy of 10 μV and hence the lowest acceptable value to be measured is 100 times larger, or 1 mV. When the gain setting is 1000 the output voltage is 1 V and this value compared with the allowed 10 V maximum for input and output delivers a dynamic range for the amplifier's stage of merely $10/1 = 10$. When the gain setting is only 1 we have a dynamic range of $10/10^{-3} = 10^4$.

The bridge rectifier in the second stage has a dead zone of 0.1 V, and if here again a 1% inaccuracy is claimed, this dead zone equals the 1% inaccuracy and the lowest acceptable value therefore is 10 V. The maximum allowed reverse voltage is 100 V and so the dynamic range of this stage is $100/10 = 10$.

As can be seen already from these two stages the voltage matching fails to be completely correct. This is illustrated in Fig. 4.8. To solve this matching problem a second amplifier stage might be introduced, but this requires a supply voltage of 100 V and so is not very attractive. A better solution is to diminish the rectifier's dead zone by a factor of 10. This can be realized by introducing a small voltage source of 0.1 V series-connected with the rectifier's output. The new threshold value of the rectifier becomes 0.01 V which then forms the 1% inaccuracy. This delivers a new acceptable lowest value of 1 V and so a better voltage matching is created.

The third stage offers no problems at all and is merely required to match the moving-coil meter. The lowest value will be determined by thermal voltages which have an order of magnitude of 1 μV, hence 100 μV is the lowest acceptable value with an inaccuracy of 1%. The upper limit can be claimed to be 100 V, then the dynamic range is $100/(100 \times 10^{-6}) = 10^6$. In this stage, complete voltage matching is easily obtained.

The last stage is the moving-coil meter and will be the most crucial one. An

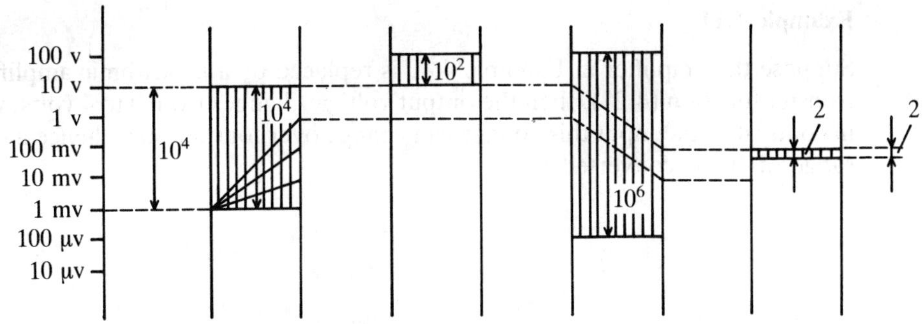

Figure 4.8 Schematic diagram of the dynamic range for the four stages

error class of 0.5 implies an absolute inaccuracy of 0.5% of full-scale deflection, which equals here 0.5 mV. Then the lowest acceptable value within the claimed inaccuracy is 100 times larger, or 50 mV. This means a dynamic range of only 100/50 = 2. It will be obvious that the last stage determines the overall accuracy. This is also shown in Fig. 4.8.

For convenience the results are summarized in Table 4.1.

Table 4.1

	Error 1%	*Lowest acceptable value (= 100 × error)*	*Upper acceptable value*	*Dynamic range*
amplifier	10 µV	1 mV	10 V	$10 - 10^4$
rectifier	0.1 V	10 V	100 V	10
attenuator	1 µV	100 µV	100 V	10^6
m.coil meter	0.5 V	50 mV	100 mV	2

* * *

Of course we can allow greater inaccuracy to improve the dynamic range, but that is not the right way to handle things. To solve this problem, logarithmic compression can be applied, which requires a type of logarithmic amplifier. The principle of operation is based on diminishing the ratio of the maximum and minimum available signal, as discussed in Chapter 3.

For instance the transfer function of a logarithmic amplifier is described by

$$V_o = \log^{10}|V_{in}/V_1| \tag{4.21}$$

in which V_{in} is the input voltage and V_1 is the lowest acceptable input voltage for accuracy reasons. Then it can easily be seen that the ratio of maximum and minimum voltage is logarithmically reduced.

* * *

Example 4.11

Suppose the amplifier in Example 4.10 is replaced by a logarithmic amplifier with transfer function (4.21), then the output voltage changes from 0 to 4 volts, whereas in contrast to the input this voltage can change over four decades, hence a dynamic range of 10^4 is converted into 4.

* * *

4.2.9 The number of measurements per unit time

This topic provides the main reason why it is becoming common to implement computing power in measurement systems. In recent processes hundreds of measurements often have to be executed in a very short time. Collected data have to be edited, calculated, stored, compared with set-points and, after conversion, have to be sent out again. Finally calibration, automatic nulling and autoranging are often highly desired features to implement in a measurement system. It will be quite clear that this is impossible to perform by hand. There is an increasing need for fully automated multi-channel measurement systems. Aspects of these systems are discussed in Chapters 2 and 3.

To emphasize the power of a modern measurement system we compare a classical configuration as illustrated in Fig. 4.9 with a sophisticated system as depicted in Fig. 4.10. The differences are summarized in Table 4.2.

The classical measurement system consists of a one-channel configuration and is composed of at least three stages. The first stage represents a type of transducer; the second stage consists of a signal modifier or conditioner which for instance implies a filter, and an amplifier; the third stage is a type of display.

In Fig. 4.10 the multi-channel arrangement shows n different input transducers n amplifiers, an interface, a multiplexer, or scanner, a low-pass filter, a sample and hold function, an A/D converter, a hard or soft disk-drive unit, a printer and plotter, a keyboard, a video-display and a computer. The amplifiers can be made of so-called modular blocks and are easily adapted to the quantity to be measured. The amplifiers and filters can be made software programmable.

A good example of a multi-channel computer-powered measuring system is the flight computer of an aircraft in which all of the above mentioned features can be found.

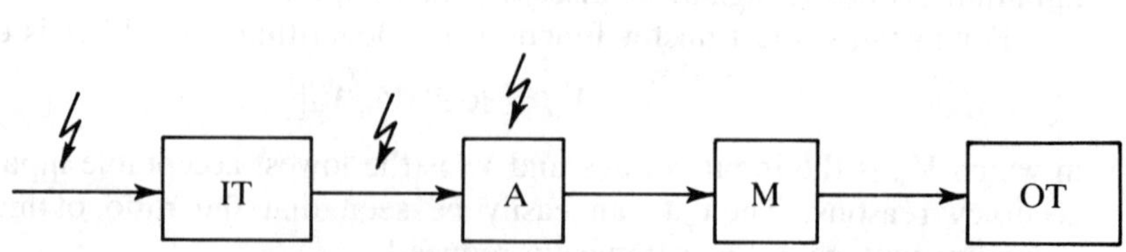

Figure 4.9 Classical measurement set-up (arrows represent interference)

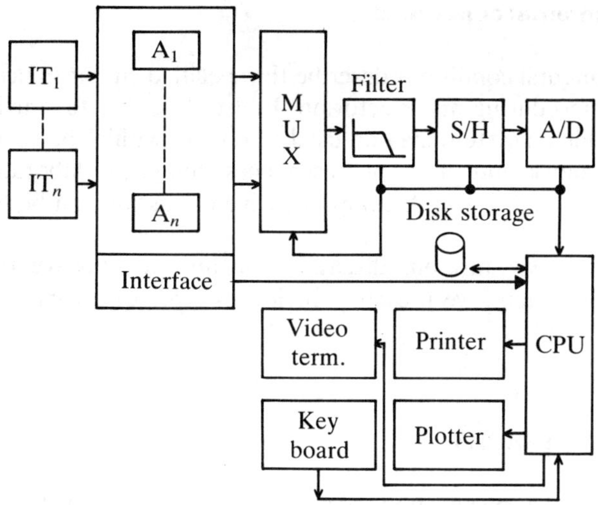

Figure 4.10 Sophisticated measurement configuration for *n* different input signals

Table 4.2 Comparison of conventional and computer-based systems

Item	*Conventional system*	*Computer based system*	*Comments*
input transducer	1	1–*n*	may be different
transducer supply voltage	fixed	var.	programmable
scanner/mux	none	1	
amplifier	1	1–*n*	programmable
signal conditioner	–	1–*n*	programmable
modifier	1	1–*n*	programmable
filtering	1	1–*n*	programmable
sample & hold	–	1–*n*	
A/D and D/A conversion	–	1–*n*	8–16 bits
storage	–	1–*n*	hard and soft
interfacing	–	1	1–*n*
output transducer	1	1–*n*	all types
multiplier	–	1	1
calibration	–	*n*	programmable
communication	–	1	serial/par.
auto ranging	–	1–*n*	all channels
auto-nulling	–	1–*n*	all channels
rectifier	1	1–*n*	
calculations	–	any	
number of meas./unit time	1	1–*n*	clock-freq. dep.

4.2.10 Environmental conditions

Environmental conditions describe the specified circumstances to which the system is subjected during manufacturing. It also describes the environmental conditions under which a system may be used to remain within the specified regions: aspects such as shock and vibration resistances, humidity, laboratory or missile launch conditions, and so on. A comprehensive discussion can be found in Chapter 5.

Before we start our discussion on how to improve the specification of an electronic system, we present a review of types of errors and their causes.

4.3 TYPES OF ERRORS

Related to the specifications of a system are of course its limits, so we can say every system has a limiting accuracy due to certain errors. Every stage will introduce some errors and to classify these errors three types can be distinguished:

1 systematic errors
2 conditional errors
3 stochastic errors

4.3.1 Systematic errors

A systematic error is one that has a constant character and is system- or process-dependent. Good examples of systematic errors are

(a) the offset of an amplifier;
(b) a non-linear behavior of the process;
(c) an undesired but yet required fixed scaling factor.

Such errors can be eliminated by an automatic nulling, or a calibration procedure applying, for instance, a digital memory. Sometimes a ratio measurement, as discussed in Chapter 2, can offer good results. Systematic errors are not very difficult to eliminate. Offset errors will be discussed in detail in Chapter 7.

4.3.2 Conditional errors

Conditional errors are those which are caused by external influences such as electromagnetic interference (EMI) and electromagnetic pulses (EMP). Conditional errors are strongly dependent on the environmental and operational conditions the system is operating in. Measures to eliminate this type of error are:

1 The choice of components which are less sensitive to the interfering source to which the system is subjected.
2 Isolation, guarding, and stabilizing the system from the interfering source.
3 Compensation of the interference source by an equal yet reverse signal. Applying a two-sensor configuration in one system can be considered as a compensation technique.

Generally the operational conditions have to be specified. Some types of conditional errors and their elimination will be discussed in more detail in Chapter 8.

4.3.3 Stochastic errors

Stochastic errors are caused by a stochastic process which basically has to be considered as a fundamental phenomenon in applied materials and components. Noise is one of the best examples of a stochastic process; others are the temperature of a material, the pressure in a gas volume and the current flow in a material. Some measures to reduce the noise influence are:

1 Reducing drastically the temperature of the noise-sensitive stages.
2 Applying a cross-correlation technique using the following expression:

$$R_{xy} = \lim_{T\to\infty} \frac{1}{T} \int_0^T x(t)y(t-\tau)\,\mathrm{d}t \tag{4.22}$$

where

$x(t)$ is the received signal
$y(t-\tau)$ is a known signal shifted over a time τ

This technique is a so-called *posterior* technique, because it is performed after the signal detection; this is in contrast to a *prior* technique which can be executed beforehand, as stated below.
3 Applying a microprocessor-based sampling technique in some circumstances can reduce noise behavior considerably. This method can be called a prior technique because it is performed before every measurement is realized. An example will be described in Section 4.5.
4 A frequency shift to a higher frequency band where less $1/f$ noise is present.
5 Another sophisticated noise-reduction technique might be the application of optical recording techniques. The application has yet to be developed because up to now only 'write once read only' (WROM) disks are available. The storage capacity is enormous. For instance an 8-inch video long-play record can store 4000 images, which is the equivalent of several gigabits of information.

Finally, in Fig. 4.11 five different representations of errors are shown, which can be assigned to one of the above categories.

In the following sections we shall discuss some widespread improving measures for electronic systems.

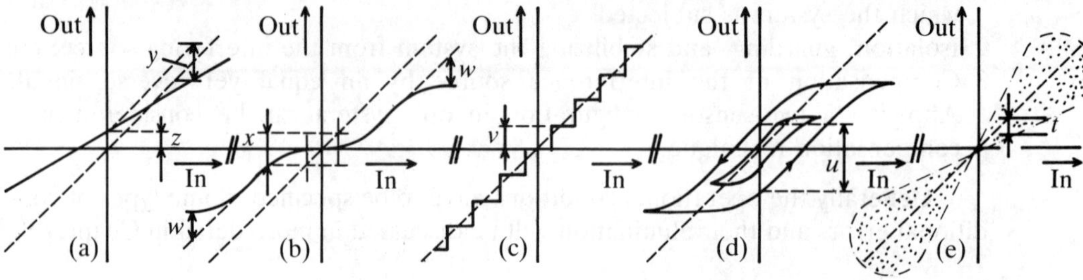

Figure 4.11 (a) Zero error *z* with scaling error *y*; (b) dead zone area *x* with saturation area *w*; (c) quantization error *v*; (d) hysteresis *u*; (e) stochastic error *t*

4.4 IMPROVING MEASURES FOR ELECTRONIC SYSTEMS

In this section we will discuss some improving measures for electronic systems. The errors that occur are common to all instruments, but it depends on the signal-to-noise ratio and other required specifications as to which special precautions must be taken to reduce these errors. This is application-dependent. Even costs aspects can play a role here, because every precaution will raise the costs of a system.

4.4.1 The feedback principle

In Chapter 3 we have discussed the principle of feedback. This is again illustrated in Fig. 4.12. Assume a linear active element has a transfer function $H(j\omega)_1$ and a time constant τ_1. In the feedback loop an attenuator K is implemented. The input signal s_i is compared with the output signal $V_f = Ks_o$, thus supplying a fraction of the output signal to the input of the amplifier. In the following discussion we further write $H(j\omega)_1 = H_1$. The resulting signal is called the error signal s_f and equals

$$s_f = s_i - Ks_o \tag{4.23}$$

and arrives at the amplifier's input; then the resulting output signal s_o equals

$$s_o = H_1(s_i - Ks_o) \tag{4.24}$$

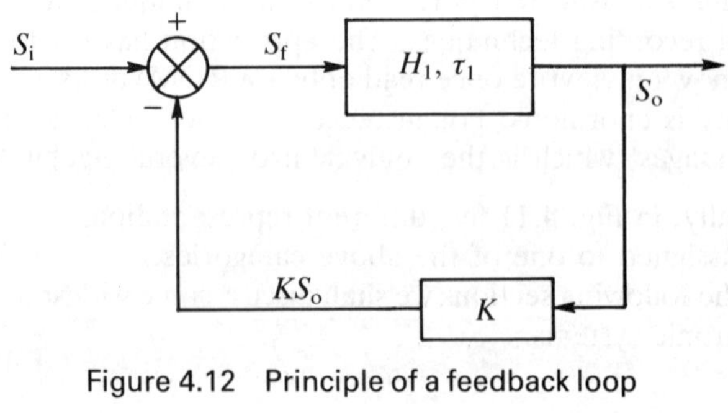

Figure 4.12 Principle of a feedback loop

and the transfer function H_t becomes

$$H_t = \frac{s_o}{s_i} = \frac{H_1}{1 + H_1 K} \tag{4.25}$$

or

$$H_t = \frac{1}{K} \frac{1}{1 + 1/KH_1} \tag{4.26}$$

From this expression some important conclusions can be drawn:

1 If H_1 is a very large value then H_t will approach $1/K$ and if H_1 is large enough, H_t will become independent of the parameters of H_1 itself.

2 The relative sensitivity for changes in H_1 is defined as

$$S_{H1} = \frac{\delta H_t/\delta H_1}{H_t/H_1} = \frac{1/(1 + H_1 K)^2}{1/(1 + H_1 K)} \tag{4.27}$$

or

$$S_{H1} = \frac{1}{1 + H_1 K} \tag{4.28}$$

As can be seen from the expression for the sensitivity a considerable decrease for changes in H_1 can be realized. The consequence of this feedback principle is that a very accurate transfer function can be realized which is completely determined by K, often composed of passive components as resistors.

3 The third conclusion which may be drawn will become clear when in (4.25) a frequency-dependent first-order transfer function is substituted for H_1. Assume H_1 is given by

$$H_1 = \frac{A_0}{1 + j\omega\tau_1} \tag{4.29}$$

then it can be found that the Bode plot for the amplitude versus frequency of the total transfer function H_t compared with the transfer function of H_1 will appear to have an increased bandwidth. This effect is illustrated in Fig. 4.13.

* * *

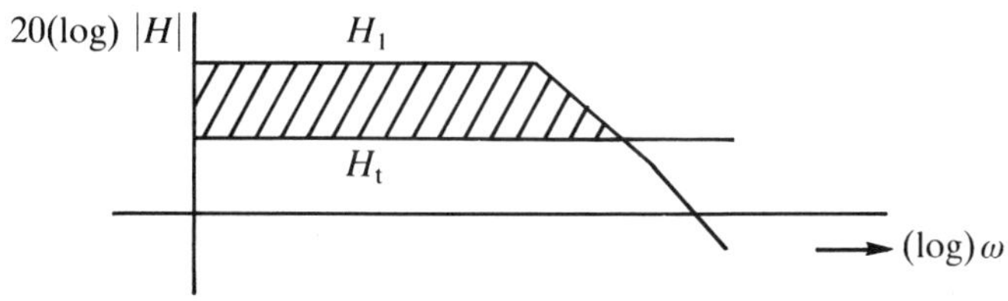

Figure 4.13 Bode plot versus frequency for open-loop gain H_1 and for the closed-loop feedback-controlled loop H_t

Example 4.12

Suppose a feedback loop is composed of an amplifier and a resistive attenuator consisting of two resistors connected in parallel to the output. The fraction of the output signal which is supplied to the input is 10^{-3}. The amplifier has a gain factor of 10^6; then applying (4.25) we find for the total transfer

$$H_t = \frac{1}{10^{-3}} \frac{1}{1 + 1/(10^6 \times 10^{-3})} = 999$$

Now assume the gain factor of the amplifier decreases by 50% owing to thermal effects and deterioration, then the gain factor becomes 5×10^5 and the transfer function H_t yields

$$H_t = \frac{1}{10^{-3}} \frac{1}{1 + 1/(5 \times 10^5 \times 10^{-3})} = 998$$

which means a reduction of only 0.2%. The transfer function, applying feedback, has become dependent on external elements, and thus almost completely independent of the amplifier's characteristics.

* * *

The above discussed type of correction applying feedback improving the accuracy is only true for *multiplicative* errors and is not valid for *additive* errors. This is clarified in Figs. 4.14 and 4.15.

In Fig. 4.14 an additive disturbance x_D is acting on the input of an active element with the same transfer function H_1. By superposition for the total output signal S_o is found to be

$$S_o = S_i \frac{H_1}{1 + H_1 K} + x_D \frac{H_1}{1 + H_1 K} \tag{4.30}$$

or

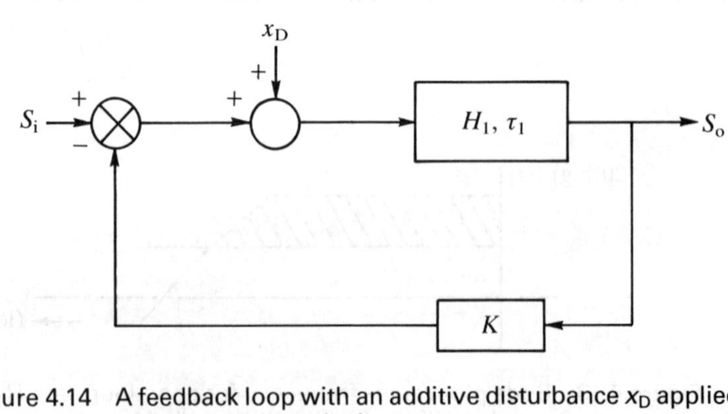

Figure 4.14 A feedback loop with an additive disturbance x_D applied at the input

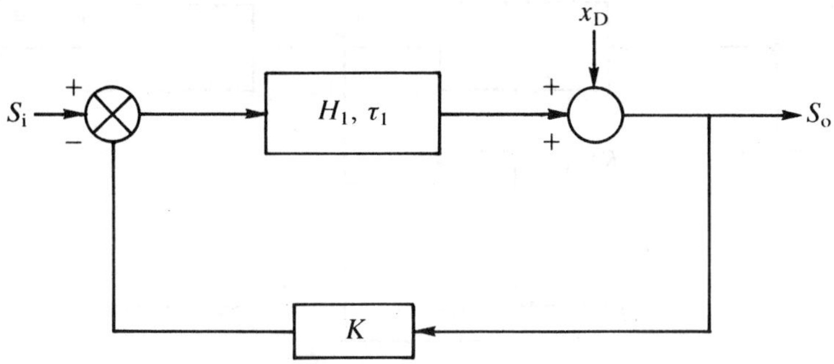

Figure 4.15 Interfering signal acting on the output

$$S_o = \frac{1}{K}\frac{1}{1 + 1/H_1K}[S_i + x_D] \tag{4.31}$$

and the signal-to-noise ratio is not improved.

In Fig. 4.15 an interfering signal x_D is applied at the output of the active element. Then considering the effect of x_D only and supposing $S_i = 0$,

$$x_D - H_1 K s_o = S_o \tag{4.32}$$

or

$$S_o = \frac{x_D}{1 + H_1 K} \approx \frac{x_D}{H_1 K} \tag{4.33}$$

This is true as long as $H_1 K \gg 1$ and compared with the original signal-to-noise ratio, the reduction will appear to have a maximum value of $1/H_1$, which is again very small.

4.4.2 Feedforward coupling

Feedforward coupling is another possible measure for improving the accuracy in a linear amplifier stage. The principle of operation is illustrated in Fig. 4.16, in which two amplifiers with transfer functions H_1 and H_2 and time constants τ_1 and τ_2 are shown. At both amplifiers a feedback coupling is applied and these are supposed to be equal with a transfer ratio of $1/K$, but in reality they will differ slightly. Suppose the transfer rates are respectively

$$H_1 = \frac{1}{K}(1 - \delta) \tag{4.34}$$

and

$$H_2 = \frac{1}{K}(1 - \theta) \tag{4.35}$$

From Fig. 4.16 going from right to left we can derive the following equations:

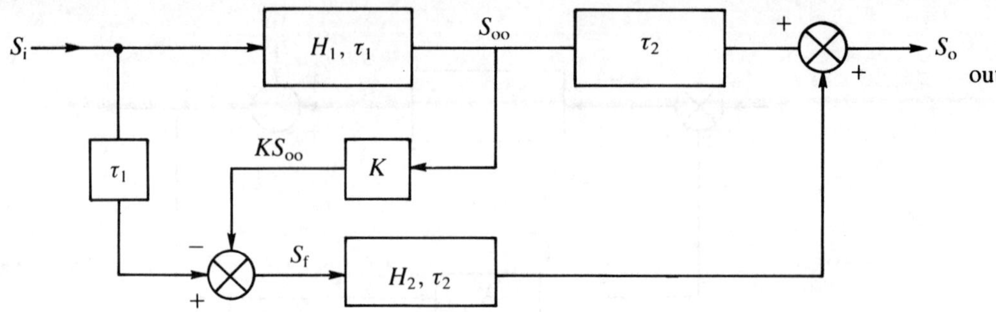

Figure 4.16 Principle of operation of feedforward coupling

$$S_o = S_f H_2 + S_{oo} \tag{4.36}$$

$$S_f = S_i - KS_{oo} \tag{4.37}$$

$$S_{oo} = S_i H_1 \tag{4.38}$$

Substituting (4.37) and (4.38) in (4.36) we find

$$S_o = (S_i - KS_{oo})H_2 + S_i H_1 \tag{4.39}$$

Again substituting (4.38) in (4.39) and rearranging gives

$$S_o = S_i(H_1 + H_2 - KH_1H_2) \tag{4.40}$$

then the total transfer function substituting (4.34) and (4.35) in (4.40) gives

$$H_t = \frac{S_o}{S_i} = \frac{1}{K}[1 - \delta + 1 - \theta - 1 + \delta\theta] \tag{4.41}$$

or the total transfer function H_t is

$$H_t = \frac{1}{K}(1 - \delta\theta) \tag{4.42}$$

If both fractions δ and θ are in the order of 1% then the total error is 10^{-4}, but it is not very difficult to achieve an accuracy of 10^{-6}.

There are some restrictions to be mentioned:

1 The accuracy of the subtraction or comparison and additive errors at the input of the second amplier.
2 The accuracy of the function of K.
3 In practical circumstances the accuracy of the addition at the output will form the most crucial restriction.
4 The high-frequency accuracy is restricted by the differences in the time delay τ which will result in adding at the wrong moments. These time delay differences can be compensated for by the depicted time delays τ_1 and τ_2 in both parallel loops.

Typical applications of this principle can be found in the field of gain improvements in very high-frequency circuits and controlled processes with dead time zones.

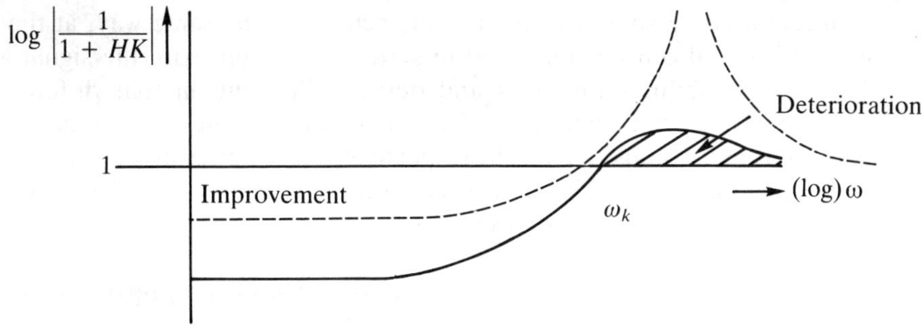

Figure 4.17 Illustration of the Westcott theorem for a second- or higher-order transfer function

It is interesting to note here that it can be proven that, for systems of second and higher order, applying feedback, apart from a frequency range where the accuracy improves, there also exists a frequency range where the accuracy deteriorates. This mechanism is expressed in the so-called *Westcott theorem* which says that integrating the logarithm of the absolute value of the frequency-dependent factor of the transfer function over the complete frequency range equals zero, or

$$\int_0^{\infty} \log|1/(1 + H(\omega)K(\omega))|\,d\omega = 0 \tag{4.43}$$

Preferably, the system will be designed maintaining the total signal range in the negative part of $\log|1/(1 + HK)|$. This is illustrated in Fig. 4.17.

Finally we make some general remarks concerning feedback.

1 Total accuracy in the subtraction of the input signal and the feedback signal is required; any fault in this part will not be eliminated.
2 In electronic systems a stable feedback can easily be realized, whereas this is not as easily applicable in non-electrical systems. The stability criterion for electrical systems requires that the loop gain $|KH_1|$ should remain always less than 1 when the argument $\phi\ (KH_1)$ equals π radians.
3 Sometimes an undesired feedback is realized especially in the case of high-frequency behavior. This can occur via a capacitive coupling between input and output and via the supply connections. We will return to this Chapter 8.

4.4.3 The influence of feedback on input and output impedance

In this section we shall discuss the influence of the applied feedback configuration on input and output impedances. The principle of feedback is that a part of the output signal is coupled back into the input. The way the coupling is realized at the input as well as at the output determines the final impedance which will be seen at the input and output.

The signal at the output can be sampled in two ways, namely by a series connection and a shunt or parallel connection. In the same way, at the input the feedback signal can be connected in series or in shunt with the signal generator. The two possibilities for input and output will result in four different possible configurations, with their specific characteristics. Some authors prefer to speak of current and voltage feedback, respectively for a series and a shunt connection. This is summarized in Table 4.3, and Fig. 4.18 illustrates all four possibilities, realized with passive impedances only.

Table 4.3 Survey of the four possibilities of applying feedback

series input–series output
series input–shunt output
shunt input–series output
shunt input–shunt output

A general treatment requires a calculation of the input voltage V_{it} in all four cases. This voltage is offered to the input terminals AB of the amplifier and can be

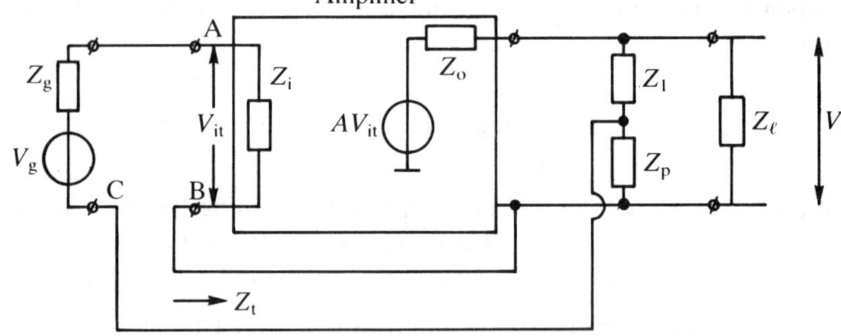

(a)

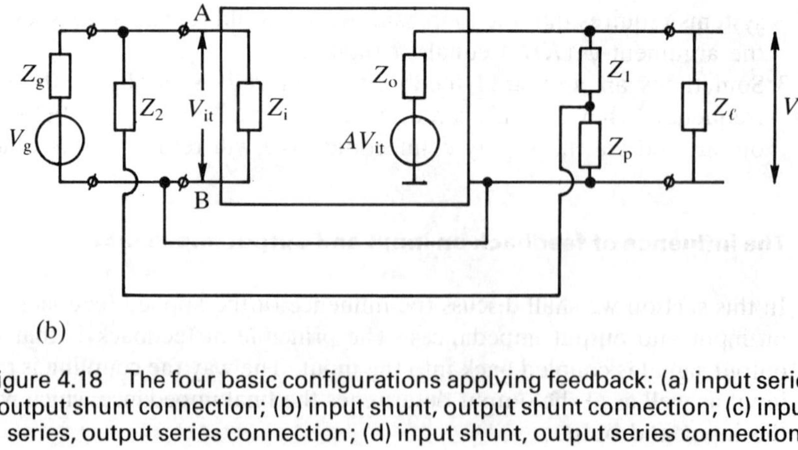

(b)

Figure 4.18 The four basic configurations applying feedback: (a) input series, output shunt connection; (b) input shunt, output shunt connection; (c) input series, output series connection; (d) input shunt, output series connection

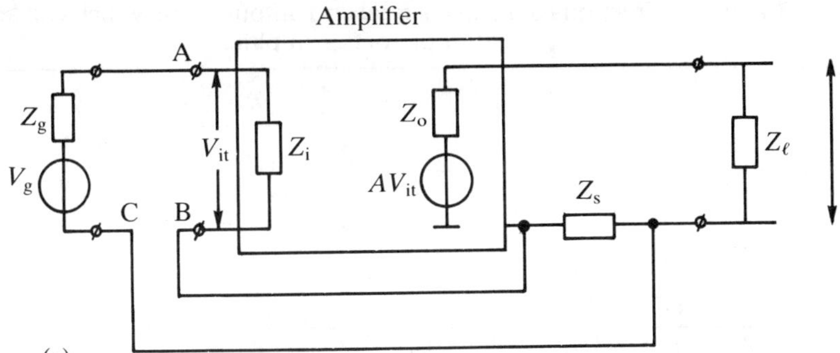

(c)

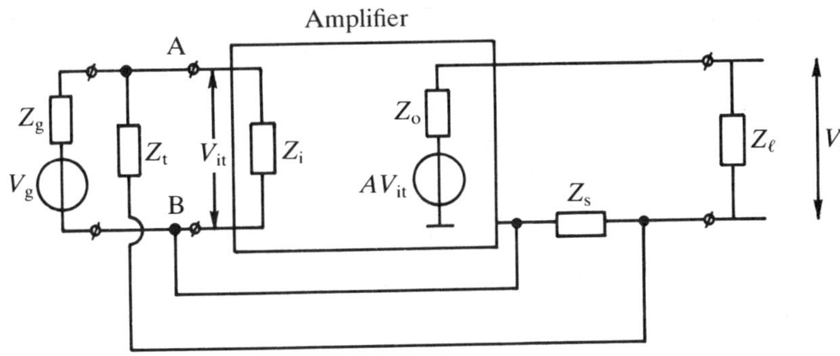

(d)

considered as the superposition of the fraction of the output voltage V_o and the fraction of the signal source V_g, or

$$V_{it} = kV_g + \beta V_o \tag{4.44}$$

where

- k describes the fraction of V_g arriving at the input terminals AB of the amplifier,
- β describes the fraction of the output voltage arriving at the input terminals AB.

It will be clear that β is composed of two fractions, one fraction originating from the output connection and one fraction developed in the input connection, each time dependent on the applied configuration. These fractions are often denoted by δ for the output and τ for the input, hence $\beta = \tau\delta$. We will provide τ and δ with indices s for series and p for a shunt configuration. Further we will assume that Z_t represents the impedance seen at the input nodes AB and CB if we look at the output. A summary is given in Table 4.4. The derivation of these expressions will be omitted.

For V_{it} the expression

$$V_{it} = V_o/A \tag{4.45}$$

Table 4.4 Fractions of source signal and output signal which will arrive at the input of the amplifier

k	τ	δ	β
$k_s = \frac{Z_i}{(Z_i + Z_g + Z_t)}$	$\tau_s = \frac{Z_i}{Z_g + Z_t + Z_i}$	$\delta_s = Z_s/Z_1$	$\tau_s\delta_s$
$k_s = \frac{Z_i}{(Z_i + Z_g + Z_t)}$	$\tau_s = \frac{Z_i}{Z_g + Z_t + Z_i}$	$\delta_p = Z_p/(Z_p + Z_1)$	$\tau_s\delta_p$
$k_p = \frac{Z_i // (Z_2 + Z_t)}{Z_g + Z_i // (Z_2 + Z_t)}$	$\tau_p = \frac{Z_i // Z_g}{Z_i // Z_g + Z_2 + Z_t}$	$\delta_s = Z_s/Z_1$	$\tau_p\delta_s$
$k_p = \frac{Z_i // (Z_2 + Z_t)}{Z_g + Z_i // (Z_2 + Z_t)}$	$\tau_p = \frac{Z_i // Z_g}{Z_i // Z_g + Z_2 + Z_t}$	$\delta_p = Z_p/(Z_p + Z_1)$	$\tau_p\delta_p$

always holds, in which A is the gain of the amplifier. If we substitute this in equation (4.44), the result is

$$V_o/A = kV_g + \beta V_o \tag{4.46}$$

Rearranging (4.46) gives the well-known expression for the amplifier with feedback, except that k expresses the fraction of V_g which arrives at the input; hence

$$V_o = V_g \frac{kA}{1 - \beta A} \tag{4.47}$$

In practice it is difficult to recognize which part of β is τ and which is δ, because they can be completely interwoven, and often the distinction is not relevant. The value of this approach is to get an insight into the fact that every feedback loop can be rearranged into one of these four basic configurations.

It will be clear that the applied configuration will determine the input and output impedance and that again four different cases can be distinguished. We will not treat the total derivation but will confine ourselves to summarizing the results in Table 4.5. As can be seen from this table, a series connection will increase the impedance at the input and at the output. In its turn a shunt connection will decrease the input and the output impedance. Note that a factor of roughly $(1 - A\beta)$ is involved each time. For clarity the following expressions are used in the table:

$$Z_n = \frac{Z_o + Z_1}{Z_1} \tag{4.48}$$

in which Z_o is the output impedance of the amplifier without feedback, and

$$Z' = Z_n \frac{Z_s}{Z_o} \tag{4.49}$$

Z_i' is the new input impedance and Z_o' is the new output impedance obtained with feedback. The factor β is as defined previously.

Table 4.5 The input and output impedance of the feedback amplifier in four basic configurations.

β	Z'_i	Z'_o
$\tau_s\delta_s$	$(1 - A\delta_s)Z_i$	$(1 - A\tau_s Z')Z_o$
$\tau_s\delta_p$	$(1 - A\delta_p)Z_i$	$Z_o/(1 - AZ_n\tau_s\delta_p)$
$\tau_p\delta_s$	$\dfrac{(Z_2 + Z_t)Z_i}{Z_2 + Z_t - A\delta_p Z_i}$	$(1 - AZ'\tau_p)Z_o$
$\tau_p\delta_p$	$\dfrac{(Z_2 + Z_t)Z_i}{Z_2 + Z_t - A\delta_p Z_i}$	$Z_o/(1 - AZ_n\tau_p\delta_p)$

In most cases, approximations can be made which simplify these expressions drastically, but it is sometimes worthwhile to have a general layout available. The above description is not exclusive; sometimes it is more convenient to express all quantities in admittances.

To conclude this discussion of system specifications we will present an example of a microprocessor-based measuring system.

4.5 A MICROPROCESSOR-BASED MEASURING SYSTEM

To demonstrate the capabilities of a microprocessor-based measuring system, a high-speed spectral density measurement set-up can deliver an output signal whereas a normal standard measuring system would only produce noise. Often spectral measurements of plasmas produce a very weak signal which is completely immersed in noise. A plasma is an ionized gas which can conduct an electrical current. For instance, the retrieved signal from the process will have the form

$$A(t) = S(t) + R(t) \tag{4.50}$$

in which $S(t)$ is the desired signal and $R(t)$ is the undesired noise contribution. If it is known that $S(t)$ shows a periodic behavior with a constant time interval T then

$$S(t) = S(t + mT) \tag{4.51}$$

with m ranging from 0 to M, where M denotes the total number of measurements.

The noise signal will show a completely random stochastic behavior and will approach an average of zero if t goes to infinity. This is realized when the number of measurements M increases to a large value. Via a processor, the signal's average, $\langle A(\mathrm{t})\rangle$, can be calculated, or

$$\langle A(t)\rangle = \frac{1}{M}\sum_{m=0}^{M-1}[R(t + mT) + S(t)] \tag{4.52}$$

and as can be seen from expression (4.52) the noise term will diminish as m increases. The applied circuitry is illustrated in Fig. 4.19 in which two micro-

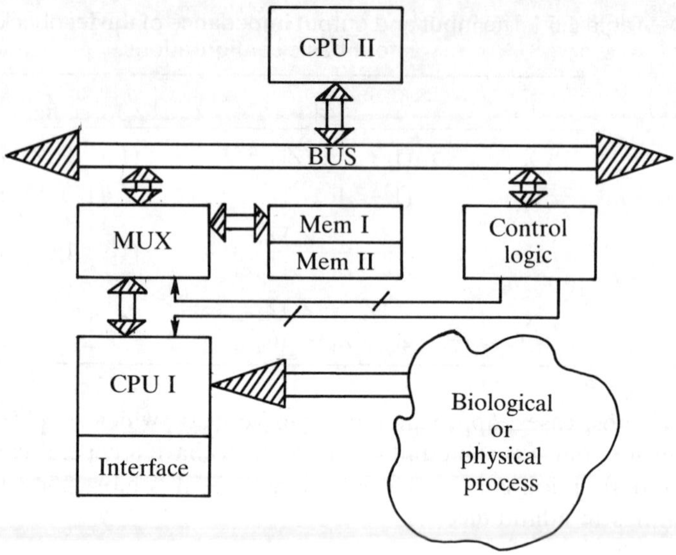

Figure 4.19 A two-microprocessor-powered measuring system

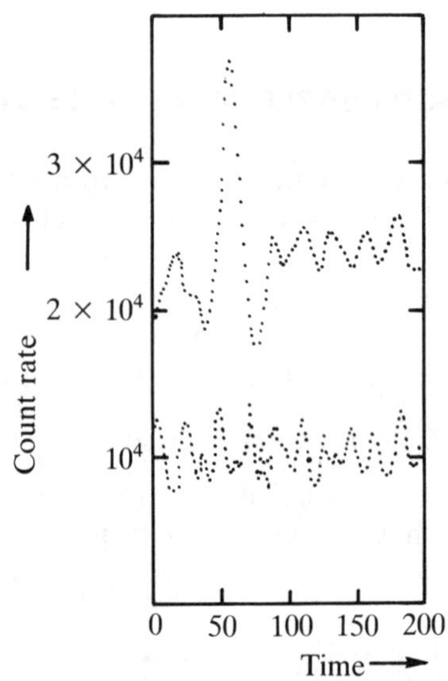

Figure 4.20 Comparison of the results of a conventional and a microprocessor-powered measuring system of a completely noise-drowned signal

processors are installed, one is only used for data collection and the other is used for numerical calculations. Both microprocessors have access to the same memory.

This principle is sometimes called a *dual ported memory* because the two synchronized microprocessors have access to the same memory and the required tasks are strictly separated between both processors. In Fig. 4.20, an example of a measuring result is given. In the lower part of the graph, the completely noise-immersed signal is depicted, measured with a conventional system.

The upper part of the graph shows the result obtained by the dual ported memory system, and may be considered as an impressive result. To obtain this result the number of summations is less than 25.

4.6 PROBLEMS

4.1 To improve the accuracy in a linear amplifier a feedforward technique can be applied. What are the main characteristics of this technique and why is no stability problem involved?

4.2 An amplifier has a gain setting of 10–1000 times. The equivalent noise voltage over the whole frequency bandwidth is 15 μV and the maximum allowed voltage range for input and output is 15 volts. Determine the dynamic range for this amplifier with a required inaccuracy of 0.5%.

4.3 The inaccuracy of a DC voltage measuring system with a 4-digit display is ±0.05% fsd plus ± 0.1% ± ½ digit of indicated value. The full-scale deflection is 1.999 mV. Determine the absolute and the fractional error in an indicated value of 1.000 V.

4.4 In Fig. 4.21 a measuring configuration for temperatures is shown. In this circuitry $R(T)$ represents a calibrated platina resistor of which the transfer function can be described by

$$R(T) = R_0(1 + aT + bT^2)$$

where

$R_0 = 100\ \Omega$
$a = 4.9 \times 10^{-3}/\mathrm{K}$
$b = -5.9 \times 10^{-7}/\mathrm{K}$
T = temperature in degrees Celsius

The resistance of the conductors is r and V_{ref} represents a reference voltage source of 6.4 V. R_1 is 1.0 kΩ and the amplifier is supposed to be ideal.

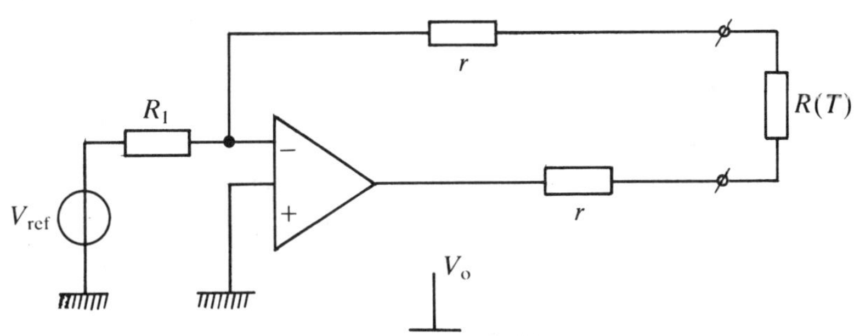

Figure 4.21 Temperature measurement configuration

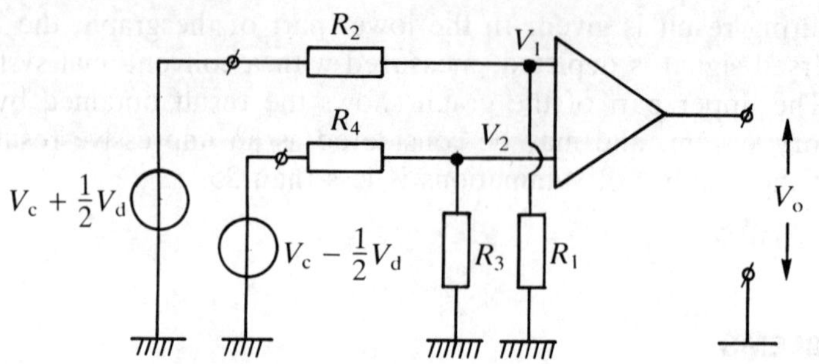

Figure 4.22 Circuit for Problem 4.5

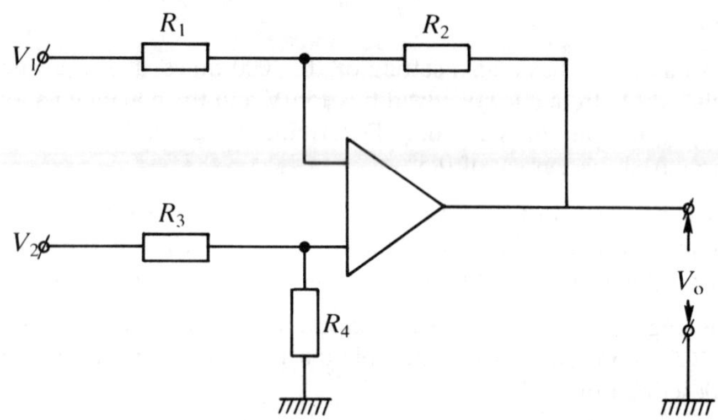

Figure 4.23 Circuit for Problem 4.6

(a) Determine the maximum allowed resistance of the conductors for an inaccuracy smaller than 0.1°C owing to r.
(b) The current through $R(T)$ will raise the temperature of the sensor with 10°C/watt dissipated in $R(T)$. Determine the fault signal in °C due to the self-heating effect of $R(T)$ at $T = 0$°C.
(c) Determine the non-linearity as a percentage value with respect to the case where b is supposed to be zero over a temperature range from -50 to $+100$°C.
(d) If $b = 0$ prove that the sensitivity of this configuration is 2.5 mV/K.
(e) If $b = 0$ determine the maximum allowed offset voltage V_{off} of the amplifier when at $T = 0$°C the fault signal is smaller than 0.1°C owing to V_{off}.
(f) Answer the same question as in (e) for the maximum allowed bias current I_{bias} of the amplifier.
(g) Determine the absolute error in the reference voltage source V_{ref}, for an inaccuracy of ±0.1°C at $T = 0$°C.
(h) Answer the same question as in (g) for an inaccuracy in R_1.

4.5 Derive an expression in β for the CMRR and the discrimination factor F for the attenuator configuration connected to an amplifier as drawn in Fig. 4.22. Introduce β_1 and β_2 for the respective fractions arriving at the input terminals of the amplifier.

4.6 For the differential amplifier circuit depicted in Fig. 4.23 derive an expression for the CMRR. If the fractional error of the resistors is $\delta = 1\%$ and the gain A is 100, which

CMRR can be guaranteed? What important conclusions can be drawn from this example?

4.7 A strain gauge has a sensitivity of 10 mV/mm and a temperature sensitivity of 10 μV/K; then a variation of for instance 0.1 mV can be caused either by a temperature variation or by a mechanical load variation. Determine the parasitic sensitivities in each case.

Five

RELIABILITY

5.1 INTRODUCTION

The formal study of reliability dates from the early 1930s and hence can be considered as a rather young science. The need for reliability predictions about systems and equipment has grown with the need for controlling society and with the still increasing complexity of the applied systems. Reliability first came into scope in equipment for military purposes; where survival conditions are basically involved reliability of equipment and systems plays a most important role. In the early 1970s the importance of reliability became accepted worldwide. In a great variety of disciplines such as transport, medicine, and even in software development and tools, the concepts of reliability became well established. Due to the nature of electronics the first major applications were found in this discipline.

Reliability is based on the quality of the different components of which every system is composed. In its turn, quality of components is based on the applied quality-control procedures in the manufacturing process. This strong relationship between reliability and quality will be discussed in more detail in a later section.

In the manufacturing process there is a strong relationship between human behavior and the effectiveness of reliability. In an organization a consensus between all departments is a basic requirement to make reliability a success. This has been proven over and over again.

As will be shown, two types of reliability can be distinguished: technical reliability and economic reliability. The maxim 'pay now, save later' has been demonstrated dramatically many times by the damage and loss of reputation caused when this rule is neglected. One other important question must be kept in mind: 'what are the consequences if a system fails?'

Ultimately we want to be able to make predictions about systems, about their lifetime and reliability under known conditions expressed in numbers. The treatment will have an introductory character but an elementary discussion of all relevant reliability parameters and aspects will be presented.

Because it is in the very nature of this discipline to *predict* specifications of components and systems of samples of a population, instantaneously and in the long run, a knowledge of probability-theory is a basic requirement.

5.2 RELIABILITY CONCEPTS

We begin this section by presenting the required definitions and concepts of reliability, about which a lot of confusion often arises.

5.2.1 Definitions

1 The *quality* of a product is defined as the instantaneous sample of technical specifications with which that product complies.

Note the word 'instantaneous' in this definition because here the specifications are determined under standardized conditions and at a certain time.

2 The *lifecycle* or *lifetime* of a system is defined as the total time between 'birth' and 'death' of that system under the condition, if applicable, of an *a priori* fixed specified maintenance scheme.

We will return to the phrase 'an *a priori* specified maintenance scheme' later on.

3 The *reliability* of a system is defined as the probability that a system will still perform successfully the functions associated with its predetermined specifications, after a certain operational period.

Note in this statement the words 'probability' and 'after a certain period'. It will be clear from this statement that in reliability time is an inherent parameter. This is the main difference from quality, but if we look for a relationship between quality and reliability the following statement holds also.

4 *Reliability* is quality over time, which can be expressed as

$$R(t) = Q/t \tag{5.1}$$

where $R(t)$ denotes the reliability as a function of time

A few examples can further stress the importance of the above statements. Suppose the calculated economic lifetime of a large mainframe computer is 10 years. The maintenance scheme is fixed at twice a year, and after 10 years the computer will be considered as economically worn out because maintenance costs exceed the accepted operational cost level. From a technical point of view the computer can still perform its tasks but it is no longer acceptable with respect to the operational cost level. In the aircraft industry it will be obvious that we want to know the reliability of every system involved in an aircraft. The same philosophy is applicable as in the example of the mainframe computer, but neglecting maintenance for reasons of cost savings may result in disasters. Here again the principle of pay now and save later is very well applicable.

5.2.2 Stochastic variables, probability distributions, functions and reliability

As stated already, probability is required to describe the reliability process. Two main reasons can be mentioned why this is true: first we want to predict the behavior of a system over a certain period of time, second the systems have become too complex to make absolute statements possible, so in every reliability statement always an amount of uncertainty will be implied. But as will be seen, the task of the reliability engineer is to minimize this uncertainty.

It is beyond the scope of this textbook to present a comprehensive discussion of probability theory, but for convenience the relevant parameters of this theory are summarized. In probability theory two types of stochastic variables can be distinguished, i.e., discrete and continuous variables, of which the parameters are defined in Table 5.1.

Throwing a dice is an example of a discrete variable; noise is an example of a continuous variable.

It has been proven that the normal lifetime of a system can be considered as a continuous variable. The lifecycle of most systems can be described by a limited number of probability density functions and the exponential distribution is often the most likely one and is the most widely used in electronic systems. We shall illustrate most concepts of reliability with the exponential distribution function instead of a more general treatment.

Table 5.1 Parameter review of stochastic variables

Discrete stochastic variable	*Continuous stochastic variable*
Probability function of x $p(x) = P(x_i)$ for $i = 1, 2, 3, \ldots$ $= 0$ elsewhere	$f(x)$ probability density function and $f(x)\Delta x$ is called the probability element for the event Δx
To every possible event a number is assigned	
Distribution function of x $p(x \leq k) = \sum_i P(x = x_i) = F(x)$ for $x_i \leq k$	$F(x) = P(x \leq k) = \int_{-\infty}^{k} f(t)\,dt$
The probability of the occurence of a maximum of k events, sometimes called *cumulative distribution.*	
Expectation of x $\mu = E(x) = x_i P(x = x_i)$	$\mu = E(x) = \int_{-\infty}^{\infty} x f(x)\,dx$
Often called *expectation of the first moment*	
Variance of x $\mathrm{Var}(x) = \sigma^2{}_x = \sum_i (x_i - \mu)^2 P(x)$	$\mathrm{Var}(x) = \sigma^2{}_x = \int_{-\infty}^{\infty} (x - \mu)^2 f(x)\,dx$
The square root of the variance is called the deviation	

Suppose the probability that a system will fail is p, where p is small, for instance 0.01, then the probability that the system will not fail is $1 - p = q$ by definition. Suppose further that np is large, where n is the number of systems to be considered, and x is the event that a system will fail, then the probability that one or more systems will fail may be calculated with the Poisson distribution:

$$f(x) = \frac{\exp(-\mu)\mu^x}{x!} = \frac{\exp(-np)(np)^x}{x!} \tag{5.2}$$

where $\mu = np$ is the mean of the Poisson distribution.

Now suppose the event of failing or not failing is a function of time and that the system fails with an average of λ times during its lifetime, then the average number of systems that will fail in a given time t is λt, with the condition that the total number of working systems remains constant. Then the probability that x systems will fail in a given time t can again be described by the Poisson distribution. The equation becomes

$$f(x) = \frac{\exp(-\lambda t)(\lambda t)^x}{x!} \tag{5.3}$$

Finally if the probability is calculated that *no* system will fail then $x = 0$. Substituting this value in equation (5.3) yields

$$f(0) = \exp(-\lambda t) \tag{5.4}$$

The two expressions (5.3) and (5.4) are depicted in Fig. 5.1, from which the relationship between the two expressions can be understood. It can be seen that for $x = 0$, the graph $f(0)$, still as a function of t, lies in front of the figure along the t-axis. The function (5.4) can be interpreted as the probability of survival for the considered system and for that reason it is called the *survival function* or *reliability* of that system, often denoted by $R(t)$. Thus $R(t)$ is the probability that the system does not fail during the interval 0 to t or is still functioning at the time t:

$$R(t) = f(0) = \exp(-\lambda t) \tag{5.5}$$

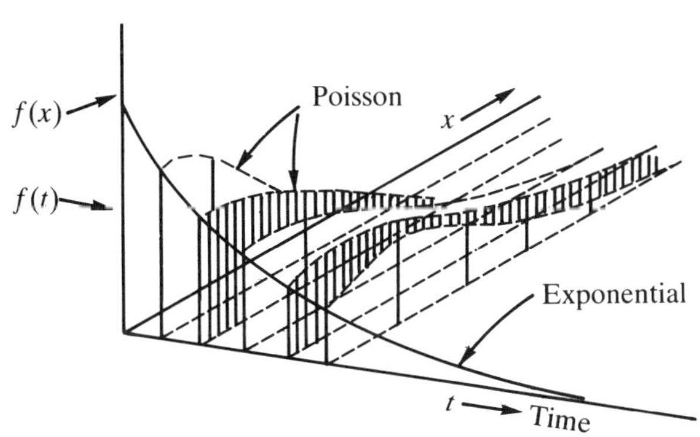

Figure 5.1 The relationship between the survival function and the Poisson function (reprinted by permission of John Wiley & Sons Inc.)

It can be seen from this graph that at $t = 0$ the probability that no system will fail is 1, as is correct of course, because all systems will work at $t = 0$. If t goes to infinity it will be clear that all systems will have failed. So the graph of the expression for $R(t)$ shows the probability that no system will fail in the interval 0 to t.

An alternative representation for the survival function may be obtained with the aid of the *cumulative distribution function* of t and is defined as

$$R(t) = 1 - F(t) = \int_t^\infty f(t)\,\mathrm{d}t \qquad \text{for } F(t) < 1 \tag{5.6}$$

and where

$$F(t) = \int_0^t f(t)\,\mathrm{d}t \tag{5.7}$$

and $f(t)$ is again the considered probability density function. Note that the limits in the integral expressions complement each other. $F(t)$ is called the cumulative failure distribution function, or failure probability, or the unreliability function. For the exponential distribution it can easily be calculated that $F(t)$ is $1 - \exp(-\lambda t)$.

5.2.3 Reliability parameters

In the application of probability distributions to reliability assessment studies of systems, more basic concepts are needed. So here are some others. If the derivative of expression (5.6) is determined and we change sign in order to get a positive expression for $f(t)$ this yields

$$-\mathrm{d}R(t)/\mathrm{d}t = f(t) = \begin{cases} \lambda \mathrm{e}^{-\lambda t} & \text{for } t \geq 0 \\ 0 & \text{for } t < 0 \end{cases} \tag{5.8}$$

For negative values of t this expression makes no sense so $f(t) = 0$ by definition for $t < 0$.

The expression (5.8) is called the *failure probability density function*, often denoted by $f(t)$ only, and can be interpreted as the degree to which the reliability decreases at time t.

For λ the following definition holds: λ is the number of failures per unit of time and per system and is called the *failure rate*. The dimension of λ is per unit of time as can be seen from expression (5.8). The applied unit of time is often denoted per 1000 hours or even per 10^6 hours.

* * *

Example 5.1

If for a certain system it is known that in 1000 hours 14 failures occur, then the failure rate equals 14/1000 = 0.014 or 1.4% per hour. The statements 'a system fails

λ times' or 'λ systems fail' in a given time t are interchangeable. No statement is made about when these failures occur or how these failures are distributed in time.

* * *

Another important parameter of reliability assessment studies is the mean of $R(t)$; this yields for the exponential distribution

$$E(t) = \int_0^\infty R(t)\,\mathrm{d}t = \int_0^\infty \exp(-\lambda t)\,\mathrm{d}t \qquad (5.9)$$

It is easily seen here that the result for the mean will be $1/\lambda$ with the dimension of time. How can this relationship be interpreted? The mean denotes an expectation, hence the calculated value for the mean of $R(t)$ describes the mean time that a system will function or not fail. This expression therefore is called the *mean time between failure* or *MTBF* and describes exactly what it says, but as you will see later on, one further distinction must be added.

* * *

Example 5.2

The reciprocal value of the failure rate λ is equal to the MTBF, so for instance suppose the failure rate λ of a component is 0.1×10^{-4}/hour, then the MTBF is $1/\lambda = 1/(0.1 \times 10^{-4}) = 10^5$ hour.

* * *

So far λ has been considered as a constant but sometimes the failure rate may be a function of time. This is especially the case when a system is brand new or almost worn out. In those circumstances we have to write $\lambda(t)$ instead of λ only. If this is true λ is called the failure rate density or *hazard rate*, or force of mortality, and is denoted by $h(t)$ or sometimes by $z(t)$. The hazard rate is defined as the ratio of the failure probability density function $f(t)$ and the reliability $R(t)$, thus

$$h(t) = \lambda(t) = \frac{f(t)}{R(t)} = \frac{-\mathrm{d}R(t)/\mathrm{d}t}{R(t)} \qquad (5.10)$$

Solving this first-order differential equation for $R(t)$ yields

$$R(t) = \exp\left\{-\int_0^t h(t)\,\mathrm{d}t\right\} \qquad (5.11)$$

which is another expression for $R(t)$ as a function of the hazard rate. If we apply both expressions (5.10) and (5.11) to the exponential failure probability density function $f(t) = \lambda\exp(-\lambda t)$ it is easily seen that the result for the hazard rate is λ and hence is a constant, and $R(t) = \exp(-\lambda t)$.

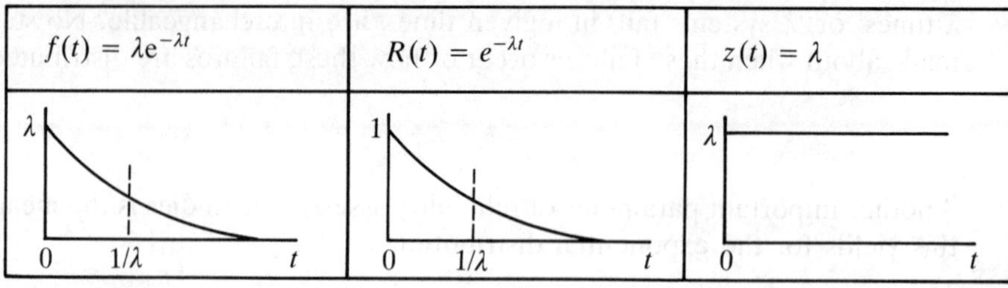

Figure 5.2 Graphical representation of three different parameters for the exponential failure distribution function

For an exponential failure distribution function the results obtained so far are illustrated in three different graphs in Fig. 5.2.

In the following paragraphs a separate discussion is devoted to MTBF, because a more detailed evaluation is required where repairable and non-repairable systems come into scope. This is needed because the mathematical description of a repairable system differs from that of a non-repairable system.

5.2.4 Mean time between failure (MTBF)

In equation (5.9) a first definition for the mean time between failure is given, but as stated already some distinction has to be added. In general, two classes of systems can be distinguished, which differ considerably in treatment and in principle of design. The two classes of systems are the repairable and non-repairable systems. A few examples will illustrate this distinction. A launched satellite is an example of a non-repairable system in most cases and its design is characterized by the diagnosed and required longevity under extremely severe operating conditions. An automobile is an example of a repairable system and its design is largely based on possible replacement of components which are worn out. This distinction can be found in all kinds of technical disciplines. Some authors prefer to speak of maintainable and non-maintainable systems which is in principle the same.

MTBF is defined as the mean over all possible time intervals in which the system has run out from its specifications when t goes to infinity. MTBF is expressed as a time and is sometimes denoted by m, hence by definition

$$\text{MTBF} = \int_0^\infty t f(t)\,\mathrm{d}t = \int_0^\infty R(t)\,\mathrm{d}t \qquad (5.12)$$

where $f(t)$ is the considered failure probability density function. As will be clear from the definition, the MTBF is always related to a repairable system. For a non-repairable system the lifetime can be measured by the *mean time to failure* (MTTF), which then exactly represents the span of its life.

Sometimes a comparison is made between repairable and non-repairable systems, and then the time to the first failure of the repairable system can be

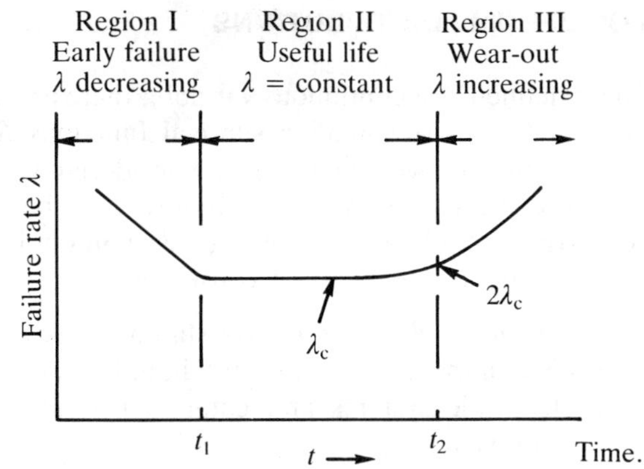

Figure 5.3 'Bath tub curve' or failure rate distribution as a function of time

measured and is expressed as a number called the *mean time to first failure* (MTTFF). We will discuss the related aspects of MTBF later on.

It has been verified experimentally that the failure rate distribution as a function of time often shows a so-called 'bath tub curve' as is shown in Fig. 5.3.

In this graph three different regions can be distinguished. In region I, λ is a function of t and will decrease with time, because in this region, the so-called 'burn in' period, a lot of failures can occur. This is often the case when a system is a totally new design and just manufactured. This region is also called the 'infant mortality' or 'early-failure' region. After a certain time, t_1, λ appears to be a constant and is no longer a function of time, and region II commences.

In region II, λ is a constant for a long period of time and is therefore called the 'useful life' period, or normal operating region. Finally, region III, the so-called 'wear-out' period, starting at time t_2, is the region where λ now is an increasing function of time. The wear-out region is defined as beginning when the failure rate is twice the failure rate of the useful-life region. In this respect an interesting analogy can be made. In living systems, the same bath tub curve can be recognized, but with large differences of shapes due to large differences in living circumstances and work loads.

We conclude this section with another simple example, which stresses the concept of failure rate.

* * *

Example 5.3

Suppose a system has a failure rate of 1.4% per hour operating time. What is the average number of failures that could have occurred after 500 operating hours? From the data it can be concluded that the failure rate per hour equals 0.014 and this equals the number of failures per hour and per system. Then with 500 operating hours it may be expected that $500 \times 0.014 = 7$ failures have occurred.

* * *

5.3 FAMILIES OF LIFETIME DISTRIBUTIONS

Since lifetime is defined as a continuous variable, there are a number of probability density functions $f(t)$, corresponding survival functions $R(t)$ and instantaneous failure rate functions $\lambda(t)$, which have been considered as well suited to described possible lifetime distributions. We shall briefly describe some of them but will not go into further detail. The following distribution functions are applicable depending on the circumstances to be considered.

1 *The exponential failure distribution.* As stated already the exponential failure distribution is one of the most important failure laws, the most widely used, very suitable for electronic systems and components, and applicable for normal operating conditions.
2 *The normal failure distribution.* The normal failure distribution is one of the best known and represents an adequate model for systems and components in which the failure is caused by some wear-out phenomena, so in region III is as illustrated in Fig. 5.3.
3 *The Rayleigh failure distribution.* The Rayleigh failure distribution is applicable when increased stress and accelerated wear-out are involved.
4 *The gamma failure distribution.* The gamma failure distribution (Γ) is a type of distribution function and is applicable when burn-in, wearout and redundancy are involved. The concept of redundancy will be discussed later on.
5 *The Weibull failure distribution.* The Weibull distribution is applicable when wear-out, aging and overloading are involved simultaneously. The Weibull law basically contains all the above mentioned distributions.

In Table 5.2 a summary is given for the reliability concepts discussed so far, with the results for the exponential failure distribution in the last column.

Derivations of expressions for reliability with other failure distribution laws than the exponential failure law incline to become complicated rapidly and are not very practical for an introductory chapter on reliability. Before continuing we present two more examples.

* * *

Example 5.4

The useful lifetime of a component, with a failure rate of 0.0001 per hour, is 1000 hours. If it may be presumed that the failure distribution is exponential, calculate the MTBF and the reliability of that component for 10 100, and 1000 hours.

From the data given it may be concluded immediately that the MTBF is $1/10^{-4} = 10\,000$ hours. The numerical value for the reliability or survival function $R(t)$ can be found by substituting the respective values for t in the exponential function $\exp(-\lambda t)$. This yields

$$R(10) = \exp(-10^{-4} \times 10) = \exp(-0.001) = 0.999$$
$$R(100) = \exp(-10^{-4} \times 100) = \exp(-0.01) = 0.990$$
$$R(1000) = \exp(-10^{-4} \times 1000) = \exp(-0.1) = 0.905$$

Table 5.2 Survey of reliability parameters

Parameter	*Mathematical expression or definition*	*Result for exponential distribution*
Survival function or reliability	$R(t) = 1 - F(t) = \int_t^\infty f(t)\,dt = \exp\left(-\int_0^t h(t)\,dt\right)$	$e^{-\lambda t}$
Failure rate	$\lambda = 1/\text{MTBF}$	λ
Failure density function	$f(t) = -\dfrac{dR(t)}{dt}$	$\lambda_e^{-\lambda t}$
Cumulative distribution or unreliability function	$F(t) = \int_0^t f(t)\,dt = 1 - R(t)$	$1 - e^{-\lambda t}$
Hazard rate	$\lambda(t) = h(t) = z(t) = -\dfrac{dR(t)/dt}{R(t)} = \dfrac{f(t)}{R(t)}$	λ
MTBF = MTTFF	$1/\lambda = \int_0^\infty R(t)\,dt = \int_0^\infty tf(t)\,dt$	$1/\lambda$

As can be seen from this example, when time increases, the reliability decreases as may be expected.

* * *

Example 5.5

A circuit consists of 20 resistors with a failure rate of 10×10^{-9}/hour/component and 30 soldered connections with a failure rate of 100×10^{-9}/hour/connection. If it may be presumed that the distribution is exponential, calculate the total failure rate and the MTBF.

Here we first have to assume that the system is no longer operational when one of the components fails, including the soldering connections. The total failure rate can be found by simply summing all components with their respective failure rates, thus

$$\lambda_{tot} = 20(10 \times 10^{-9}) + 30(100 \times 10^{-9}) = 3200 \times 10^{-9}/\text{hour}$$

Then the MTBF is

$$1/\lambda = 1/(3200 \times 10^{-9}) = 312\,500 \text{ hours} = 35.67 \text{ years.}$$

Here the strong influence of the soldered connections may be recognized.

* * *

5.4 RELIABILITY OF PARTS, COMPONENTS, EQUIPMENT AND SYSTEMS

In this section we shall present a few more definitions which are frequently used in the wrong context. We have used the concept of component and system several times without making any distinction between them; to avoid further confusion here are the correct definitions.

A *part* is the smallest building block of which a system is composed. Examples are resistors, transistors, integrated circuits (ICs), capacitors, and so on. Parts are characterized by the fact that they are non-repairable and cannot perform a function on their own.

A *component* is a small structure composed of several parts which can perform a certain task, such as a power supply, a keyboard, or a display. A component might be repairable, but a completely moulded power supply is an example of a component which is non-repairable.

Equipment is composed of several components and can perform a function or task as part of a larger system. In general, equipment is repairable. Examples are a temperature control, a fuel control system or a video terminal.

A *system* is defined as a structure of different pieces of equipment such as a large mainframe computer, an aircraft or a telephone station, which can perform large specific tasks.

It will be clear that the reliability of parts are the cornerstone of reliability of systems. In addition to the reliability of parts, a lot of other factors are involved which finally determine the overall reliability of the system. We will discuss several of these factors in the following paragraphs, such as:

1 the required MTBF;
2 the operational conditions;
3 the purpose of use;
4 the demanded reliability;
5 the applied configuration;
6 the maintenance protocol;
7 the required availability.

From the viewpoint of system reliability the reliability engineer is concerned with the identification and selection of an appropriate set of measures of reliability effectiveness and with defining the suitable reliability objectives. Very often the reliability engineer wants to approach the derived optimum lifecycle as a function of cost or vice versa as close as possible. An example of such a lifecycle versus cost function is depicted in Fig. 5.4. Here, reliability is the chosen independent variable and cost the dependent variable. Graph (a) represents the manufacturing costs plus the cost of preventive maintenance of the system under development. Graph (b) represents the cost of non-estimated maintenance, and graph (c) represents the sum of (a) and (b). Finally, graph (d) gives an impression of the total to which is added the loss on production rate during non-estimated repairing time of the system.

A poorly designed system means a cheap system with a poor reliability and with very high maintenance cost. A highly sophisticated designed system can

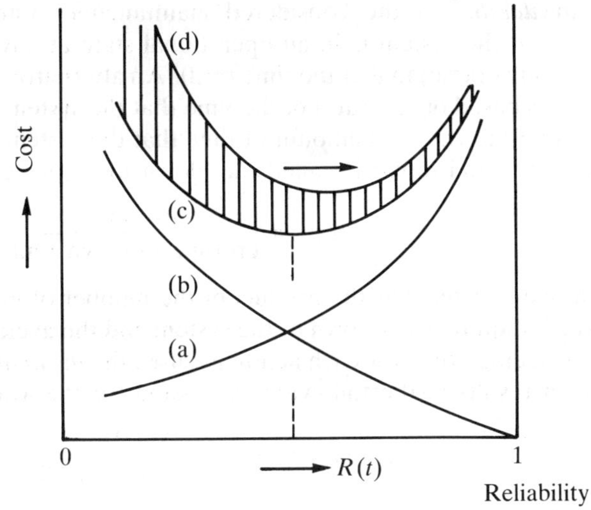

Figure 5.4 Cost versus reliability for a given system (courtesy Delft University Press)

provide a high reliability with very modest maintenance cost, but an extremely high sales price as a consequence. Somewhere there is an optimum which shall be approached as closely as possible by the manufacturer.

5.5 AVAILABILITY, REPAIRABLE AND NON-REPAIRABLE SYSTEMS

In this section we shall go into some more detail concerning repairable and non-repairable systems. The idea of repairable and non-repairable systems implies that the system can be maintained, and so is synonymous with the concept of maintainable and non-maintainable systems, although a repair action differs from a maintenance action. The problem of developing mathematical tools for the reliability of systems that can be maintained while in use, can be approached applying Markovian techniques. Markovian techniques are especially useful for describing the stochastic behavior of systems under a great variety of failure and repair conditions. This means that Markovian techniques are able to take account of repair actions, which of course require a finite period of time to restore a system to an operational state. In short, the probability of every possible transfer state has to be defined and calculated, so, for instance, given the system is working normally in a certain state, the question 'what is the probability that the system will achieve another state' has to be answered. So forward and backward differential equations in the time domain have to be developed, whereas in non-maintained systems only forward transitions have to be considered. This approach results in an additional figure of merit for describing the reliability of maintained systems, the so-called availability, often denoted by $A(t)$. In general, availability is a function of time.

The *availability* of the considered maintainable system is defined as the probability that the system is in an operational state at any time t given that the system was fully operational at the time $t = 0$. An alternative definition more easily understood, is based on the ratio of the time that the system is operational, the so-called 'up time', to the total amount of time that the system is needed. If the time for maintenance and repair is called the 'down time', we can write

$$\text{Availability} = \frac{\text{up time}}{\text{up time} + \text{down time}} \tag{5.13}$$

The downtime is defined as the product of the number of maintenance and repair actions which require a shutdown of the system and the average time for each such action. The average time for each action is called the *mean time to repair* (MTTR). This concept results in the following expression for the availability:

$$\text{Availability} = \frac{\text{MTBF}}{\text{MTBF} + \text{MTTR}} \tag{5.14}$$

If the MTBF and thus the reliability increases, the MTTR can also increase; if the MTBF decreases, the average time for a repair action must also decrease in order to maintain the same availability. In a great majority of cases, involving many different types of components, equipment and systems, here again an exponential distribution of the repair time can be justified.

As stated already, to be able to make comparisons possible between maintained and non-maintained systems, the concept of MTTFF can be introduced for maintained systems. We have to introduce here another definition to make calculations more accessible. The *repair rate* is defined as the frequency per unit of time with which repair actions are effected and is denoted by μ, often expressed in numbers per hour.

You will be not surprised that the reciprocal value of μ is the MTTR, which is in complete analogy with the failure rate and the MTBF.

* * *

Example 5.6

Suppose the repair rate of a system is $\mu = 0.1$/hour, then the MTTR is $1/\mu = 10$ hours.

* * *

If the distribution is exponential and the MTBF $= 1/\lambda$ and the MTTR $= 1/\mu$, then by substitution of these values in the expression for the availability (5.14) this results

$$\text{Availability} = \frac{1/\lambda}{1/\lambda + 1/\mu} = \frac{\mu}{\mu + \lambda} \tag{5.15}$$

* * *

Example 5.7

If the MTTR of a computer system equals 5 hours, and the MTBF is 5000 hours, then by substitution in expression (5.15) the availability equals

$$A = \frac{1/5}{1/5 + 1/5000} = 0.9990$$

* * *

If in expression (5.15) $\mu = 0$ is substituted it may be expected that the expression reduces to an expression for non-maintained systems, but this is wrong and the result appears to be zero. As stated already availability is, in general, a function of time and expression (5.15) is called the stationary or *steady-state availability* when t goes to infinity. The general expression for $A(t)$ is called the *momentary availability* and results in a more complicated expression requiring Markovian techniques for derivation. For instance, here is an example for the expression of $A(t)$ for a single system with single repair actions:

$$A(t) = \frac{\mu}{\lambda + \mu} + \frac{\lambda}{\lambda + \mu}\exp[-(\mu + \lambda)t] \tag{5.16}$$

If now $\mu = 0$ is substituted we get the well-known expression (5.5) for the reliability $R(t) = \exp(-\lambda t)$ for a single system, and it may be concluded that the availability is a generalization of the concept of reliability. The exponential term is called the 'transient term' and as t goes to infinity it can be seen from (5.16) that this term vanishes and the steady-state availability results.

Summarizing, the following availability concepts are used:

1 The momentary availability $A(t)$. The momentary availability $A(t)$ is defined as the probability that the system is in an operational state at any time t given that the system was operational at the time $t = 0$. The momentary availability is well suited to describe systems which must be available at any time.

2 The steady-state availability A_{ss} is defined as the mean of $A(t)$ if $t \to \infty$, hence

$$A_{ss} = \lim_{t \to \infty} \frac{1}{t} \int_0^t A(t)\,dt \tag{5.17}$$

3 The mission availability $A_m(t)$ is defined as the mean for a certain time interval t_1 t_2; hence

$$A_m(t) = \frac{1}{t_2 - t_1} \int_{t_1}^{t_2} A(t)\,dt \tag{5.18}$$

In Fig. 5.5 the availability as a function of time is depicted from which it can be seen that for $t \to \infty$ the three kinds of availability tend to coincide with the steady-state availability A_{ss}.

* * *

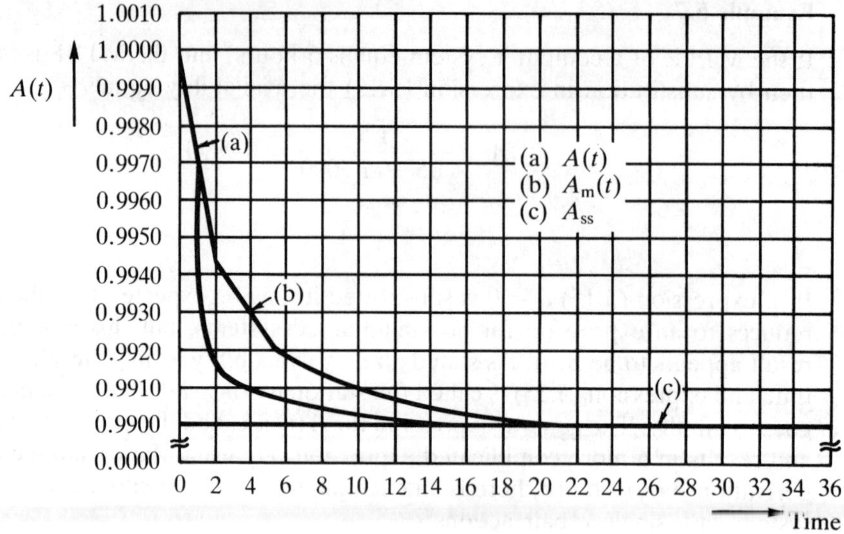

Figure 5.5 Availability as a function of time (reprinted by permission of John Wiley & Sons Inc.)

Example 5.8

For a single subsystem having the parameter MTTR $= 1/\mu = 100$ hours, the proportion spent in an operational state is 0.9901 and that spent in a failed state is 0.0099. These figures indicate that over a period of for instance 10 000 hours the system is operable for 9901 hours and occupies a failed state for 99 hours.

* * *

5.6 BOOLEAN ALGEBRA AND RELIABILITY

This section considers the application of Boolean algebra to find the reliability of a system which is composed of multiple components or subsystems. In most circumstances we have to deal with such composite systems and other sophisticated tools are required to calculate reliability. In many applications failing or not failing can be characterized by two states, i.e. the systems works, or it does not work. This statement asks for a Boolean approach, that is to say we can apply Boolean algebra to calculate the reliability of composite systems.

Suppose a system C consists of two subsystems A and B and A, B and C denote the state of the respective systems. The following statement:

$$\mathrm{C} = A + B \tag{5.19}$$

can be read as: the system C works if either the system A or B or both systems are in an operational state, which represents an OR function.

Consider now a situation in which the system C works when and only when both A and B are in an operational state, which represents an AND function:

$$C = AB \tag{5.20}$$

Finally if we define a working state with 1 and a non-working state with 0 it is easily seen that all basic rules of Boolean algebra can be applied to composite systems for reliability calculations. For convenience, here are the basic rules of Boolean algebra without any derivation:

$$\begin{aligned} A + 0 &= 0 \qquad & A0 &= 0 \\ A + 1 &= 1 \qquad & A1 &= A \\ A + A &= A \qquad & AA &= A \end{aligned} \tag{5.21}$$

$$\begin{aligned} A + B &= B + A \qquad & A + \bar{A} &= 1 \\ (A + B) + C &= \mathrm{A} + (B + C) \qquad & A\bar{A} &= 0 \\ AB + AC &= A(B + C) \qquad & A\bar{\bar{A}} &= \mathrm{A} \\ (AB)C &= A(BC) & & \end{aligned} \tag{5.22}$$

And finally De Morgan's theorem:

$$\overline{AB} = \bar{A} + \bar{B} \quad \text{and} \quad \overline{A + B} = \overline{AB} \tag{5.23}$$

We will demonstrate this with an example.

* * *

Example 5.9

Suppose we want to compare the reliability of the two systems depicted in Fig. 5.6, we have only to determine the Boolean expression for the working state of each of the system configurations.

For a series-connected system, consisting of two or more subsystems, it will be clear that the system will function if both subsystems are functioning; on the other hand, for a parallel-connected system, the system will function if one of the subsystems is functioning. It is further assumed that all subsystems are indepen-

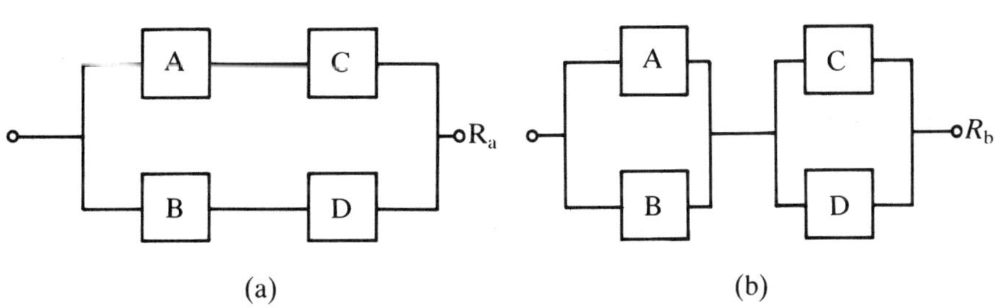

Figure 5.6 Two systems composed of the same four subsystems performing the same function, but with different reliability

dent, and that the respective states are A, B, C and D, then for the state working of the upper branch in Fig. 5.6a the following expression holds:

$$W_1 = AC$$

and for the lower branch

$$W_2 = BD$$

and hence for the complete system

$$W_{sys\cdot a} = W_1 + W_2 = AC + BD \tag{5.24}$$

In Fig. 5.6b the Boolean expression becomes

$$\begin{aligned} W_{sys\cdot b} &= (A + B)(C + D) \\ &= AC + AD + BC + BD \\ &= W_{sys\cdot a} + AD + BC \end{aligned} \tag{5.25}$$

By comparison of (5.24) and (5.25) the conclusion will be justified that the configuration of Fig. 5.6b will represent a larger reliability than that of Fig. 5.6a. This result can easily be understood by inspection of Fig. 5.6 because system (b) still functions if either A and D or B and C are failing.

* * *

So far only the Boolean functions of both systems for the working state are determined. If it is required to calculate the reliability then we have to transfer the Boolean expression into a probability function as in the following example, where it is presumed again that all the subsystems are independent.

* * *

Example 5.10

Suppose the reliability of every subsystem in Fig. 5.6 is 0.1, then the reliability for system (a) can be calculated as follows.

$$\begin{aligned} P(W_{sys\cdot a}) &= P(AC + DB) \\ &= P(AC) + P(DB) - P(ABCD) \\ &= P(A)P(C) + P(D)P(B) - P(A)P(B)P(C)P(D) \end{aligned}$$

and by substitution of $P(A) = P(B) = P(C) = P(D) = 0.1$ the reliability is

$$P(W_{sys.a}) = 0.1 \times 0.1 + 0.1 \times 0.1 - (0.1)^4 = 0.0199$$

For system (b) we find for the reliability function

$$\begin{aligned} P(W_{sys.b})\ & P(A + B)P(C + D) \\ &= [P(A) + P(B) - P(AB)][P(C) + P(D) - P(CD)] \end{aligned}$$

and by substitution

$$P(W_{\text{sys.b}}) = (0.2 - 0.01)(0.2 - 0.01) = 0.0361$$

The reliability of the configuration of Fig. 5.6b is 1.8 times larger than that of Fig. 5.6a. So a well chosen configuration can improve reliability considerably.

* * *

5.7 THE RELIABILITY AND MTBF OF SYSTEMS IN SERIES

Here we consider a system composed of components and subsystems connected in series and forming one complete system. At first no repair action will be taken into account, hence only the failure rate of the respective components and subsystems is involved. The system fails when any of the subsystems attains a failed state. We will start with an example.

* * *

Example 5.11

Of an X_R missile type, applied to launch telecommunication satellites, it is known that 7 out of 10 launches are successful. Of the communication satellite X_S it is known that after launching 95 out of 100 are functioning in accordance with their predetermined specified characteristics. The applied launching system X_L fails in 5 out of 250 launching operations, so we can conclude that the reliability of the launching system is

$$R(X_L) = 225/250 = 0.90$$

Where the reliability of the whole system is a combination of all items involved, supposing every system is independent we find for the total reliability

$$R(X_R)R(X_S)R(X_L) = (0.7)(0.95)(0.90) = 0.5985$$

* * *

It is not very difficult to apply the same philosophy to a type of vital measuring system monitoring a critical temperature in a nuclear reactor or in an aircraft engine. This concept is illustrated in Fig. 5.7 and can be generalized. Suppose a system is composed of n subsystems then for the reliability of the complete system the following multiplying rule can be applied:

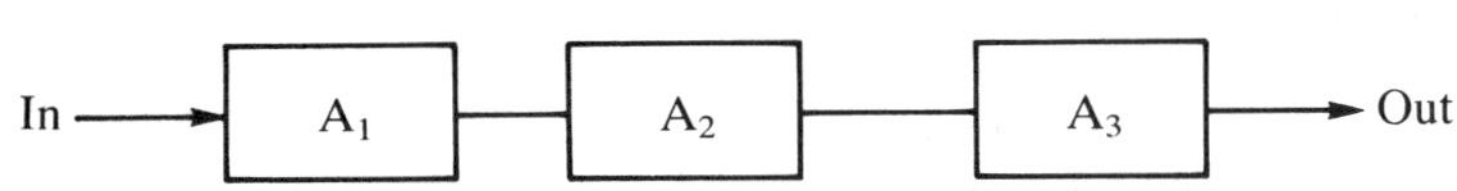

Figure 5.7 Series system consisting of three subsystems performing one function

$$R_{(\text{system})} = \prod_{i=1}^{n} R_i \tag{5.26}$$

where R_i is the reliability of the subsystem i.

If the reliability of every independent subsystem may be denoted by the exponential time function $\exp(-\lambda_i t)$, which is often the case, then expression (5.26) can be rewritten as

$$R(t) = \prod_{i=1}^{n} R_i(t) = \prod_{i=1}^{n} \exp(-\lambda_i t) = \exp(-\lambda_s t) \tag{5.27}$$

with

$$\lambda_s = \sum_{i=1}^{n} \lambda_i = \lambda_1 + \lambda_2 + \ldots + \lambda_n \tag{5.28}$$

If all n subsystems are identical, i.e. $\lambda_1 = \lambda_2 = \ldots \lambda_n = \lambda$, then

$$R(t) = \exp(-n\lambda t) \tag{5.29}$$

The MTBF of a series system can be calculated as follows:

$$\text{MTBF} = \int_0^{\infty} \exp\left(-\sum_{i=1}^{n} \lambda_i t\right) \mathrm{d}t = \frac{1}{\sum_{i=1}^{n} \lambda_i} \tag{5.30}$$

where again it is presumed the systems follow an exponential distribution and are independent; if the systems are not independent a conditional reliability must be included which complicates calculations considerably.

For n identical subsystems the MTBF becomes

$$\text{MTBF} = 1/n\lambda \tag{5.31}$$

Fig. 5.8 is a graphical representation of the reliability and the MTBF of a single system and a system composed of two subsystems in series with a failure rate of 0.25 per hour for each system. Note that the reliability and MTBF of a single system are higher than those of a series system.

As we have seen already, talking about the MTBF of a series system involves repair and maintenance actions to be taken into account, but this will be discussed later on. The expressions rapidly increase in complexity because the repair rate μ is involved.

5.8 THE RELIABILITY AND MTBF OF SYSTEMS IN PARALLEL

Systems are often connected in parallel to improve reliability. Especially when a vital system must be monitored and controlled, there is a demand for more than one system to be installed. For example, in an aircraft, besides a lot of other doubled-installed equipment such as the two flight computers as a basic pre-

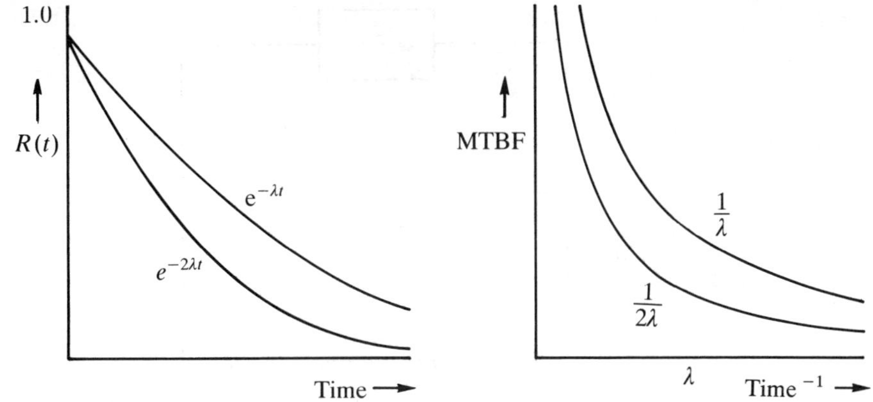

Figure 5.8 Comparison of the reliability and the MTBF of a single system and a series system with $n = 2$ with exponential distribution and failure rate of 0.25 per hour

requisite, sometimes the fly by wire system is installed in four-fold. Very often power supplies are installed twice, and in power stations all vital equipment is installed more than once. The same idea can be found in computers and memories and a large variety of sophisticated measuring instruments.

Every piece of equipment connected in parallel is able to execute the complete task on its own, independent of the other pieces of equipment. The background of this philosophy can be described as *redundancy*, so if a system retains redundancy, more than one identical system is installed and, as we will see, reliability is increased considerably.

In parallel systems we have to be more specific because the three following possibilities frequently occur, each with their own characteristics:

1 n systems are connected in parallel of which all n systems are active. This is called an *active on-line standby system.*
2 n systems are connected in parallel of which one system is active only. This is called a *cold on-line standby system*, or sometimes a passive on-line system. It is assumed here that if one system fails, a new system is switched on immediately and the failed system is switched off.
3 n systems are connected in parallel but, to be operational, only k out of n systems have to be active, with $k < n$, sometimes called a *parallel k out of n system.*
4 Mixed configurations in which series and parallel connected systems are combined.

We first discuss the reliability and MTBF of these systems without considering repair and maintenance actions.

5.8.1 The active on-line standby system

In this mode of operation n systems are provided operating simultaneously, performing one function only and with equal failure rates. It can be seen that the

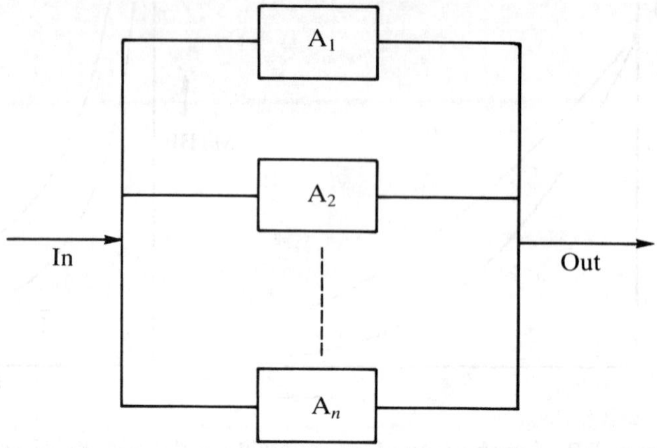

Figure 5.9 Configuration of an active on-line standby system composed of n identical systems

system is still in an operational state when $n - 1$ systems have reached a failed state since any of the systems can execute the correct function. This is illustrated in Fig. 5.9.

In this system $n - 1$ systems are redundant and the system still works when one system works, hence we can describe the reliability with the Boolean expression for an OR function of n systems:

$$R = P(A_{sys}) = P(A_1 + A_2 + \cdots + A_n) \tag{5.32}$$

This expression can be rewritten with the basic rules of probability for independent systems, which is really a cumbersome affair. It is much easier to convert the problem, calculating the probability the system will fail and then taking the complement. The probability the system will fail is

$$F(A_{sys}) = \prod_{i=1}^{n} [1 - P(A_i)] \tag{5.33}$$

then the probability that the system will not fail equals

$$R(_{sys}) = 1 - F(_{sys}) = 1 - \prod_{i=1}^{n} [1 - P(A_i)] \tag{5.34}$$

If again an exponential distribution is assumed then by substitution for n identical systems expression (5.34) yields

$$R(_{sys}) = 1 - \prod_{i=1}^{n} [1 - \exp(-\lambda_i t)] \tag{5.35}$$

Suppose $n = 2$ and the systems are identical with a failure rate of λ, then expression (5.35) yields

$$R(t, 2) = 2\exp(-\lambda t) - \exp(-2\lambda t) \tag{5.36}$$

You can try to find for yourself an expression when the failure rates are unequal. To find an expression for the MTBF we have to apply the basic definition for the MTBF, hence

$$\text{MTBF} = \int_0^\infty R_S \, \mathrm{d}t = \int_0^\infty \left[1 - \prod_{i=1}^{n} (1 - \exp(-\lambda_i t) \right] \mathrm{d}t \qquad (5.37)$$

To execute a general calculation requires quite a lot of effort, therefore we shall confine the calculation a little and perform the calculation for two systems with unequal failure rates.

$$\text{MTBF}(t, 2) = \int_0^\infty [1 - (1 - \exp(-\lambda_1 t))(1 - \exp(-\lambda_2 t))] \mathrm{d}t$$

$$= \int_0^\infty [\exp(-\lambda_1 t) + \exp(-\lambda_2 t) - \exp\{-(\lambda_1 + \lambda_2)t\}] \mathrm{d}t$$

Solving this multiple integral gives

$$\text{MTBF}(t, 2) = 1/\lambda_1 + 1/\lambda_2 - 1/(\lambda_1 + \lambda_2) \qquad (5.38)$$

If the failure rates are identical then the MTBF becomes

$$\text{MTBF}(t, 2) = \frac{3}{2\lambda} \qquad (5.39)$$

and in the same fashion for a triple identical system it can be found that the MTBF is

$$\text{MTBF}(t, 3) = \frac{11}{6\lambda} \qquad (5.40)$$

A graphical representation of these results can clarify the consequences for the reliability and MTBF of a parallel active on-line system. As can be seen from Fig. 5.10, the reliability and the MTBF will increase considerably in comparison with a single system. As stated already we have neglected any repair action, so in the following example the MTTF is calculated.

* * *

Example 5.12

Suppose the failure rate of a non-repairable system equals 0.0015/hour, then the MTTF of a single system is simply 1/0.0015 = 667 hours. For an active on-line standby system consisting of two such systems the MTTF equals, applying (5.39),

MTTF = 3/2 × 1/0.0015 = 1000 hours.

* * *

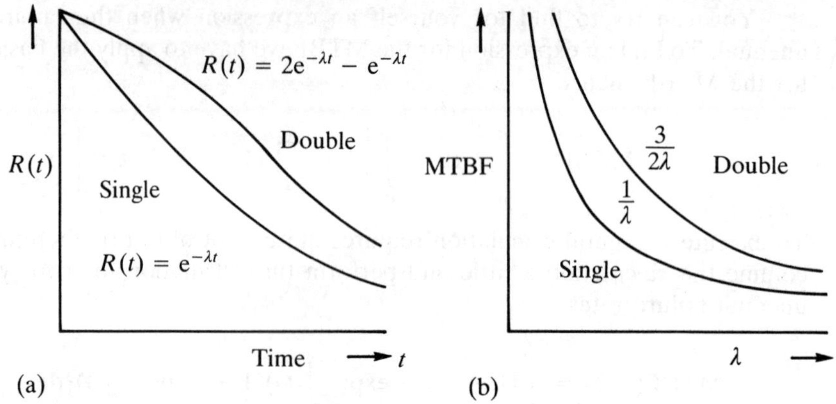

Figure 5.10 The reliability and MTBF of an active on-line standby system composed of two identical systems compared with a single system

It can be shown that for an n-fold active on-line standby system when no repair action is involved, the MTTF is

$$\text{MTTF} = \sum_{i=1}^{n} \frac{1}{i\lambda} = \frac{1}{\lambda} + \frac{1}{2\lambda} + \cdots + \frac{1}{n\lambda} \tag{5.41}$$

5.8.2 The cold on-line standby system

In this configuration only one system is on standby; the other systems will be switched on when a system fails. No repair actions are performed and maintenance is executed when all systems have failed. This is called a non-maintained cold on-line standby system. The basic configuration is depicted in Fig. 5.11 for which the following assumptions have to be made:

1 The system fails when all subsystems have failed.
2 The failure probabilities of all systems are independent.
3 Cold systems not in operation cannot fail.
4 The exponential distribution is applicable for all systems.
5 The reliability of switching a system off and on is 100%.

So it can be said that the system is still working when no system has failed, or if one system has failed, or two systems, etc., up to $n - 1$ systems. This refers to a Poisson distribution which can be described as follows:

$$\begin{aligned} R(t) &= P[x \leq (n-1)] = \sum_{k=0}^{n-1} \frac{(\lambda t)^k \exp(-\lambda t)}{k!} \\ &= \exp(-\lambda t)\left[1 + \lambda t + \frac{(\lambda t)^2}{2!} + \cdots + \frac{(\lambda t)^{n-1}}{(n-1)!}\right] \end{aligned} \tag{5.42}$$

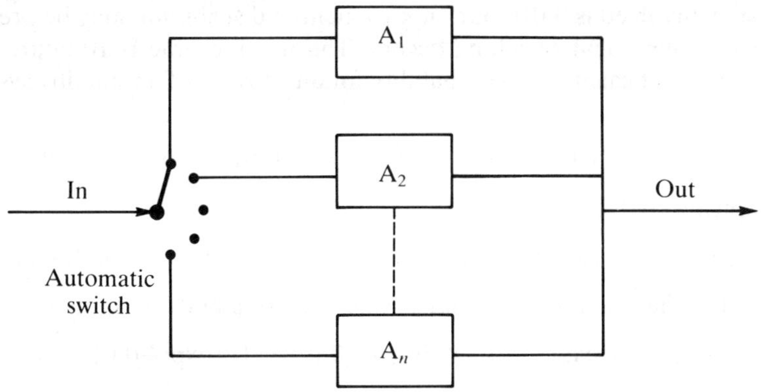

Figure 5.11 A parallel cold on-line standby system configuration of *n* identical systems

It is easy now to derive an expression for a cold on-line standby system composed of two identical systems. Substituting $n = 2$ in expression (5.42) gives

$$R(t) = \exp(-\lambda t) + \lambda t \exp(-\lambda t) \tag{5.43}$$

Again, applying the basic definition for the MTTF, it can be found that for the two-system configuration the MTTF is

$$\text{MTTF} = \int_0^\infty R(t)\,\mathrm{d}t = \int_0^\infty [(\exp(-\lambda t) + \lambda t \exp(-\lambda t)]\,\mathrm{d}t \tag{5.44}$$

and by partial integration for the second term the result is

$$\text{MTTF} = 1/\lambda + 1/\lambda = 2/\lambda \tag{5.45}$$

If the same calculation is executed for n systems then we find

$$\text{MTTF} = n/\lambda \tag{5.46}$$

Note that, here again, repair actions are omitted so it makes more sense to talk about MTTF instead of MTBF, of which the definition implies that repair actions can be performed.

In the following example we will compare both types of parallel configurations. It will be seen that a parallel cold on-line system delivers a slight advantage above a parallel active on-line system. However, it must be borne in mind that for the cold on-line system it is assumed that no switching problems exist, which in practice is not always the case.

* * *

Example 5.13

A two-system parallel configuration is considered of which the active and passive configuration must be compared on reliability aspects. The failure rate of each system involved is 0.01/hour, an exponential distribution may be presumed and if a system fails, ideal switching occurs. The mission time is 10 hours.

We first calculate the reliability for an active on-line standby system. Applying (5.35),

$$R(t, 2) = 1 - (1 - \exp(-0.01 \times 10))^2 = 0.9909$$

Applying (5.39),

$$\text{MTTF}(t, 2) = 3/2\lambda = 3/(2 \times 0.01) = 150 \text{ hours}$$

For the cold on-line standby system the reliability is, applying equation (5.43),

$$R(t, 2) = \exp(-0.01 \times 10) + (0.01 \times 10)\exp(-0.01 \times 10) = 0.9953$$

and applying (5.45), the MTTF becomes

$$\text{MTTF}(t, 2) = 2/0.01 = 200 \text{ hours}$$

which is an improvement of 25% compared with the active on-line standby system.

* * *

5.8.3 A parallel *k* out of *n* system

Here again, a distinction must be made between active and cold on-line standby systems. For the *active* on-line standby configuration composed of n identical systems of which k out of n systems are active, a binomial distribution is applicable. Then it can be shown that the reliability, when an exponential distribution is assumed, equals

$$R(n, k, p) = \sum_{i=k}^{n} \binom{n}{i} [\exp(-\lambda t)]^i [1 - \exp(-\lambda t)]^{n-i} \tag{5.47a}$$

and for the MTTF,

$$\text{MTTF}(n, k, p) = \frac{1}{\lambda} \sum_{i=0}^{n-k} \frac{1}{(n - i)} \tag{5.47b}$$

If $k = 1$ then a cold on-line standby configuration is present, as discussed already in Section 5.8.2, because only one system must be operational. If $k = n$ then by definition a series system is implied, because all systems must be active.

For a *cold* on-line standby system composed of n identical systems of which k must be operational, the reliability for an exponential distribution is given by

$$R(n, k, p_{\text{cold}}) = \sum_{i=0}^{n-k} [\exp(-\lambda t)]^k \frac{(k\lambda t)^i}{i!} \tag{5.48}$$

and for the MTTF it can be shown that

$$\text{MTTF}(n, k, p_{\text{cold}}) = \frac{n - k + 1}{k\lambda} \tag{5.49}$$

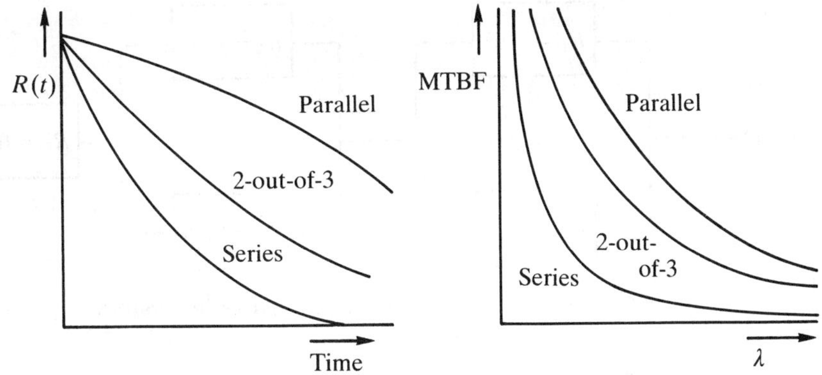

Figure 5.12 Graphs of the reliability and MTTF versus time for a series system, an active on-line standby system and a cold on-line k out of n system

From the above it can be seen that reliability increases as n increases. As claimed already, all systems are considered to be independent of each other. For comparison, Fig. 5.12 illustrates the reliability and MTTF behavior versus time for a series system, an active on-line standby parallel system, and a k out of n system.

5.9 METHODS TO DETERMINE THE RELIABILITY OF SYSTEMS

There are numerous methods available to determine the reliability of systems, some more powerful than others. We shall discuss five methods by illustration with examples. Four of these methods have one thing in common, that is how to simplify a complex system to a subset of known systems to make reliability calculations possible. The fifth method is called the Markov technique and is the most powerful one.

5.9.1 The network reduction method

This method implies a straightforward reduction of the system to simpler systems until just one single system remains.

* * *

Example 5.14

We will determine the reliability of the mixed parallel–series system depicted in Fig. 5.13. The failure distribution is presumed to be exponential and the parallel subsystems are of the active standby type.

The two parallel subsystems can be substituted by a single system as in Fig. 5.14, hence the result is a series system of three subsystems. Finally one system will remain when this is replaced again as illustrated in Fig. 5.15.

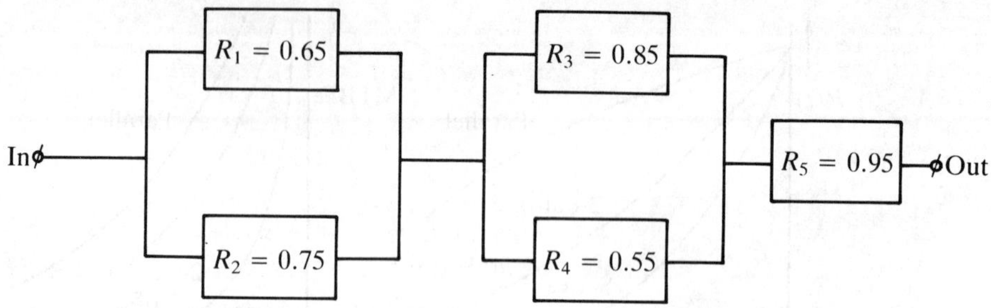

Figure 5.13 A mixed parallel-series system

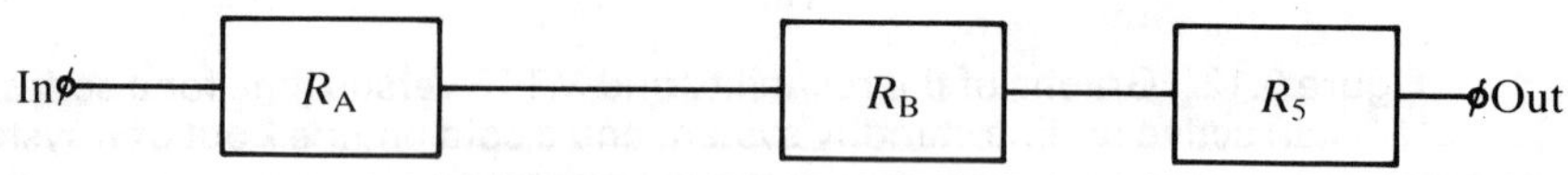

Figure 5.14 Reduction of the mixed system of Fig. 5.13 to a series system only

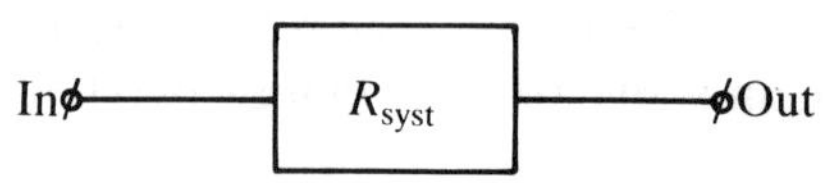

Figure 5.15 Complete reduced system of Fig. 5.13 representing the original system with the same function

The calculation can be executed by first applying equation (5.35) for both parallel subsystems, hence

$$R(\text{syst}) = 1 - \prod_{i=1}^{2} [1 - \exp(-\lambda_i)t] \tag{5.50}$$

then

$$R_A = 1 - (1 - 0.65)(1 - 0.75)$$
$$= 1 - 0.0875 = 0.9125$$

and

$$R_B = 1 - (1 - 0.85)(1 - 0.55)$$
$$= 1 - 0.0675 = 0.9325$$

Then the system reliability can be found by multiplication:

$$\begin{aligned} R_{syst} &= R_A R_B R_5 \\ &= (0.9125)(0.9325)(0.95) = 0.8084 \end{aligned} \tag{5.51}$$

In the same way the MTTF can be calculated, resulting in the following expression for the failure rate:

$$\lambda_{sys} = \lambda_A + \lambda_B + \lambda_5 \tag{5.52}$$

If λ is a constant and identical for all subsystems then the MTTF equals

$$\lambda_{sys} = 2\lambda/3 + 2\lambda/3 + \lambda = 7\lambda/3 \qquad (5.53)$$

Taking the reciprocal value of (5.53) delivers the MTTF,

$$MTTF_{sys} = 1/\lambda_{sys} = 3/7\lambda$$

Note that for parallel systems the failure rate decreases and for series system it increases. If in this example five different failure rates were involved all expressions would be considerably more complicated.

* * *

5.9.2 The path-tracing technique

This method considers the different ways a system can function satisfactorily. The systems are supposed to be independent. The method will be illustrated by the following example. The method is a rather simple one and is well suited for systems that are not too complicated.

* * *

Example 5.15

Applying Boolean algebra and probability theory on the mixed independent parallel–series system depicted in Fig. 5.16, we can write the following expression for the reliability, summing both paths AC and BC for a successful operational state:

$$\begin{aligned} R_{sys} &= P(AC + BC) = P(AC) + P(BC) - P(ABC) \\ &= P(A)P(C) + P(B)P(C) - P(A)P(B)P(C) \end{aligned} \qquad (5.54)$$

If the values for the reliability of the independent systems are known, the total reliability of the system can be found by substitution. For instance, if $P(A) = 0.95$, $P(B) = 0.85$ and $P(C) = 0.80$, then $R_{sys} = 0.794$.

* * *

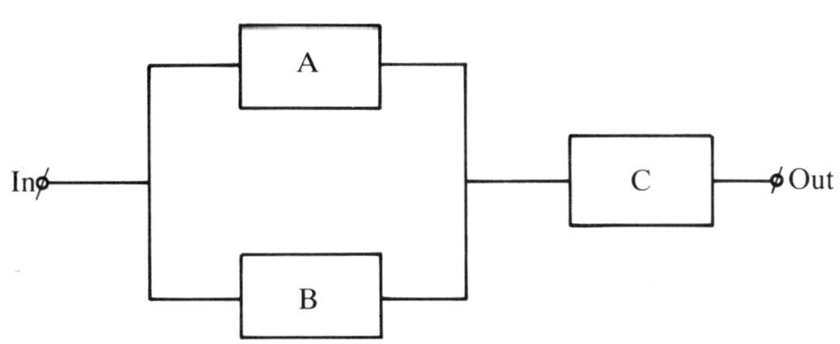

Figure 5.16 Parallel-series configuration of three independent systems

5.9.3 The decomposition technique

The decomposition technique is well suited to the determination of the reliability of more complex systems. A good example of such a system is the bridge configuration depicted in Fig. 5.17. The aim of this method is to reduce the system to a configuration of less complicated subsystems to make reliability calculations possible with reduced effort.

Here we have to follow a quite different strategy, but suppose first that we try to apply the path-tracing technique, then the following expression can be found for the reliability of the complete system, because five different paths can be found for successful operation:

$$R_{\text{sys}} = P(AD + BK + ACK + BCD) \tag{5.55}$$

If you try to convert this expression, over 30 terms are found. The result can be obtained after a tremendous effort with a great chance of making mistakes.

To solve this problem we use the following theorem, sometimes improperly called the Bayes theorem, but this is only part of it.

$$P(E) = \sum_i P(E/B_i)P(\mathrm{B_i}) \tag{5.56}$$

with

$$\sum_i P(B_i) = 1$$

This theorem says that if a certain event E depends on a number of mutually exclusive events B_i, then the probability that E occurs equals the product of the probability of event B_i and the conditional probability $P(E/B_i)$, summed over all possible events B_i. This theorem will be applied on the bridge configuration by making a special choice of events! Here is how it works.

* * *

Example 5.16

1 Remove a well-chosen subsystem from the configuration under consideration.

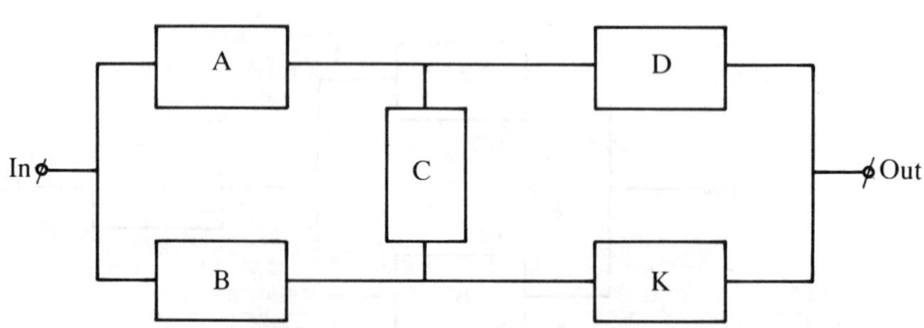

Figure 5.17 Bridge configuration of five independent systems

This removed system is sometimes called the 'key system'. In this example C will be removed.

2 Replace the removed system by another virtual subsystem V which can never fail, and calculate the conditional reliability.
3 Replace the removed system by another virtual system $\bar{V}$ which can never work, and calculate the conditional reliability.
4 Find the reliability of the sub-system E applying equation (5.56) combined with the above statements 1, 2 and 3, then

$$R_{sys} = P(E) = P(\text{E works/V works always})\, P(\text{V works always}) + P(\text{E works}/\bar{\text{V}} \text{ never works})\, P(\bar{\text{V}} \text{ never works}) \quad (5.57)$$

You can see from this equation that for V and $\bar{V}$ two conditional probabilities for the given events are introduced in accordance with equation (5.56), hence two new systems can be drawn, R_{E1} and R_{E2}, of which the reliability can be calculated by standard equations. These two newly introduced systems are depicted in Figs 5.18 and 5.19. For the system depicted in Fig. 5.18 the reliability is (with abbreviated notation)

$$R_{E1} = P(E/V) \quad (5.58)$$

and for the system depicted in Fig. 5.19 the reliability is

$$R_{E2} = P(E/\bar{V}) \quad (5.59)$$

then the reliability of the total system E is

$$R(E) = R_{E1}P(V) + R_{E2}P(\bar{V}) \quad (5.60)$$

or

$$R(E) = P(E/V)\mathrm{P}(V) + P(E/\bar{V})P(\bar{V}) \quad (5.61)$$

For the parallel–series system of Fig. 5.18 we can write

$$R_{E1} = [1 - (1 - P(A))(1 - P(B))][1 - (1 - P(D))(1 - P(K)] \quad (5.62)$$

and for the parallel system of Fig. 5.19 we can write

$$R_{E2} = 1 - [(1 - P(A)P(D)][1 - P(B)P(K)] \quad (5.63)$$

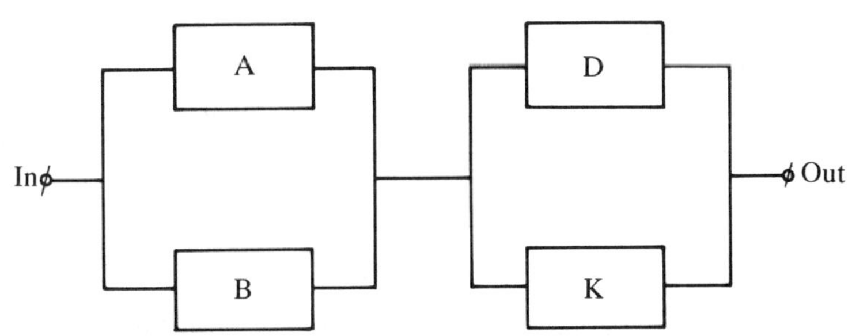

Figure 5.18 Bridge configuration of which system C is replaced by a system V which never fails

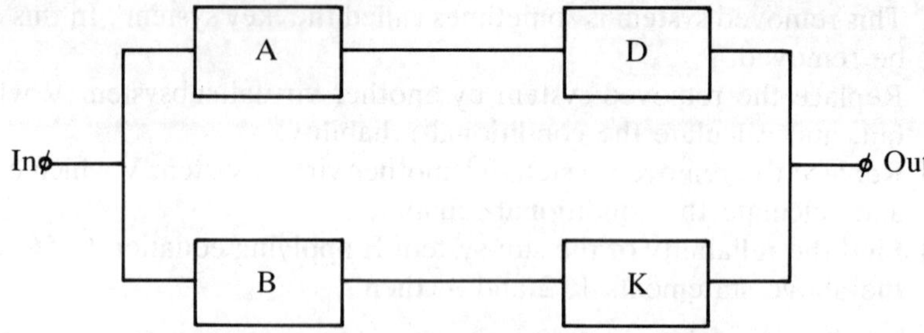

Figure 5.19 Bridge configuration of which system C is replaced by a system V which never works

Now one final step must be set for the virtual element V: if we equate the reliability of our key element C to the reliability of V then

$$\begin{aligned} P(V) &= P(C) \\ P(\bar{V}) &= 1 - P(V) = 1 - P(C) \end{aligned} \tag{5.64}$$

and the reliability of the bridge configuration becomes

$$R_{sys} = R_{E1} P(C) + R_{E2}[1 - P(C)] \tag{5.65}$$

All systems are presumed to be independent. As you can see, it is not that easy but it is the shortest way to find the reliability for a more complex system such as a bridge configuration. If all systems are identical and have the same reliability R you can verify that the reliability of the bridge configuration is

$$R_{sys} = 2R^2 + 2R^3 - 5R^4 + 2R^5 \tag{5.66}$$

When the reliability is found, the MTTF can be determined by applying the proper definition for the MTTF.

* * *

5.9.4 The minimal cut set technique

The minimal cut set technique is a method well suited to computer calculations. The method relies on the fact that in the system under consideration one searches for the minimum collection of elements that must be removed from the system so that the system fails. In general more than one such a minimum cut set can be found for which the system fails, but every minimum cut set cannot be reduced to more simply cut sets. If all possible minimum cut sets $A_1, A_2, \ldots, A_n$ are determined then the reliability is

$$R_{sys} = 1 - P[\bar{A}_1 + \bar{A}_2 + \cdots + \bar{A}_n] \tag{5.67}$$

where $\bar{A}_1, \bar{A}_2, \ldots, \bar{A}_n$ is the event the cut set will fail.

* * *

Example 5.17

We will demonstrate this method by the example illustrated in Fig. 5.20. By investigation of the system the number of cut sets A so that the system fails is five. They are

$$A_1 = BC \quad A_2 = BD \quad A_3 = CD \quad A_4 = D \quad \text{and} \quad A_5 = BCD$$

but these are not all minimum cut sets because A_4 is involved in A_5 and so are A_3 and A_2. The minimum cut set are A_1 and A_4 only, so the reliability of the system is

$$R_{\text{sys}} = 1 - P(\bar{A}_1 + \bar{A}_4) \tag{5.68}$$

$$\begin{aligned} R_{\text{sys}} &= 1 - P(B\bar{C} + \bar{D}) \\ &= 1 - [P(\bar{B})P(\bar{C}) + P(\bar{D}) - P(\bar{B})P(\bar{C})\ P(\bar{D})] \end{aligned} \tag{5.69}$$

in which it is again implicitly assumed that the systems are independent.

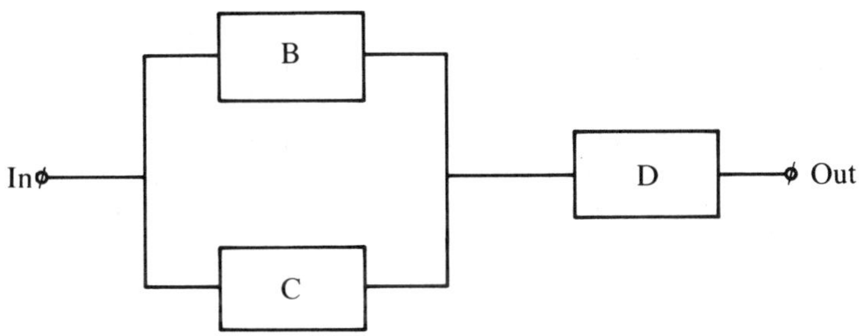

Figure 5.20 Minimum cut set technique applied to a parallel–series system

* * *

5.9.5 Markov techniques

Markov techniques are the most powerful methods available for reliability calculations. Applying this method makes it possible to account for dependent and independent systems, time dependent failure and repair rates which are not necessarily equal. Besides these aspects, systems with more than two possible states can be taken into consideration. This makes it possible to take into account possible transitions between different states. We first have to introduce some necessary definitions before demonstrating this method with an example.

A Markov state model of order k and with m possible events for every stochastic variable x_i, can be in m^k possible states. The number of symbols with which every state must be described is called the *order* of the Markov chain.

Suppose there is a system with $m = 2$ and the two stochastic variables $x_i = 1$ or 0, then the number of possible states equals 4 and these can be denoted $S_{0,0}$, $S_{0,1}$, $S_{1,0}$ and $S_{1,1}$. In Table 5.3 this is illustrated with still another example.

Table 5.3 Possible states related to the order of a Markov chain

	S	S_m	$S_{m,m}$	$S_{m,m,m}$
Order k	0	1	2	3
Number of states	1	2	4	8
States	S	S_0, S_1	$S_{0,0}; S_{0,1}$ $S_{1,0}; S_{1,1}$	$S_{0,0,0}; S_{0,0,1}$ $S_{0,1,0}; S_{0,1,1}$ $S_{1,0,0}; S_{1,0,1}$ $S_{1,1,0}; S_{1,1,1}$

If a Markov chain is in a certain state, a transition can occur from one state to another. In particular, if a possible state is left, another state can occur, but the original state can be obtained again. Thus after a state is left m different events can occur; therefore the number of transitions is determined by the expression m^{k+1}.

If for instance $k = 1$ and $m = 2$ with $x_i = 0$ or 1, then the following four transitions can occur: $0 \rightarrow 0$; $0 \rightarrow 1$; $1 \rightarrow 0$; and $1 \rightarrow 1$.

As can be seen from Table 5.3, for $m = 2$ and $k = 3$ we find 8 different states and 16 possible transitions. The next step is to assign a probability to every *transition*. This means that given a certain state, we require the probability that a possible transition will occur. This describes precisely the definition of a conditional probability. The collection of transitions and states is often depicted in a so-called *state diagram*. An example of such a diagram is given in Fig. 5.21.

As can be seen from this diagram, the indices of S denote the sequence numbers of the states and after the equals sign the value of the used symbol is denoted, so for instance $S_2 = 1$ is the second possible state with value 1. A transition $P(1/0)$ should be read as the probability that state 1 will occur given the system is in state 0.

For this state diagram the following two equations will hold:

$$P(0/0) + P(1/0) = 1 \tag{5.70}$$

$$P(0/1) + P(1/1) = 1 \tag{5.71}$$

You can check for yourself that this has to be true. In general for a Markov chain the statement will hold that given a certain state S_a the sum of all possible transition probabilities is 1, or

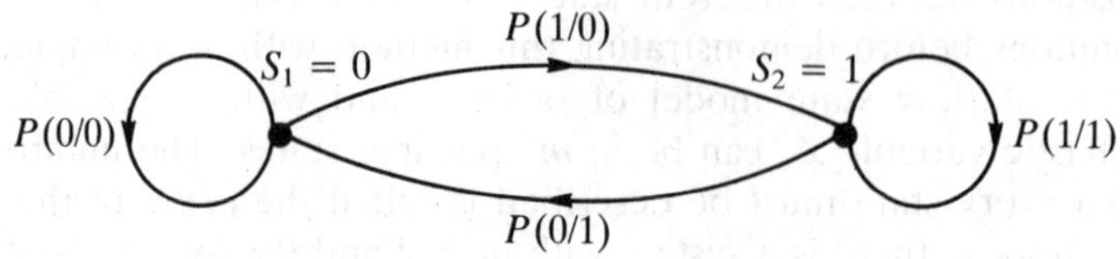

Figure 5.21 State diagram for a Markov chain with two states of order 1 and $m = 2$

$$\sum_i P(S_i/S_a) = 1 \tag{5.72}$$

where S_i denotes another possible state to be reached.

Finally, to apply Markov techniques some other assumptions and definitions have to be made:

1 Every state is mutually exclusive and states are independent of each other.
2 The sum of probabilities for every possible state is 1.
3 In the time interval Δt only one event can occur.
4 Failure rates and repair rates are supposed to be constant.
5 The probability a system will fail in the time interval Δt equals $\Delta t\lambda$.
6 The probability a system will not fail in the time interval Δt equals $1 - \Delta t\lambda$.
7 The probability of repairing a system in the interval Δt is $\Delta t\mu$.
8 The probability a system will not be repaired in the time interval Δt equals $1 - \Delta t\mu$.
9 The state in which all systems have failed will be denoted by $S_i = 0$; this is called an absorbing state, because certainly $P(0/0) = 1$!
10 One of the systems is operational, the other systems are to be repaired or are waiting for repair actions. This is denoted by $S_2 = 1$.
11 All systems are working, a state often called the initial state and denoted by $S_3 = 2$.
12 An exponential distribution will be assumed.

Example 5.18

We will start with the most simple example possible. Consider a system which consists of one system only and cannot be repaired. Then $S_2 = 1$ cannot occur, hence the state diagram is as depicted in Fig. 5.22.

This is a Markov chain of order 1 with two symbols, the number of states equals $m^k = 2$ and the number of transitions is $m^{k+1} = 4$, of which one transition cannot occur, because repair is excluded. This is depicted by the dashed line in Fig. 5.22.

Of both possible states S_1 and S_3 the probability has to be determined at the time $t + \Delta t$. Applying equation (5.56) we can write the following two equations for the two states to be considered:

$$S_1(t + \Delta t) = P(0)_{t+\Delta t} = P(0/0)\,P(0) + P(0/2)\,P(2) \tag{5.73}$$

$$S_3(t + \Delta t) = P(2)_{t+\Delta t} = P(2/0)\,P(0) + P(2/2)\,P(2) \tag{5.74}$$

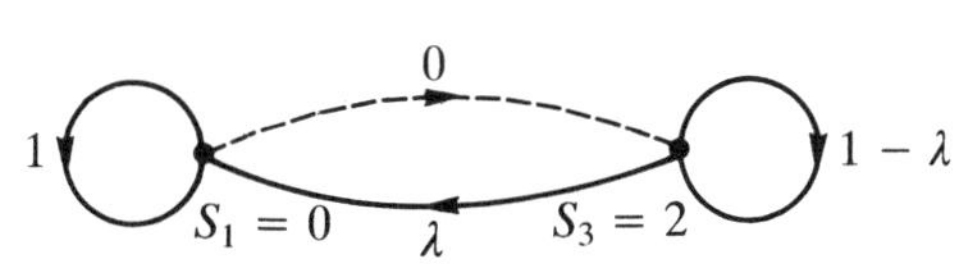

Figure 5.22 State diagram of a single system without repair actions

It is important now to recognize that if S_3 is calculated, the reliability of the system is found, because this is the probability the system will work at the time $t + \Delta t$ by definition of the above statements.

From the state diagram the transitions with their respective (conditional) probabilities can be read and can be written as a matrix:

$$\begin{array}{cc} & \begin{array}{cc} 2 & 0 \leftarrow \text{at time } t + \Delta t \end{array} \\ P = \begin{array}{c} 2 \\ 0 \\ \uparrow \\ t = t \end{array} & \begin{pmatrix} 1 - \lambda & \lambda \\ 0 & 1 \end{pmatrix} \end{array} \tag{5.75}$$

At the top of the matrix the states at time $t + \Delta t$, called the final state, are depicted, and to the left of the matrix the states at time t, called the initial state, are written. Within the matrix we write the transition probabilities for $t + \Delta t$. For instance, $P(2/2) = 1 - \lambda$ is the probability that state 2 will occur at $t + \Delta t$ given the system is in state 2 at t, etc. By substitution of these values in equations (5.73) and (5.74) the following difference equations are found, in accordance with statements 5 and 6:

$$P(0)_{t+\Delta t} = 1P(0)_t + \Delta t \lambda P(2)_t \tag{5.76}$$

$$P(2)_{t+\Delta t} = 0P(0)_t + (1 - \Delta t \lambda) P(2)_t \tag{5.77}$$

Rearrangement of these equations yields two differential equations

$$\lim_{\Delta t \to 0} \frac{P(0)_{t+\Delta t} - P(0)_t}{\Delta t} = P'(0)_t = \lambda P(2)_t \tag{5.78}$$

$$\lim_{\Delta t \to 0} \frac{P(2)_{t+\Delta t} - P(2)_t}{\Delta t} = P'(2)_t = -\lambda P(2)_t \tag{5.79}$$

or

$$P'(0)_t - \lambda P(2)_t = 0 \tag{5.80}$$

$$P'(2)_t + \lambda P(2)_t = 0 \tag{5.81}$$

To solve these equations two boundary conditions must be known, namely at time $t = 0$ the system works so $P(2)_{t=0} = 1$; and at $t = 0$ the system cannot fail, thus $P(0)_{t=0} = 0$. Applying Laplace transforms gives

$$sP(0) - P(0)_0 = \lambda P(2) \tag{5.82}$$

and

$$sP(2) - P(2)_0 = -\lambda P(2) \tag{5.83}$$

Substituting the boundary conditions gives

$$sP(0) - 0 = \lambda P(2) \tag{5.84}$$

$$sP(2) - 1 = -\lambda P(2) \tag{5.85}$$

Solving (5.84) and (5.85) yields

$$P(2) = \frac{1}{s + \lambda} \tag{5.86}$$

and

$$P(0) = \frac{\lambda}{s(s + \lambda)} \tag{5.87}$$

Inverse transformation delivers

$$S_3 = P(2) = \exp(-\lambda t) \tag{5.88}$$

and

$$S_1 = P(0) = 1 - \exp(-\lambda t) \tag{5.89}$$

As is stated already the reliability $R(t)$ is represented by S_3 and is in accordance with equation (5.5).

* * *

We will give now a more complicated example of an active on-line standby parallel system.

* * *

Example 5.19

The system to be considered is depicted in Fig. 5.23 and can be interpreted as a measurement and control system in an active on-line standby configuration monitoring and controlling a process. The failure rate and repair rate are supposed to be constant.

Both systems are active and controlling each other's operational behavior; when one of the two systems is operational the systems works. The reliability can be described as

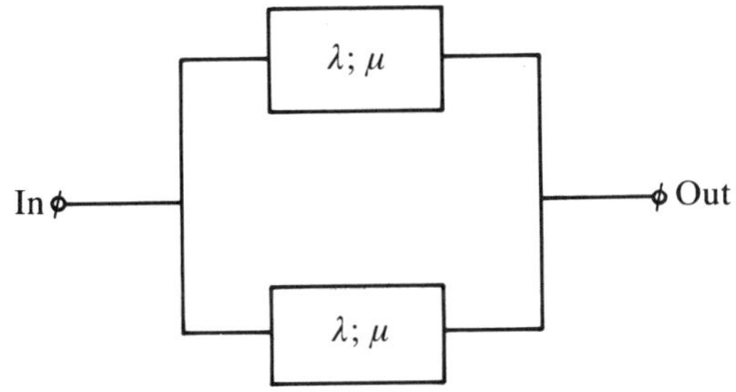

Figure 5.23 Measurement and control systems of two identical active on-line standby systems

$$R(t) = 1 - P(0)_t \tag{5.90}$$

The system can be interpreted as a Markov chain of order 1 with three state symbols, namely

$S_1 = 0$ both systems have failed
$S_2 = 1$ one system is operational and the other is in repair
$S_3 = 2$ both systems are active

Hence the number of states is three and the number of transitions is $m^{k+1} = 9$. Some transitions cannot occur. The state diagram is depicted in Fig. 5.24, from which several important conclusions may be drawn.

1 The probability the system will come into state S_1 given that the system is in state S_1 is a sure event, thus $P(0/0) = 1$. In other words, a failed system cannot become operational.
2 The probability the system will come into action again when both systems have failed is 0, hence $P(2/0) = 0$.
3 The probability the system will fail when both systems are active is 0, hence $P(0/2) = 0$.

The state diagram can be represented as a matrix:

$$\begin{array}{cc} & \begin{array}{ccc} 2 & 1 & 0 \leftarrow \text{state at } t = t + \Delta t \end{array} \\ P = \begin{array}{c} 2 \\ 1 \\ 0 \\ \uparrow \\ t = t \end{array} & \begin{pmatrix} 1 - 2\lambda & 2\lambda & 0 \\ \mu & 1 - \mu - \lambda & \lambda \\ 0 & 0 & 1 \end{pmatrix} \end{array} \tag{5.91}$$

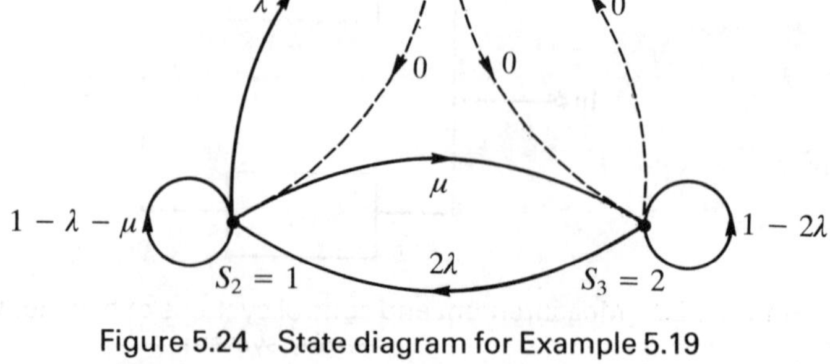

Figure 5.24 State diagram for Example 5.19

Note that the sum of the conditional probabilities in every row of the matrix is 1; this is a check for the validity of the matrix.

The appropriate equations relating to this matrix are

$$\begin{aligned} P(0)_{t+\Delta t} &= P(0/0)P(0)_t + P(0/1)P(1)_t + P(0/2)P(2)_t \\ P(1)_{t+\Delta t} &= P(1/0)P(0)_t + P(1/1)P(1)_t + P(1/2)P(2)_t \\ P(2)_{t+\Delta t} &= P(2/0)P(0)_t + P(2/1)P(1)_t + P(2/2)P(2)_t \end{aligned} \tag{5.92}$$

Substituting the appropriate values gives a set of difference equations

$$\begin{aligned} P(0)_{t+\Delta t} &= 1P(0)_t + \Delta t\lambda P(1)_t + 0P(2)_t \\ P(1)_{t+\Delta t} &= 0P(0)_t + (1 - \Delta t\lambda - \Delta t\mu)P(1)_t + 2\lambda\Delta tP(2)_t \\ P(2)_{t+\Delta t} &= 0P(0)_t + \Delta t\mu P(1)_t + (1 - \Delta t2\lambda)P(2)_t \end{aligned} \tag{5.93}$$

By rearranging and taking the limit as $\Delta t \to 0$, the set of difference equations can be transformed into a set of differential equations:

$$\left.\begin{aligned} P'(0)_t - \lambda P(1)_t &= 0 \\ P'(1)_t + (\lambda + \mu)P(1)_t - 2\lambda P(2)_t &= 0 \\ P'(2)_t - \mu P(1)_t + 2\lambda P(2)_t &= 0 \end{aligned}\right\} \tag{5.94}$$

Applying the Laplace transform of (5.94), by definition the following boundary conditions at the moment $t = 0$ are valid, namely $P(2)_{t=0} = 1$, because the system will work at that time, and $P(1)_{t=0} = P(0)_{t=0} = 0$, for the reason both systems are operational at that time. We find

$$sP(0)_s - \lambda P(1)_s = 0 \tag{5.95}$$

$$-(\mu + \lambda + s)P(1)_s + 2\lambda P(2)_s = 0 \tag{5.96}$$

$$-\mu P(1)_s + (2\lambda + s)P(2)_s = 0 \tag{5.97}$$

If we solve this set of equations for $P(1)_s$ only, the other quantities are also known. Combining (5.96) and (5.97) gives

$$P(1)_s = \frac{2\lambda}{2\lambda^2 + (3\lambda + \mu)s + s^2} \tag{5.98}$$

This equation can be written as

$$P(1)_s = \frac{2\lambda}{(s - s_1)(s - s_2)} \tag{5.99}$$

where s_1, s_2 are the two values of

$$\tfrac{1}{2}[-(3\lambda + \mu) \pm \sqrt{\lambda_2 + 6\lambda\mu + \mu^2}] \tag{5.100}$$

Fractional splitting of (5.99) delivers

$$P(1)_s = \frac{2\lambda}{s_1 - s_2}\left[\frac{1}{s - s_1} - \frac{1}{s - s_2}\right] \tag{5.101}$$

Taking the inverse Laplace transform of (5.101) gives $P(1)$ in the time domain:

$$P(1)_t = \frac{2\lambda}{s_1 - s_2}[\exp(s_1 t) - \exp(s_2 t)] \tag{5.102}$$

We are not yet finished because we have to find $P(0)_t$ before ultimately the reliability can be found. Rearranging the first equation of (5.94) yields

$$P'(0)_t = \lambda P(1)_t \tag{5.103}$$

then $P(0)_t$ becomes

$$P(0)_t = \int_0^T P'(0)\,\mathrm{d}t = \int_0^T \lambda P(1)\,\mathrm{d}t \tag{5.104}$$

or after integration

$$P(0)_t = 1 + \frac{s_2 \exp(s_1 t) - s_1 \exp(s_2 t)}{s_1 - s_2} \tag{5.105}$$

and finally applying equation (5.90) and substituting (5.105) the reliability becomes

$$\begin{aligned} R(t) &= 1 - P(0)_t \\ &= \frac{s_1 \exp(s_2 t) - s_2 \exp(s_1 t)}{s_1 - s_2} \end{aligned} \tag{5.106}$$

where s_1 and s_2 are as given in equation (5.100).

* * *

This method requires some effort! But once the Markov state diagram is determined for a given system, the reliability can be found by a systematic application of the concepts given above. The elegance and the power of this method lie in the fact that we can take into account time- and state-dependent states and transitions.

5.10 DESIGN ASPECTS

The design process of a system is very complex because numerous aspects have to be taken into account before the right decisions can be made. Some basic rules are available and a very systematic approach is needed. The best way to go ahead is to apply a type of checklist and to investigate every part of it. To give you an impression of such a checklist which can be used in a design procedure, here is an example.

1 Define the application and purpose of the system.
2 Define the required reliability with respect to the environmental conditions the system will be used in.
3 Use a type of index system to describe the effectiveness of the reliability.
4 Define the system and divide the system into pieces which are independent of each other.

5 Find for every subsystem the optimum configuration with respect to costs and reliability.
6 Determine the costs of manufacturing for every subsystem.
7 Define a complete system configuration, and a maintenance philosophy relating to all other factors and conditions.
8 Assign failure rates and repair rates to every part of the system.
9 Predict the reliability of every subsystem and the whole system. Use for this part of the design procedure the latest published data and failure rate handbooks. The most reliable sources on data for components and systems are issued by the Department of Defense, DOD (Washington, DC).
10 Compare the calculated results with the predicted values. Repeat the procedure where required to determine best possible strategy.

An important note can be made here. As we have seen in Fig. 5.4, costs versus reliability always play a vital role in the design procedure. But every time a disaster has occurred, apparently basic human factors were involved. It is interesting to mention that some of these human factors are very often responsible for great disasters. Some factors are: running out of time, hurry, arrogance, greed, commercial stress, negligence, wrong or failing communication procedures in the organization, etc. That this is true can easily be verified by analyzing disasters such as a space shuttle failure and a nuclear powerplant accident. No system is only technique; human factors are always involved and these are the most difficult to grasp. Therefore, despite all possible implemented precautions, 100% reliability cannot exist.

There are some other points of view to be considered here. It will be clear that in the design procedure the manufacturer will use different arguments than the consumer. Last but not least safety aspects can be important, as is the case in the aircraft industry, in the medical field and when the environment is involved. These aspects can drastically increase the costs of design.

It is beyond the scope of this textbook to present a detailed discussion of all types of design procedures. We will confine the treatment to some relevant aspects to stress their importance.

In general in the design stage eight different quality classes can be distinguished ranging from the highest military standard class through industrial class, to consumer class with the least severe requirements. To each class a so-called quality factor B_Q can be assigned, as summarized in Table 5.4.

Concerning the environmental conditions, 11 different environmental classes are distinguished. These classes range from benign laboratory conditions to severe environmental conditions during a missile launch with large shock and vibration levels. The environmental classes with their respective applied symbols are summarized in Table 5.5.

An example for integrated circuits is given in Table 5.6. In this table the circuit complexity is given in number of gates per circuit. In each column the failure rate can be read as a function of the environmental conditions and circuit complexity in failure rates per 10^6 hr.

Table 5.4 Summary of quality classes with assigned quality factors

Quality class	*Description*	B_Q
S	Procured in full accordance with MIL-M-38510, Class S requirements.	1
B	Procured in full accordance with MIL-M-38510, Class B requirements	2
B-1	Procured to screening requirements of MIL-STD-883, Method 5004, Class B, and in accordance with the electrical requirements of MIL-M-38510 slash sheet or vendor or contractor electrical parameters. The device must be qualified to requirements of MIL-STD-883, Method 5005, Class B. No waivers are allowed.	5
B-2	Procured to vendor's equivalent of screening requirements of MIL-STD-883, Method 5004, Class B, and in accordance with vendor's electrical parameters. Vendor waives certain requirements of MIL-STD-883, Method 5004, Class B.	10
C	Procured in full accordance with MIL-M-38510, Class C requirements.	16
C-1	Procured to screening requirements of MIL-STD-883, Method 5004, Class C and in accordance with the electrical requirements of MIL-M-38510 slash sheet or vendor or contractor electrical specification. The device must be qualified to requirements of MIL-STD-883, Method 5005, Class C. No waivers are allowed.	90
D	Commercial (or non-mil standard) part, hermetically sealed, with no screening beyond the manufacturer's regular quality assurance practices.	150
D-1	Commercial (or non-mil standard) part, packaged or sealed with organic materials (e.g., epoxy, silicone, or phenolic).	300

Table 5.5 Summary of the 11 environmental classes with applied symbols

Environment	*Symbol* B_E	*Nominal environmental conditions*	*No.*
Ground, Benign	G_B	Nearly zero environmental stress with optimum engineering operation and maintenance.	1
Space, Flight	S_F	Earth orbital. Approaches Ground, Benign conditions without access for maintenance. Vehicle neither under powered flight nor in atmospheric re-entry.	2
Ground, Fixed	G_F	Conditions less than ideal to include installation in permanent racks with adequate cooling air, maintenance by military personnel and possible installation in unheated buildings.	3

Table 5.5 Continued

Environment	*Symbol* B_E	*Nominal environmental conditions*	*No.*
Ground, Mobile	G_M	Conditions more severe than those for G_F, mostly for vibration and shock. Cooling air supply may also be more limited, and maintenance less uniform.	9
Naval, Sheltered	N_S	Surface ship conditions similar to G_F but subject to occasional high shock and vibration.	5
Naval, Unsheltered	N_U	Nominal surface shipborne conditions but with repetitive high levels of shock and vibration.	6
Airborne, Inhabited, Transport	A_{IT}	Typical conditions in transport or bomber compartments occupied by aircrew without environmental extremes of pressure, temperature, shock and vibration, and installed on long mission aircraft such as transports and bombers.	7
Airborne, Inhabited, Fighter	A_{IF}	Same as A_{IT} but installed on high-performance aircraft such as fighters and interceptors.	8
Airborne, Uninhabited, Transport	A_{UT}	Bomb bay, equipment bay, tail, or wing installations where extreme pressure, vibration, and temperature cycling may be aggravated by contamination from oil, hydraulic fluid and engine exhaust. Installed on long mission aircraft such as transports and bombers.	9
Airborne, Uninhabited, Fighter	A_{UF}	Same as A_{UT} but installed on high-performance aircraft such as fighters and interceptors.	10
Missile Launch	M_L	Severe conditions of noise, vibration, and other environments related to missile launch, and space vehicle boost into orbit, vehicle re-entry and landing by parachute. Conditions may also apply to installation near main rocket engines during launch operations.	11

For instance, to find the failure rate of a system the following formula can be used:

$$\lambda_{BE} = \sum_{i=1}^{n} N_i(\lambda_G B_Q)_i B_L \qquad (5.107)$$

Table 5.6 Example of failure rates per 10^6 hr for integrated circuits as a function of circuit complexity and environmental conditions

	Environmental conditions									
Circuit complexity	G_B & S_F	G_F	A_{IT}	A_{IF}	N_S	G_M	A_{UT}	A_{UF}	N_U	M_L
1–20 gates	0.010	0.048	0.099	0.16	0.14	0.12	0.21	0.30	0.25	0.24
21–50 gates	0.048	0.19	0.31	0.40	0.43	0.34	0.73	0.86	0.92	0.52
51–100 gates	0.076	0.31	0.48	0.59	0.68	0.54	1.2	1.3	1.5	0.78
101–500 gates	0.19	0.82	1.2	1.4	1.7	1.3	3.1	3.4	3.9	1.7
501–1000 gates	0.32	1.4	2.0	2.3	2.8	2.1	5.1	5.5	6.4	2.6
1001–2000 gates	0.74	3.1	4.6	5.2	6.4	4.8	12.0	13.0	15.0	6.0
2001–3000 gates	2.0	8.4	13.0	14.0	17.0	13.0	33.0	35.0	41.0	16.0
3001–4000 gates	5.4	23.0	35.0	39.0	47.0	36.0	90.0	96.0	111.0	44.0
4001–5000 gates	15.0	62.0	94.0	105.0	128.0	97.0	241.0	258.0	299.0	121.0
ROM** ≤ 320 bits	0.021	0.087	0.13	0.15	0.18	0.14	0.33	0.36	0.42	0.17
ROM** 321–576 bits	0.031	0.13	0.19	0.22	0.27	0.20	0.49	0.53	0.62	0.26
ROM** 577–1120 bits	0.048	0.20	0.31	0.35	0.42	0.32	0.78	0.84	0.98	0.41
ROM** 1121–2240 bits	0.072	0.30	0.45	0.52	0.63	0.48	1.2	1.3	1.5	0.61
ROM** 2241–5000 bits	0.11	0.46	0.70	0.80	0.96	0.74	1.8	1.9	2.2	0.94
ROM** 5001–11000 bits	0.17	0.70	1.1	1.2	1.5	1.1	2.7	2.9	3.4	1.5
ROM** 11001–17000 bits	0.25	1.1	1.6	1.9	2.2	1.7	4.1	4.5	5.2	2.2
Linear, ≤ 32 transistors	0.011	0.052	0.12	0.20	0.16	0.15	0.22	0.35	0.27	0.33
Linear, 33–100 transistors	0.023	0.11	0.24	0.41	0.35	0.31	0.48	0.73	0.60	0.66

where

- λ_{BE} is the total failure rate in the applicable environment
- λ_G is the specific failure rate of the ith component in the applicable environment (see for instance Table 5.6)
- B_Q is a quality factor (see Table 5.4)
- B_L is a learning factor which equals 10 when a change in the design is made
- N_i is the number of components of type i
- n is the number of different categories of components

Besides the application of redundancy, one of the most effective measurements to improve the reliability or to reduce the failure rate of a system is to apply *derating*. This technique implies a reduction in workload for every component involved in the design. Successful application of this method requires a thorough knowledge of the component's failure rates. Its effect can be demonstrated by the example depicted in Fig. 5.25 in which the failure rate of a capacitor versus temperature is shown with normalized voltage load as parameter. Derating can easily deliver factors of improvement of 1000, as shown.

However, six other different environmental circumstances which influence the reliability can be distinguished, and these are summarized in Table 5.7.

If any of these conditions can occur, additional factors in the calculation of the failure rate have to be accounted for.

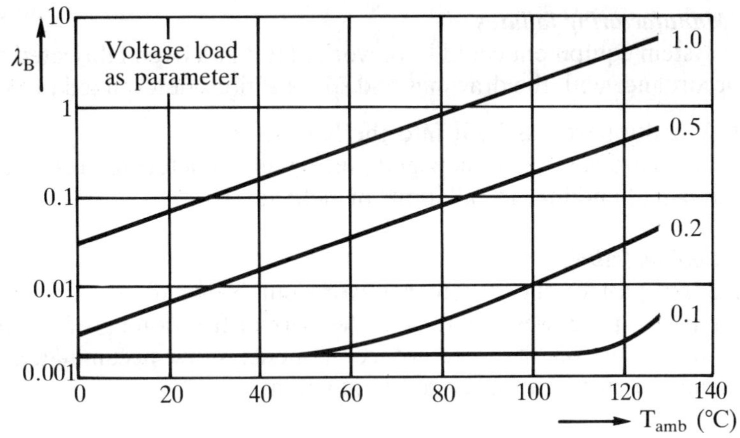

Figure 5.25 Failure rate versus temperature with voltage load as parameter (courtesy Delft University Press)

Table 5.7 Environmental circumstances

Thermal	Infrared radiation and large temperature gradients
Mechanical	Pressure, shocks and vibration
Pollution	Moisture, sand, dust, chemical contamination
Electrical	Electric magnetic interference (EMI), transients, lightning, etc.
Radiation	Electric magnetic pulses (EMP) caused by nuclear blasts, neutrons, gamma-radiation, etc.
Others	Human factors, maintenance and repair procedures

5.11 CAUSES OF FAILURE

A failure can be defined as an event where the system has failed or has run out of its specifications. The system has obtained a repairable or non-repairable state. In general, failures can be catastrophic or non-catastrophic. A catastrophic failure is one in which the system has stopped functioning; in a non-catastrophic failure the system has stopped functioning within its specifications. Failures can be divided in three main technical categories and one non-technical category.

1 Design failures

(a) The design is inherently incapable of performing the actual mission or operation, because either the system's functional logic does not correspond with the requirements or there is an unworkable combination of parts implemented.

(b) The environment is beyond the capability of the system because the environmental conditions were misestimated or the system is never qualified to that environment.

2 Manufacturing failures

The system equipment could have worked if it had been fabricated and maintained in accordance with the drawings and specifications determined in the design phase.

(a) A faulty part was built into the hardware.
(b) The hardware was damaged during manufacturing, test, reprocessing, or repair, handling, installation, or field service.

3 Random failures

(a) Some part of the system has worn out owing to the action of chemical or physical processes inherent in the normal functioning of the part.
(b) Some component or set of components show accumulated environmental damage sufficient to cause malfunction.

4 Human factors

Human factors in the causes of failure are all those not falling into one of the categories mentioned above. Examples are as mentioned earlier:

(a) Miscommunication
(b) Hurry and lack of time
(c) Failing organization structure
(d) Price versus market policies and acceptance
(e) Failing quality control
(f) System management and manufacturing philosophy

Fig. 5.26 is an example illustrating, by means of a sector diagram, the division of different system failures in ground test circumstances of a satellite.

This chapter concludes with one of the most dramatic graphs to be shown

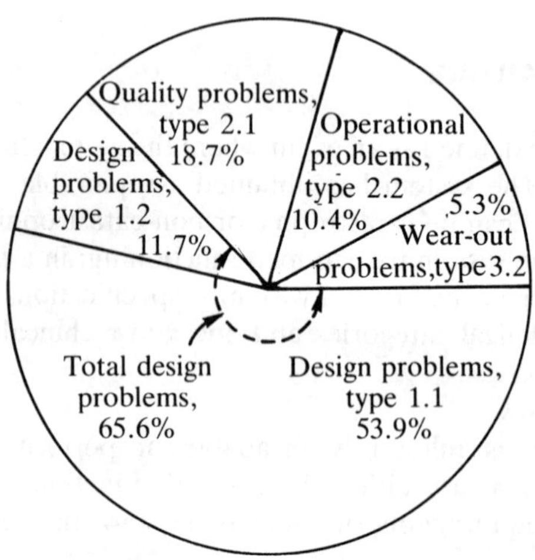

Figure 5.26 Division of system failures of a satellite in ground test circumstances (courtesy McGraw Hill Publishing Co.)

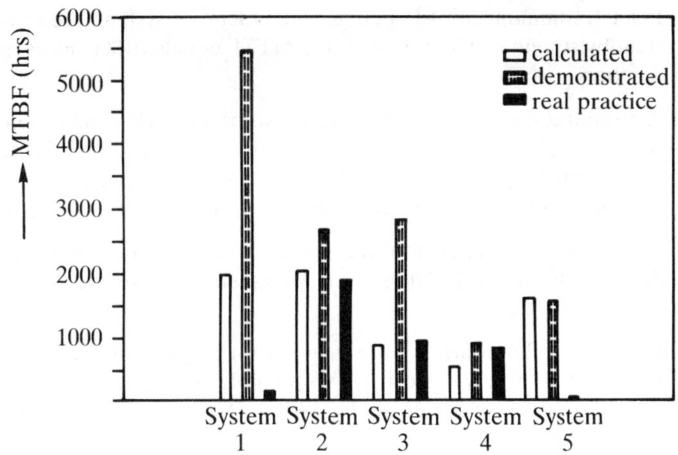

Figure 5.27 MTBFs of different systems under differing circumstances (courtesy Computer Science Press)

relating to reliability. This graph shows the large discrepancy which still exists between calculated values, measured values in laboratory circumstances, and in real practical circumstances. This graph is shown in Fig. 5.27.

A very important reason for such discrepancies is that the tools developed so far are still not satisfactory, because the real world is more complex than we can understand. In most cases all tests performed to determine the reliability of a system are single sequential tests executed one after the other. The real world does not act as a sequential test environment, but all environmental conditions influence the system in parallel and simultaneously. This makes things more complicated than we are able to understand and handle at this moment. However, reliability techniques have become a useful science and tool and are still in development.

5.12 PROBLEMS

5.1 In a manufacturing process of integrated circuits, 5 of every 100 ICs satisfy the specifications. Calculate the probability that from a sample of 10 ICs just two of them pass the specifications.

5.2 Calculate the failure rate and MTBF of a circuit consisting of 20 resistors and 30 soldered connections when the failure rates are 10×10^{-9}/hour/component and 1×10^{-9}/hour/connection respectively.

5.3 A system with a useful lifespan of 1000 hours has a failure rate of 0.0001/hour/system and is supposed to have an exponential distribution. Calculate the reliability for that system for 10, 100 and 1000 hours of operation.

5.4 The power supply unit of an aircraft consists of two identical systems with a failure rate of 350×10^{-6}/hour/system. The system configuration is an active on-line standby system with exponential distribution. Determine the reliability and the MTBF for the unit.

5.5 In a telecommunications satellite, three identical systems are installed in a cold on-line standby system configuration. If the MTTF equals 10 000 hours determine the survival function.

5.6 A ground communication system consists of three identical systems in an active on-line standby configuration. If two of the three systems are operational, the communication system works satisfactorily. The failure rate of each system is 400×10^{-6}/hour. Calculate the system reliability and the MTBF for a 24 hour period.

5.7 Apply the decomposition technique to determine the reliability of the system depicted in Fig. 5.28. The system works when one of the traces CD, GD, GF or EF works error free.

5.8 Draw a possible Markov state diagram for a system of second order with $m = 2$ and $x_i = 0$ or 1. Every state has two possible transitions. Write down for every state the appropriate transition probabilities. Determine the number of transitions which are free to be chosen.

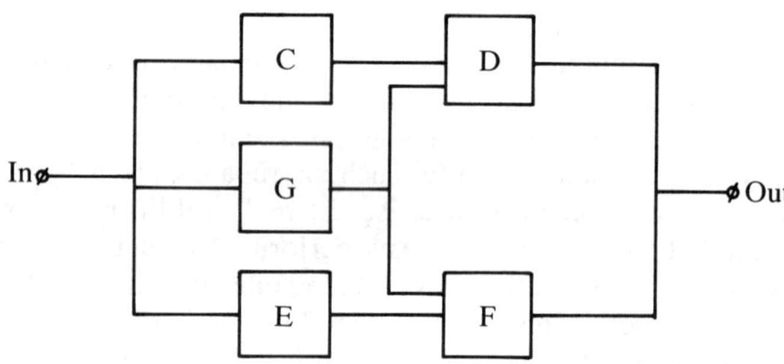

Figure 5.28 System configuration for Problem 5.7

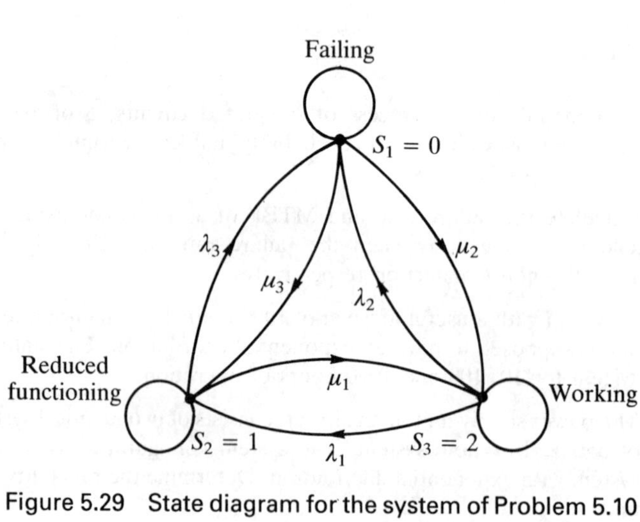

Figure 5.29 State diagram for the system of Problem 5.10

5.9 A system has two states with symbols 0 and 1 with $P(1/0) = p$ and $P(0/1) = q$. Calculate the probabilities $P(0)$ and $P(1)$. *Hint*: Draw first the state diagram.

5.10 A control and measuring system has three states: $S_1 = 0$ is the state that the system fails, $S_2 = 1$ is the state that the system works without control capacity, $S_3 = 2$ is the state that the system works normally. The appropriate state diagram is depicted in Fig. 5.29. The boundary conditions are $P(2)_{t=0} = 1$, $P(1)_{t=0} = P(0)_{t=0} = 0$. Arrange for this system the Laplace transformed differential equations.

Six

TRANSDUCERS

6.1 INTRODUCTION

It is the author's belief that transducers belong to the most intriguing and complex fields of electronics today. This is because all types of disciplines meet each other in the science, the research, the development and the manufacture of transducers. The knowledge of sensors is multidisciplinary and requires a comprehensive knowledge of general physics, solid-state physics, electronics, technology and manufacturing techniques.

Basically sensors are energy converters, but as we have seen already, energy information conversion is the objective of a sensor. The information available in one energy form must be converted into the same or another energy form, with exactly the same information content as the originating energy form. It is said that the information is carried by a certain type of energy. In Chapter 1 we have seen the great advantages of handling information in electrical form. Besides this, there is an increasing trend to use semiconductor technology in the manufacturing of transducers. If this is successful, large cost reductions may be expected as a result.

In all types of measuring systems, transducers are found at the input and at the output. In several cases energy conversion will be found within the system also. This is illustrated in Fig. 6.1.

The identification unit contains a device which reacts to the desired physical quantity and converts it into an electrical quantity. The input transducer can be found here, normally called a sensor. In the modifier stage the electrical energy is converted into another shape as discussed in Section 2.2. In the presentation unit, two possibilities arise depending on the nature of the 'system' to be served. If a machine or process must be supplied with new information, the output transducer must be able to provide it. In that case, a type of actuator may be required to execute the right action. In measurement systems to be perceived by humans, we need a type of display on which our senses react, or can take action. So if we state that the output transducer is an actuator we are always right.

In this chapter a detailed survey of possible energy conversions will be presented. But due to the great advantages of conversion into and/or from the electrical energy domain, the main emphasis is given to these conversions. In a separate section we shall discuss the possibilities of semiconductor materials used

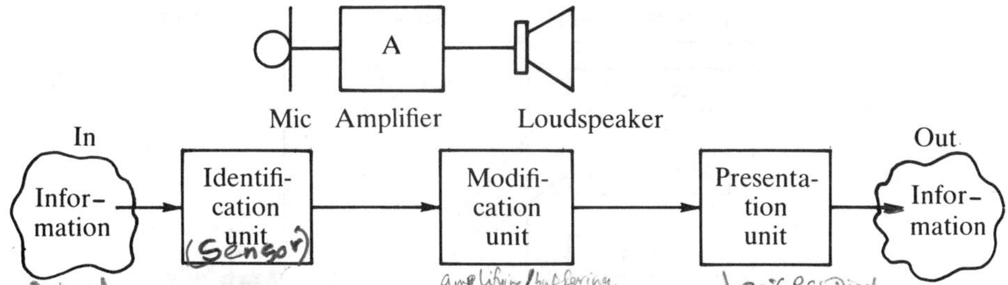

Figure 6.1 Functional block diagram of a measurement system

to develop and manufacture sensors. To avoid any confusion we will start by presenting the required definitions and terminology.

6.2 DEFINITIONS AND TERMINOLOGY

Generally speaking, the word 'transducer' can be thought to be derived from the Latin verb *traducere* which means 'to convert'. So a *transducer* can be defined as a device capable of converting energy from one form into another. Transducers can be found both at the input stage as well as at the output stage of a measuring system. The input transducer is called the *sensor*, because it senses the desired physical quantity and converts it into another energy form. The output transducer is called the *actuator* because it converts the energy into a form on which another independent system can react, whether it is a biological system or a technical system. So, for a biological system the actuator can be a numerical display, or a loudspeaker on which the visual and aural senses react respectively. For a technical system the actuator could be a recorder or a laser, producing holes in a ceramic material. The results can be interpreted by humans.

6.2.1 Types of energy forms

We can distinguish six different energy domains: radiant, mechanical, thermal, electrical, magnetic, and chemical. If certain information is already available in the electrical domain it can be claimed that it requires no energy conversion, but in general there is shape conversion left and this is just the domain which belongs to the field of electronics and electrical science and engineering. A good example of such a sensor only sensitive to electrical energy is the probe of an oscilloscope, with which a good adaptation to the signal source is realized. In the modifier stage we meet other examples of shape converters as there are for instance the A/D and D/A converters. If only the identification unit of Fig. 6.1 is considered we can draw Fig. 6.2 in which the possible energy conversions at the input are depicted schematically.

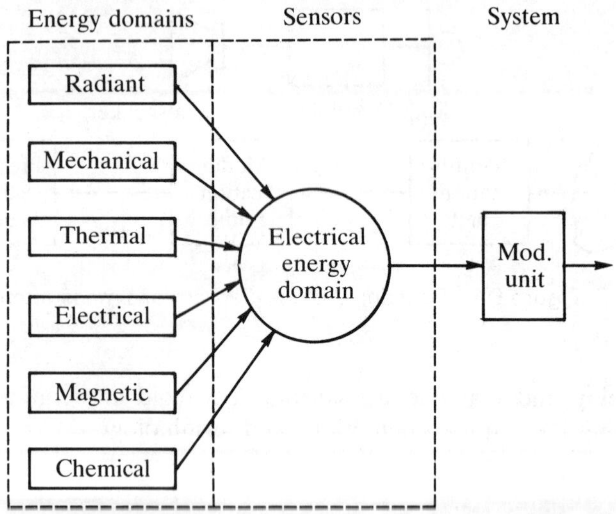

Figure 6.2 The different energy domains at the input of a measurement system

In the same way, the six different domain conversions at the output can be drawn. This is illustrated in Fig. 6.3, where compared with Fig. 6.2, the only difference is the reverse indicating arrows.

6.2.2 Types of energy sources

If the energy sources at the input are considered, or the objects the actuators are acting on at the output, we can distinguish between technical systems and biological systems. Let us consider this in some more detail. It will be clear that technical systems can produce all six energy forms. Hence at the input side the six different types of energy sources can be always recognized. For biological systems this is not so clear, but a more careful consideration will reveal the same six different types of energy forms.

1 Radiant energy is produced by biological systems. This is normally infrared radiation and can be detected although it is not visible. Radiant energy is generated by the system's proper temperature. It is well known that some kinds of fish produce radiant energy in the visible region between 400–800 nm.
2 Biological systems can also produce mechanical energy due to movements or the liquid pressure in the vessels.
3 Thermal energy is produced by all systems in which oxidation takes place.
4 Electrical energy for instance is produced by the heart muscle at a potential of several millivolts.
5 Magnetic energy is also produced by the human heart muscle and can be detected for example by squids.

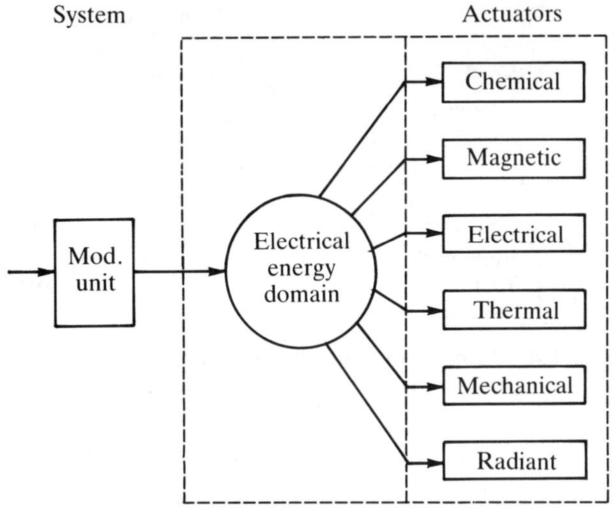

Figure 6.3 The different energy domains at the output stage

6 In biological systems chemical energy is produced in all types of processes and they can act as an energy source also.

At output we can find the same six energy domains for both types of systems. Again for technical systems this will be clear, but biological systems require more explanation.

1 Biological systems with their sense of vision can react to radiant energy as is the case for instance with the information displayed on a cathode-ray tube (CRT).
2 Biological systems are sensitive to mechanical forces and will react to them.
3 Thermal energy influences biological systems to a large extent and can for instance determine the velocity of growth.
4 Biological systems are sensitive to electrical energy and muscles for instance will react on electrical pulses.
5 It is already known that magnetic fields can cause chemical changes in biological tissues and cell structures and so change their information content.
6 There are numerous examples in which biological systems are sensitive to chemical actuators (substances). Examples are all types of medicine.

The above is summarized in a Fig. 6.4 representing the same situation as Fig. 6.1 but supplied with the two types of systems at input and output.

6.2.3 Modulating and self-generating transducers

Another important characteristic related to transducers implies the distinction of transducers into two categories. This is because some transducers require an

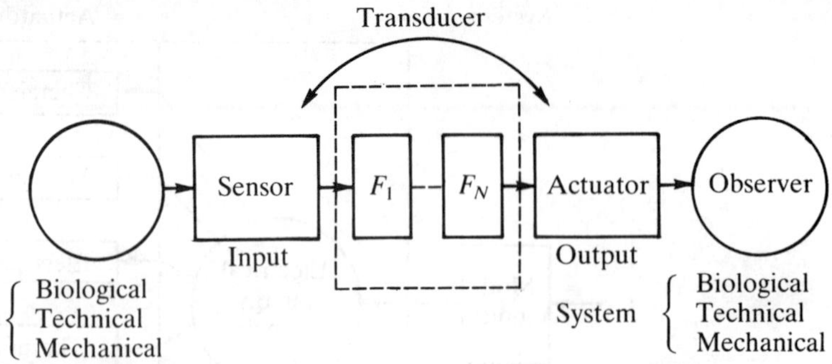

Figure 6.4 At input and output a biological or technical system can act as a source and as object for all six different energy domains

auxiliary energy source to become operational. Hence a distinction is made between modulating and self-generating transducers.

A *modulating* transducer is defined as one which requires an auxiliary energy source to convert energy from one domain into another. A good example of a modulating transducer which is found at the input of a system, is a strain gauge. This type of transducer requires an electrical energy source to become operational. The electrical current flowing in the strain gauge is modulated by a mechanical force which is converted into an electrical voltage change via a change of the resistive elements. As a consequence the input energy modulates the energy of the auxiliary energy source.

At the output of a system a liquid crystal display (LCD) is an example of a modulating transducer, because again an auxiliary energy source is required to perform the correct transduction. If no incident light is falling onto a LCD, the information cannot be detected. So the electrical modulating energy, in which the information is present, modulates the incoming auxiliary radiant energy and is presented as outcoming radiant energy.

A *self-generating* transducer is defined as a transducer which requires no auxiliary energy source to convert energy. Examples of a self-generating transducer which can be found at the input of a system are a solar cell, or a thermocouple, and at the output we can find for instance a heating element converting electrical energy into thermal energy without the need of an auxiliary energy source.

We can illustrate these concepts in two simple diagrams. In Fig. 6.5a the diagram represents a self-generating transducer and can be considered as a two-port device with respect to information (= signal) and energy. Fig. 6.5b represents a modulating transducer and can be considered as a three-port device with respect to energy and a two-port device with respect to information.

To avoid confusion at this point it is necessary to mention another classification for transducers. Sometimes the words 'active' and 'passive' are used in this respect. An active transducer is one which is capable of modulating energy. Normally the efficiency or power gain is much greater than 1. A transistor, a mechanically driven switch or a liquid crystal display (LCD) are examples of active devices. A passive transducer is defined as a device able to convert energy without an auxiliary energy

(a) Self-generating transducers

(b) Modulating transducers

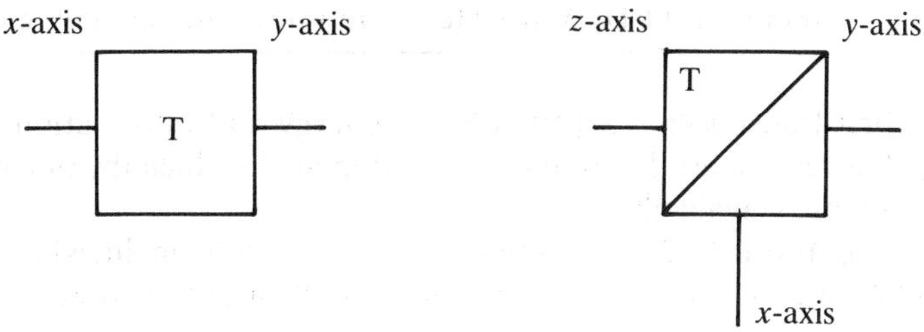

(a) No auxiliary energy source required

Examples

Input: Solar cell
Thermocouple

Output: Heating element

x	*y*	*z*
Info in	Info out	00
energy	energy	–

(b) Auxiliary energy source requred

Examples

Input: Strain Gauge

Output: LCD

x	*y*	*z*
–	Info out	Info in
energy	energy	energy

Figure 6.5 Symbolic representation of (a) self-generating transducers, (b) modulating transducers

source. Examples are a solar cell, a thermocouple or a piezo element. The efficiency of these is always smaller than 1. In the discipline of electronics we meet the same type of definitions. Components such as transistors, which are able to deliver gain, are called active components. This in contrast to resistors, coils and capacitors, etc., which are not able to deliver gain. These devices are called passive components. For convenience the definitions are summarized in Table 6.1. We will consider the first given definitions as more characteristic.

A convenient way of describing the different transduction possibilities is by applying a three-dimensional energy space diagram as is illustrated in Fig. 6.6, in which:

1 The *x*-axis represents the input energy domain. It contains the required auxiliary energy only, if a modulating transducer is involved. In the case of a self-

Table 6.1

Active transducers	*Passive transducers*
Modulating transducers	Self-generating transducers
Efficiency ⩾1	Efficiency <1
transistor, LCD, switch, etc.	resistor, coil, capacitor, etc.

generating transducer it represents the energy and information bound to it.

2 The *y*-axis represents the output energy domain in which the outcoming information content is available.

3 The *z*-axis represents the modulating energy domain in which the incoming information is available if an auxiliary energy source is required.

In the $x-y$ plane or the so-called input–output plane all 36 possible self-generating transducers can be found.

An interesting comment can be made on what is found on the main diagonal in the $x-y$ plane and the space diagonal. In the transducers found on the main diagonal no energy conversion is involved, but an energy *shape* conversion is executed only. This type of transducer is normally called a modifier, because it only modifies the available information in the same energy domain. Examples of

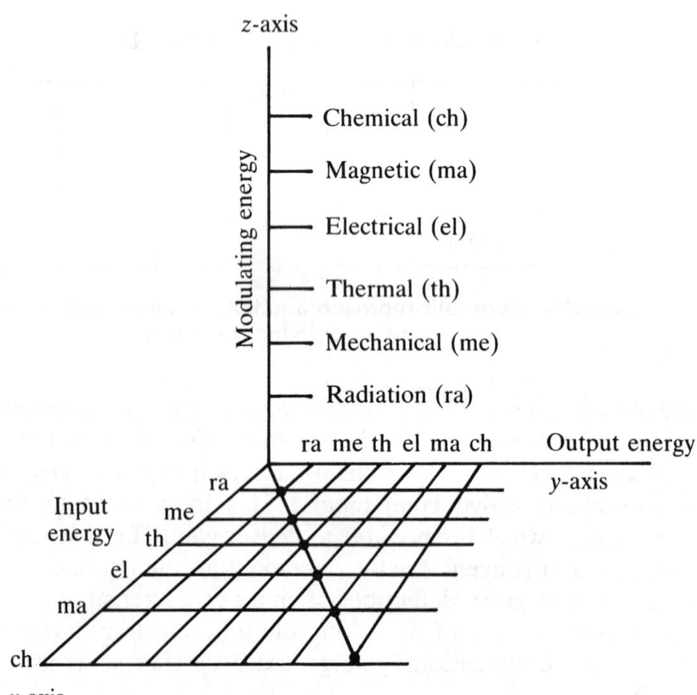

Figure 6.6 The *x-y-z* energy space diagram as a basis for describing all possible transducers

modifiers or shape transducers are the already mentioned A/D and D/A converters. So we can distinguish between the self-generating modifiers and the modulating modifiers.

In the mechanical energy domain, a gear set, with which the number of revolutions per minute can be changed, is a realization of a modulating modifier. Other examples are a transistor in the electrical energy domain and color filters in the radiant energy domain. Note that here the words 'modifier' and 'modulating' can easily lead to confusion and therefore preference is given to the name 'shape transducer'.

Based on the three-dimensional energy space diagram it is not very difficult to see that the modulating and self-generating transducers can be described by a three-dimensional vector $[x, y, z]$ in which the z component represents the modulating energy domain. For this purpose and in analogy with the so-called Miller indices, here Miller indices for energy can be applied.

This convention results then in [input energy, output energy, modulating energy]. For *signal* or *information*, for modulating transducers this results in: [00, information out, information in] and for self-generating transducers we find: [information in, information out, 00]. For both kinds of transducers the energy and information content for the x, y and z axes are depicted in Table 6.2.

Table 6.2 Energy and information content of x, y and z axes

	Modulating transducer			*Self-generating transducer*		
	x	y	z	x	y	z
Energy	in	out	in	in	out	non
Information	non	out	in	in	out	non

Table 6.3 lists several examples applying this notation.

Table 6.3 Applications of Miller indices for transducers

transistor	[el, el, el]	Modulating shape transducer
thermocouple	[th, el, 00]	Self-generating input transducer
pH-meter	[ch, el, 00]	Self-generating input transducer
LED display	[el, ra, 00]	Self-generating output transducer
LCD display	[ra, ra, el]	Modulating output transducer
coil	[el, ma, 00]	Self-generating output transducer
magnetoresistor	[el, el, ma]	Modulating input transducer
photoconductor	[el, el, ra]	Modulating input transducer
lateral photodetector	[ra, el, me]	Modulating input transducer

In this table the following abbreviations are used: el = electrical, th = thermal, ch = chemical, ra = radiation, ma = magnetic, me = mechanical. Note that the type of energy denoted in y functionally determines the kind of transducer with respect to input or output, hence if y = el then an input transducer is involved and if $y \neq$ el then an output transducer is implied.

In \ *Out*	*Radiant*	*Mechanical*	*Thermal*	*Electrical*	*Magnetic*	*Chemical*
Radiant	Filter			Photodiode		
Mechanical		Gear box		Piezo resistor		
Thermal			Heat exchanger	Seebeck effect		
Electrical	LED	Piezo electricity		Transistor	Coil	Electrical plating
Magnetic				Magneto resistor	Magnetic circuit	
Chemical				pH-meter		Chemical reaction

Figure 6.7 Matrix showing a collection of electronic self-generating transducers with examples of shape transducers on the main diagonal

Referring to the LCD again, the input energy is radiation (x-axis), the output energy is radiation (y-axis) and the modulating energy is electrical (z-axis) and, with respect to information, the z-axis represents the incoming information in electrical form and the y-axis the outcoming information in radiant energy.

In the $x-y$ plane every energy domain contains one modifier, five self-generating input transducers and five self-generating output transducers. This is illustrated in more detail in Fig. 6.7 for electronic devices only. In the same figure, the main diagonal shows the shape transducers.

To illustrate the use of the above concepts we will give the example of a digital thermometer with LCD readout as depicted in Fig. 6.8.

In a digital thermometer the thermal energy is converted into an electrical energy applying a thermocouple or a pn-junction, then the signal is modified applying an amplifier stage and an A/D converter and finally the electrical signal is converted again by a modulating output transducer to produce radiant energy to be perceived by the eye (LCD).

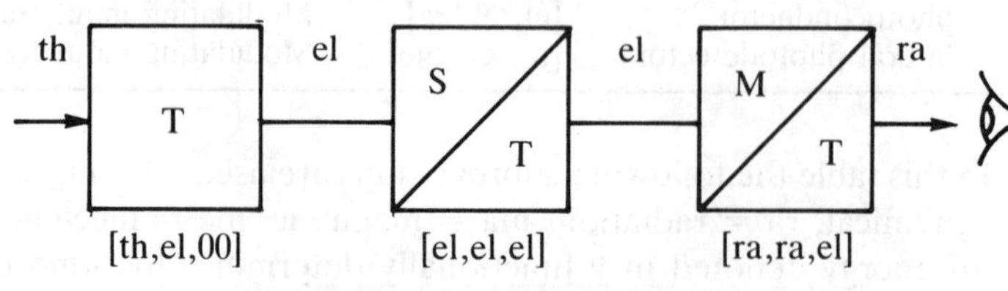

Figure 6.8 A digital thermometer with LCD display in a functional block diagram applying Miller indices

6.3 STATE DESCRIPTIONS

The state of a transducer can be described by combining physical quantities of which the product represents an energy. This state description reveals some other common characteristics of transducers. A most important characteristic to be kept in mind is that no transducer is sensitive to one physical energy only. The ideal situation exists when the transducer can be isolated from all physical influences except the desired one. One information source should be measured at a time. How well this can be realized determines to a large extent the accuracy of the transducer and the transducer is often the weakest link in a measuring chain. Besides this aspect, reliability aspects, as discussed in Chapter 5, play a vital role and even safety aspects come into scope here, when life-saving systems are involved. Excellent examples can be found in intensive-care departments where the personnel is completely dependent on the information delivered by the systems monitoring the patient.

For a transducer, two state descriptions are available, namely

1 Steady-state
2 Dynamic state

We will discuss both of them.

6.3.1 The steady-state description

Practically every transducer is sensitive to every type of energy, so if we consider a small volume $\mathrm{d}V$ in which the transducer is placed, the energy content $\mathrm{d}W$ in this volume contains the summation of all possible energies, or

$$\begin{aligned} \mathrm{d}W &= \Sigma I_i \cdot \mathrm{e}_i \\ &= \sigma \mathrm{d}l + P\mathrm{d}\delta/\delta + V\mathrm{d}q + E\mathrm{d}D + H\mathrm{d}B + T\mathrm{d}S \end{aligned} \tag{6.1}$$

where

$\mathrm{d}W$ total energy content in volume dV (J/m^3)
I_i represents an intensive quantity
e_i represents an extensive quantity
σ is a mechanical force per unit of volume (N/m^3)
$\mathrm{d}l$ is a displacement (m)
P is a pressure (Nm/m^3)
$\mathrm{d}\delta$ is a volume density of mass (kg/m^3)
V is a voltage (Nm/C)
$\mathrm{d}q$ is charge per unit of volume (C/m^3)
E is the electrical field (N/C)
$\mathrm{d}D$ is charge per unit of surface (C/m^2)
H is the magnetic field strength (A/m)
$\mathrm{d}B$ is the magnetic induction (Vs/m^2)
T is the absolute temperature (K)
$\mathrm{d}S$ is entropy per unit of volume (NmK^{-1}m^{-3})

Note that in this energy relation every product contains an intensive and an extensive quantity and that all products are expressed in energy by volume, which is as it should be.

Between a specific intensive quantity I_i and an extensive quantity e_i the following relation exists:

$$I_i = \frac{dW}{de_i} \tag{6.2}$$

with all other parameters kept constant.

For instance, the intensive quantity temperature can be defined as $T = dW/dS$ and the intensive quantity voltage as $V = dW/dq$. The steady state of matter and, hence, of a transducer, can be expressed in an extensive quantity accounting for all possible intensive quantities. When the variations are small enough a linear relation for all intensities may be assumed and when the superposition theorem is applicable we may write for a certain infinitesimal extensive quantity de_1

$$de_1 = \sum_{m=1}^{n} s_{1,m}\, dI_m \tag{6.3}$$

in which $s_{1,m}$ denotes the *sensitivity* 1 for an energy of type m.

We can express the intensive quantity dI_1 as the superposition of all types of extensive quantities, or

$$dI_1 = \Sigma M_{1,m}\, de_1 \tag{6.4}$$

in which $M_{1,m}$ denotes the *stiffness*, *rigidity* or the *moduli* for the energy of type m.

We will demonstrate these concepts with an example.

* * *

Example 6.1

Suppose a pressure transducer is subjected to and sensitive to the three types of intensive quantities, temperature T, pressure P producing a force σ on the resistive membrane elements, and electrical field strength E. Note that in this case the pressure is converted into a mechanical stress. This is sometimes called a tandem effect. We want to describe the related extensive variations in length dl of the pressure elements, the change in entropy dS and the change in electrical flux dD respectively, as a function of the mentioned intensive quantities. We can write the following equations:

$$dl = \left.\frac{\partial l}{\partial \sigma}\right|_{T,E} d\sigma + \left.\frac{\partial l}{\partial T}\right|_{\sigma,E} dT + \left.\frac{\partial l}{\partial E}\right|_{T,\sigma} dE \tag{6.5}$$

$$dS = \left.\frac{\partial S}{\partial \sigma}\right|_{T,E} d\sigma + \left.\frac{\partial S}{\partial T}\right|_{\sigma,E} dT + \left.\frac{\partial S}{\partial E}\right|_{T,\sigma} dE \tag{6.6}$$

$$dD = \left.\frac{\partial D}{\partial \sigma}\right|_{T,E} d\sigma + \left.\frac{\partial D}{\partial T}\right|_{\sigma,E} dT + \left.\frac{\partial D}{\partial E}\right|_{T,\sigma} dE \tag{6.7}$$

To get some insight, these equations can be presented in a matrix depicting all nine sensitivities.

$$\begin{pmatrix} \mathrm{d}l \\ \mathrm{d}S \\ \mathrm{d}D \end{pmatrix} = \begin{pmatrix} \left.\frac{\partial l}{\partial \sigma}\right|_{T,E} & \left.\frac{\partial l}{\partial T}\right|_{\sigma,E} & \left.\frac{\partial l}{\partial E}\right|_{T,\sigma} \\ \left.\frac{\partial S}{\partial \sigma}\right|_{T,E} & \left.\frac{\partial S}{\partial T}\right|_{\sigma,E} & \left.\frac{\partial S}{\partial E}\right|_{T,\sigma} \\ \left.\frac{\partial D}{\partial \sigma}\right|_{T,E} & \left.\frac{\partial D}{\partial T}\right|_{\sigma,E} & \left.\frac{\partial D}{\partial E}\right|_{T,\sigma} \end{pmatrix} \begin{pmatrix} \mathrm{d}\sigma \\ \mathrm{d}T \\ \mathrm{d}E \end{pmatrix} \tag{6.8}$$

On the main diagonal, the shape conversions can be recognized, because no type of energy conversion is involved. These effects are sometimes called *direct* effects. All other effects imply energy conversions combining two forms of energy and are sometimes called *cross-effects*. For instance, in this case

$$\left.\frac{\partial l}{\partial \sigma}\right|_{T,E}$$

shows a direct effect or shape conversion and describes the sensitivity for a mechanical stress and is called elasticity; in the same way

$$\left.\frac{\partial D}{\partial E}\right|_{T,E}$$

is a direct effect and is the definition of the permittivity. The following sensitivity describes a cross-effect and is called the specific thermal expansion:

$$\left.\frac{\partial l}{\partial T}\right|_{\sigma,E}$$

Of course we can ask for the expression for the intensive quantities temperature T, electric field E and mechanical force σ, and in this case the relations become as shown in (6.9). In this matrix again the direct effects are to be recognized on the main diagonal and the others are the cross-effects. Here all coefficients are called moduli. All coefficients can be given the appropriate name.

$$\begin{pmatrix} \mathrm{d}\sigma \\ \mathrm{d}T \\ \mathrm{d}E \end{pmatrix} = \begin{pmatrix} \left.\frac{\partial \sigma}{\partial l}\right|_{S,D} & \left.\frac{\partial \sigma}{\partial S}\right|_{l,D} & \left.\frac{\partial \sigma}{\partial D}\right|_{l,S} \\ \left.\frac{\partial T}{\partial l}\right|_{S,D} & \left.\frac{\partial T}{\partial S}\right|_{l,D} & \left.\frac{\partial T}{\partial D}\right|_{l,S} \\ \left.\frac{\partial E}{\partial l}\right|_{S,D} & \left.\frac{\partial E}{\partial S}\right|_{l,D} & \left.\frac{\partial E}{\partial D}\right|_{l,S} \end{pmatrix} \begin{pmatrix} \mathrm{d}l \\ \mathrm{d}S \\ \mathrm{d}D \end{pmatrix} \tag{6.9}$$

* * *

6.3.2 The dynamic state description

If we consider dynamic effects then current densities J_i are the extensive quantities and the gradients G_i are the intensive quantities. The gradients can be considered as the driving forces and the current densities can be considered as current flows apart from the type of energy involved. The product of G_i and J_i represents a heat generation per unit of time and by volume, hence the units are J sec^{-1} m^{-3}, or watts m^{-3}. So here power by volume is involved.

For an arbitrary current density this can be denoted as

$$J_i = \sum_{m=1} \sigma_{i,m} G_{i,m} \tag{6.10}$$

in which $\sigma_{i,m}$ is called the *specific conductivity* for the respective process. For the gradient we can write

$$G_i = \sum_{m=1} \beta_{i,m} J_{i,m} \tag{6.11}$$

in which $\beta_{i,m}$ is called the *specific resistivity*. For instance, a driving force for heat flow is a temperature gradient $\mathrm{d}T/\mathrm{d}x$ and for an electric current a voltage gradient $\mathrm{d}V/\mathrm{d}x$. It should be kept in mind that all dynamic state equations ought to be denoted as vector equations, and only for simplicity has this been omitted here. In the same way as before cross-effects, combining two types of energy, and direct effects can be distinguished. In the following example as an application of these equations the current flow through a pn-junction is described.

* * *

Example 6.2

In a semiconductor the flow of charge carriers in a pn-junction consists of holes and electrons, hence two equations are required to describe this process. For the flow of charge carriers two driving forces are responsible. These driving forces are a voltage gradient, or electric field, and a charge carrier concentration gradient. If only the flow of electrons and holes in the x-direction is considered, we find

$$J_n = q\mu_n n \,\mathrm{d}V/\mathrm{d}x + qD_n \mathrm{d}n/\mathrm{d}x \tag{6.12}$$

$$J_p = q\mu_p \mathrm{d}V/\mathrm{d}x - qD_p \mathrm{d}p/\mathrm{d}x \tag{6.13}$$

where

J_n	is the electron current density (A/m^2)
J_p	is the holes current density (A/m^2)
q	is the specific charge for electrons
μ_n	is the electron mobility in the semiconductor
μ_p	is the hole mobility in the semiconductor
n	is the concentration of electrons as a function of position
p	is the concentration of holes as a function of position

dV/dx is the voltage gradient in x-direction
D_n is the diffusion coefficient for electrons
D_p is the diffusion coefficient for holes
dn/dx is the concentration gradient for electrons
dp/dx is the concentration gradient for holes

For instance, the products $q\mu_n n$ and qD_n represent the specific conductivities related to the two driving forces.

In equations (6.10) and (6.11) you can easily recognize Ohm's law. We mention here two other well-known cross-effects, namely the Peltier effect, producing a heat flow due to a voltage gradient, and the Seebeck effect, producing a voltage gradient due to a temperature gradient.

6.4 TABLES OF ENERGY CONVERSIONS

In this section, an attempt is made to present in five tables a survey of most known energy conversions which can be found in solid-state transducers. The division of the tables is in accordance with the six already known energy domains and no distinction is made between sensors and actuators or between self-generating and modulating transducers, because this can easily be derived respectively from the required type of input and output energy and the applied notation. As mentioned already the coefficients respectively represent the input energy, the output energy and the modulating energy. When the electrical energy domain is taken as a reference, the sensors (input) and actuators (output) are easily recognized.

A calculation leads theoretically to 6^3 effects for modulating transducers and to 6^2 effects for self-generating transducers including cross- and direct effects. It is interesting to note that some effects are not (yet) known. Only the transducers which can be realized in solid-state technology are presented. For cross-references, abbreviations are used as for instance T3, which refers to the third transducer effect in the thermal domain, R3 refers to the third effect in the radiant energy domain, etc.

6.4.1 The radiant energy domain

Table 6.4

Name of effect	*Notation*	*Macroscopic description*
1 Photovoltaic	[ra, el, 00]	A voltage is generated by incident radiation at the junction of two dissimilar materials
2 Photo-magneto electric	[ra, el, ma]	An electrical field is generated to both a magnetic field and incident radiation
3 Photoconductivity	[el, el, ra]	The increase of electrical conductivity of a material due to incident radiation

Table 6.4 Continued

Name of effect	*Notation*	*Macroscopic description*
4 Photoelectric	[el, el, ra]	Electrons and holes are generated and separated in a junction area due to incident radiation
5 Photodielectric	[el, el, ra]	The change of a dielectric constant and loss due to incident radiation
6 Laser	[el, el, ra]	An energy level population inversion is generated due to incident radiation
7 Photoluminescence	[ra, ra, 00]	Radiant energy is emitted due to incident radiation with shorter wavelength
8 Radioluminescence	[ra, ra, 00]	Visible radiant energy is emitted due to incident X-rays or γ-rays.
9 Radiation heating	[ra, th, 00]	The increase of temperature of a material by incident radiation
10 Photomagnetic	[ra, ma, 00]	The change of magnetization by incident radiation
11 Photochemical	[ra, ch, 00]	The change of structure due to incident radiation
12 Electroluminescence (Destriau)	[el, ra, 00]	The illuminating excitation of a material due to an alternating electrical field
13 pn-luminescence (Lossev)	[el, ra, 00]	The radiation of recombination energy in a forwardbiased pn-junction
14 Incandescence	[el, ra, 00]	The emission of radiation by thermal movement of atoms activated by an electric current
15 Kerr electro-optic	[ra, ra, el]	The generation of double refraction of radiation due to an electrical field
16 Kerr magneto-optic	[ra, ra, ma]	The change of a polarization plane of polarized radiation by a magnetic field
17 Faraday	[ra, ra, ma]	See 16
18 Pockel's effect	[ra, ra, el]	The rotation of polarization of polarized radiation by an electrical field
19 Cotton–Mouton	[ra, ra, ma]	The generation of double refraction of radiation in a liquid due to a magnetic field

6.4.2 The mechanical energy domain

Table 6.5

Name of effect	*Notation*	*Macroscopic description*
1 Piezoelectric	[me, el, 00]	The generation of a surface charge due to a mechanical force

Table 6.5 Continued

Name of effect	*Notation*	*Macroscopic description*
2 Acousto-electro	[me, el, 00]	The generation of an electric current by a traveling longitudinal accoustic wave
3 Tribo-electro	[me, el, 00]	The generation of positive or negative surface charges due to rubbing of materials
4 Piezo-resistance	[el, el, me]	The change in conductivity in semiconductors due to a mechanical force
5 Dember effect	[ra, el, me]	The generation of a voltage across two regions in a semiconductor due to radiation
6 Lateral photovoltaic	[ra, el, me]	The generation of a lateral voltage across a junction by radiation
7 Lateral photoelectric	[ra, el, me]	The generation of a current division across two contacts at one side of a reverse biased pn-junction by radiation
8 Thermoelastic	[th, el, me]	The generation of a voltage in two regions of a metal due to a mechanical strain and a temperature difference in the regions
9 Magnetostriction	[me, ma, 00]	The change in magnetization by a mechanical force
10 Piezo-optic	[ra, ra, me]	The change in refractive index due to a mechanical force
11 Photoelastic	[ra, ra, me]	The generation of double refraction by a mechanical force
12 Magnetostriction	[ma, me, 00]	The mechanical deformation of a material by a magnetic field

6.4.3 The thermal energy domain

Table 6.6

Name of effect	*Notation*	*Macroscopic description*
1 Thermoelectric Seebeck	[th, el, 00]	The generation of an electric current in a closed loop of two dissimilar conductors by different junction temperatures
2 Piezoelectric	[th, el, 00]	The change in polarization due to a temperature change
3 Nernst effect	[th, el, ma]	The generation of an electric field due to a temperature gradient and a magnetic field which are mutually perpendicular
4 Thermodielectric	[el, el, th]	The change of permittivity of a ferroelectric due to temperature

Table 6.6 Continued

Name of effect	*Notation*	*Macroscopic description*
5 Thermoconductivity	[el, el, th]	The change in conductivity due to temperature
6 Superconductivity	[el, el, th]	The conductivity is infinite below a specific temperature
7 Thermoelastic	[me, el, th]	The generation of a voltage between two regions in a metal due to mechanical strain and their temperature difference
8 Incandescence	[th, ra, 00]	The emission of radiant energy when a material is heated
9 Thermoluminescence	[th, ra, 00]	The emission of radiant energy of certain crystals due to temperature
10 Curie temperature	[th, ma, 00]	The change to paramagnetism of ferromagnetic material at specified temperature
11 Néel temperature	[th, ma, 00]	Same as 10 for ferrimagnetic material
12 Thermochemical	[th, el, 00]	The change of structure due to temperature
13 Electrothermal	[el, th, 00]	The generation of heat in a conductor by electric current
14 Peltier effect	[el, th, 00]	The generation of a temperature difference between two junctions when a current passes through it

6.4.4 The magnetic energy domain

Table 6.7

Name of effect	*Notation*	*Macroscopic description*
1 Hall effect	[el, el, ma]	The generation of an electric field in a conductor due to a current and a magnetic field which are mutually perpendicular
2 Magnetoresistance	[el, el, ma]	The change in resistivity of a material by a magnetic field
3 Suhl effect	[el, el, ma]	The change in conductivity of a semiconductor surface by a magnetic field
4 Superconductivity	[el, el, ma]	The change to normal state of a superconductor at a specific magnetic field
5 Photomagneto-electric	[ra, el, ma]	The generation of an electric field due to a magnetic field and incident radiant energy
6 Nernst effect	[th, el, ma]	See T3
7 Magnetostriction	[ma, me, 00]	A mechanical deformation is generated in a ferromagnetic material by a magnetic field

Table 6.7 Continued

Name of effect	*Notation*	*Macroscopic description*
8 Faraday effect	[ra, ra, ma]	The rotation of a polarization plane of polarized radiation by a magnetic field
9 Kerr magneto-optic	[ra, ra, ma]	See R16
10 Cotton-Mouton	[ra, ra, ma]	See R19
11 Ettinghausen effect	[el, th, ma]	The generation of a temperature gradient due to an electric current and a magnetic field
12 Maggi–Righi–Leduc	[th, th, ma]	The change in thermal conductivity by a magnetic field
13 Righi–Leduc	[th, th, ma]	The generation of a temperature gradient due to a heat flow and a magnetic field
14 Electromagnetic	[el, ma, 00]	The change in magnetization due to an electric current
15 Curie and Néel	[th, ma, 00]	See T10 and T11

6.4.5 The chemical energy domain

The domain of solid-state transducers is the most difficult to characterize, because all types of aggregation states can occur and often the desired information has to be retrieved from gradients in concentrations.

Table 6.8

Name of effect	*Notation*	*Macroscopic description*
1 Volta effect	[ch, el, 00]	The generation of a voltage between two dissimilar metals
2 Galvano-electric	[ch, el, 00]	The generation of a voltage between two dissimilar metals in an electrolyte
3 Chemical-dielectric	[el, el, ch]	The change in permittivity by a gas or liquid
4 Chemical-electric	[el, el, ch]	The change in conductivity of a semiconductor surface by an electrolyte
5 Chemical-magnetic	[ch, ma, 00]	The change in magnetization of a magnetic material by absorption of a gas
6 Electrochemical	[el, ch, 00]	The change of structure by an electric current
7 Thermochemical	[th, ch, 00]	The change of structure by temperature

Some transducers show a tandem effect, which means that more than one energy conversion is involved in one transducer element. For instance, in a flow transducer the gas flow generates a temperature difference and the temperature difference generates a resistive difference which is converted into a voltage difference, so two energy conversions are involved: me–th and th–el.

6.5 INTEGRATED SEMICONDUCTORS AND TRANSDUCERS

In the early 1960s J. S. Kilby of Texas Instruments, J. A. Hoerni and R. N. Noyce of Fairchild invented planar silicon integrated technology. Nobody at that time was able to predict the powerful impact on the electronics industry and the huge social implications on daily life that this technology would have.

The relatively simple planar technology is made possible by application of a photolithographic method and the use of oxide window masking. As has been revealed, these techniques have the important characteristics that passive and active components as well as the (aluminum) interconnections can be fabricated on one side of a silicon wafer. This all requires about 40 process steps of which most are very critical for the overall performance. One of the most general critical aspects in the fabrication of semiconductors is contamination by dust. Today we can meet a great variety of applications of electronics in all types of market sectors from domestic appliances to the automobile industry and from the watch industry to the toy industry. A very futuristic outlook of a possible society is presented by Alvin Toffler in his book *The Third Wave*, in which large emphasis is made on a communication society, all made possible by modern electronics.

In the meantime IC technology has also influenced a great variety of measuring techniques. Nowadays in this process four groups of devices can be recognized:

1 Integrated circuits, fabricated with standard IC technology, whose performance can improve transducer characteristics. These are often general-purpose ICs.
2 Custom-made ICs designed for one specific task, also made in standard IC technology.
3 Transducers fabricated in standard IC technology for one specific physical quantity.
4 Integrated circuits fabricated in combination with other compatible technologies resulting in a specific transducer.

The great advantages of planar silicon technology have raised the question as to whether standard silicon technology could be used also for the manufacture of transducers. In the early 1970s S. Middelhoek and A. F. P. van Putten of the Delft University of Technology made a successful first attempt to apply *standard* IC technology for the manufacturing of flow transducers. This resulted in a silicon anemometer or gas flow velocity meter based on the principle of the generation of a temperature gradient due to a gas flow in a silicon chip, whose average temperature is maintained well above ambient temperature.

In the next sections we first discuss some important advantages of semiconductor technology, which appear also to be of importance for the manufacturing of transducers. Then the specific physical effects of silicon, which can be applied to transducers, will be discussed. It has been revealed that other semiconductor materials can be applied for the manufacture of transducers, but the most promising with respect to costs are those which are compatible with silicon technology. So let us first have a look at the great advantages of silicon technology.

6.6 CHARACTERISTICS OF SILICON TECHNOLOGY

The main characteristics of silicon technology can be summarized in the following strongly related aspects:

1 Size
2 Complexity
3 Power consumption
4 Reliability
5 Price

6.6.1 Size

Photolithographic techniques have made it possible to produce integrated circuits with very tiny details for which the micrometer is the normal measuring instrument. The ultimate resolution obtained by these techniques is determined by the applied wavelength of the source in the photolithographic process. It has been shown that line widths of less than 0.1 μm can be fabricated and the smallest transistor ever made cannot be detected with a normal light microscope with maximum magnification. This very large-scale integration (VLSI) technique is made possible by applying an electron beam or an X-ray beam as the 'light' source in the photolithographic process. A complete electronic circuit requires a few mm^2 of silicon and apart from this effect the applied chip areas are still increasing. As an example, a complete microprocessor requires a chip area of $6 \times 6\ mm^2$. This reduction in line width versus time is illustrated in Fig. 6.9.

6.6.2 Complexity

The large number of components which can be produced on a single chip have made it possible to integrate complete electronic systems on a single chip. Today a memory chip contains not only the memory cells but all the required addressing and sensing circuitry such as decoders, registers, buffer control logic and so on. Today we are confronted with 32 bit microprocessors on a single chip of $10 \times 10\ mm^2$, with all the necessary peripheral electronics such as control circuitry decoders, bus interface, register, RAM and ROM capacity, timing controller, floating point unit, etc., all on the same chip.

It will be clear that for electronic measurement systems transducing elements integrated with a large part of the modifier are a real possibility resulting in a large

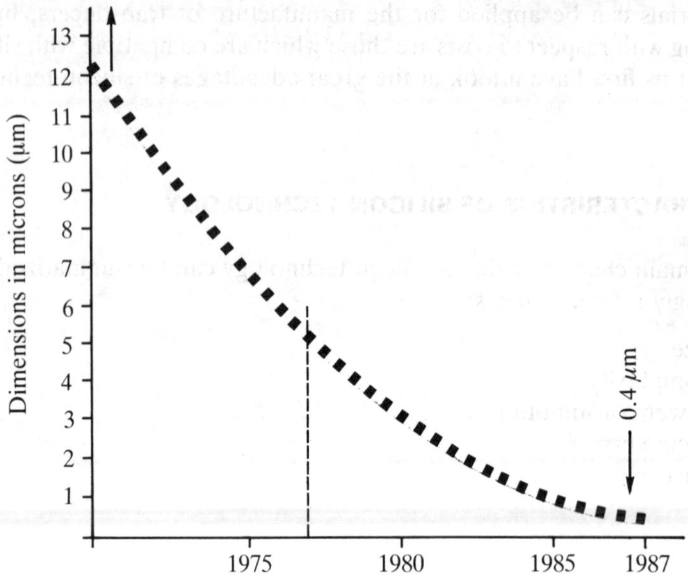

Figure 6.9 Line width trend versus time of IC technology (courtesy Van de Kooy)

cost reduction. To illustrate this increasing complexity an illustration of 'Moore's law' is depicted in Fig. 6.10.

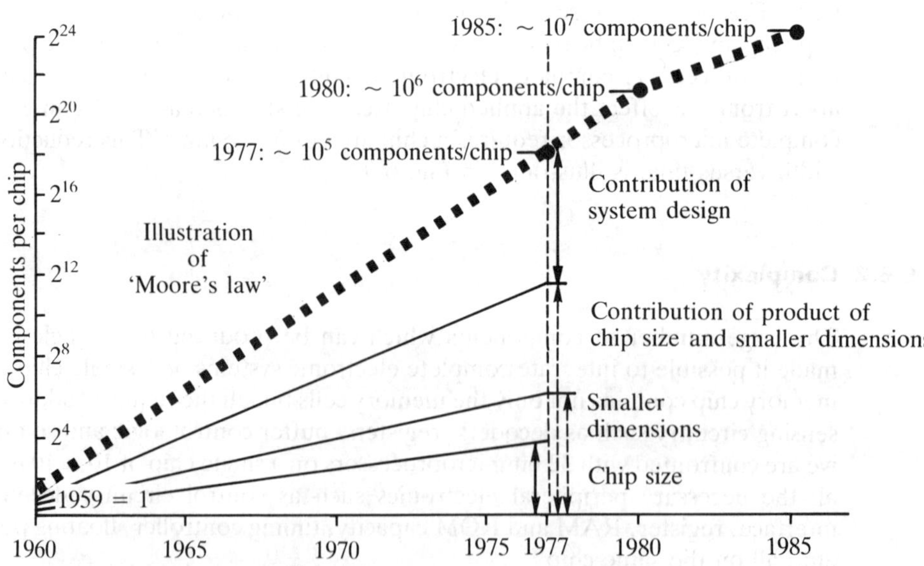

Figure 6.10 Illustration of increasing complexity according to Moore's law (courtesy Van de Kooy)

6.6.3 Power consumption

To illustrate the reduction in power consumption in electronic components and systems over the past decades a comparison may be made between the power consumption of the first computer, the ENIAC (built in 1946), and a simple handheld programmable pocket calculator. The ENIAC, built with vacuum tubes had a power consumption of about 150 kW, whereas the pocket calculator with much more computing power than the ENIAC, consumes less than 1 W. In Table 6.9; another comparison is made between the ENIAC and a more recently developed microcomputer which illustrates some striking developments.

Table 6.9 A comparison between the ENIAC computer and a 1987 microcomputer

	ENIAC	*Microcomputer*
year	1946	1987
MTBF	2 hrs	15 000 hrs
power consumption	150 kW	<40 W
clock-frequency	100 kHz	4–10 MHz
active components	18 000	±40
weight	30 000 kg	<1 kg
RAM memory	1 kbit	>1 Mbyte
price	$100 000	$1000

Other striking examples are electronic watches with LCD displays consuming less than 1 mW. The power consumption of logic gates and memory cells is often denoted in a figure of merit called the *power–delay product*, which is the product of power consumption and gate delay time and is expressed in picojoules (pJ). In Table 6.10 the power–delay product of some technologies is summarized.

Table 6.10 Power–delay product of some technologies (1986)

Technology	*Power-delay product* (pJ)
p-channel silicon gate	1
n-channel silicon gate	0.5
CMOS silicon gate	0.5
SOS CMOS	0.1
I^2L	0.01–1
Silicon (bipolar)	1
GaAs (bipolar)	0.5
Josephson gate	10^{-6}

6.6.4 Reliability

In Chapter 5 we have discussed aspects of reliability in detail. Integrated circuit technology achieves a large increase in reliability compared with vacuum tube technology. Integrated circuits can achieve an MTTF (non-repairable system) of 10^{11} hours, whereas vacuum tubes were not better that 10^5 hours, hence an improvement of 10^6 is obtained. Compared with discrete transistors the improvement is roughly a factor 10^3.

The most important parameters determining the reliability of integrated circuits are:

1 Contamination and dust in processing rooms
2 Wafer handling procedures
3 Accuracy of aligning the several masks
4 Visual, mechanical and electrical test procedures
5 Wire-bond testing
6 Packaging of the chips

Wire-bond (pull) testing is mentioned explicitly because it can make up to 30–60% of manufacturing costs. For instance, some chips require 100 wire bonds or more and to fulfill military standard specifications they should be tested all separately.

6.6.5 Price performance ratio

Today, integrated circuits are made in a fully automated production process and many different circuits can be simultaneously processed on one wafer. Of one 6 in. (150 mm) wafer the yield can be over a 1000 chips of 9 mm^2. The result is a drastic reduction in costs per component and often packaging costs surpass the chip costs. Prices of integrated circuits have fallen dramatically and for comparison purposes this is often expressed in a price performance ratio (price/gate). Fig. 6.11 illustrates

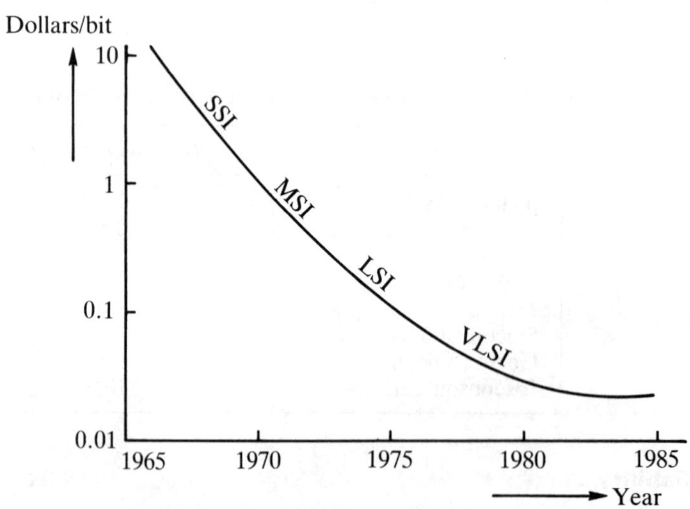

Figure 6.11 Price per gate development over two decades for a MOS circuit (courtesy Delft University Press)

Table 6.11 A hypothetical illustration of the development of semiconductors using a comparison with the automobile industry (Source Moore)

	Model T Ford	*'Rolls-Royce'*
year	1919	1979
speed	60 km/hr	160 000 km/hr
fuel economy	5 km/l	20 000 km/l
life span	10 years	centuries
price	$1000	$0.10

this development showing the development in price per gate for a circuit made in MOS technology over two decades.

Sometimes another example is given to illustrate the dramatic developments realized in semiconductor integrated circuit technology. For this purpose a comparison is made between some characteristics of a Model T Ford of 1919 and a type of hypothetical 'Rolls-Royce' if the automobile industry had made the same development and progress as the semiconductor industry. Such a comparison is presented in Table 6.11.

Considering all the above features of silicon technology, it will be clear that silicon or another silicon-compatible semiconductor can also offer great advantages for manufacturing transducers. But for that purpose, the next step is to investigate all known present physical effects in silicon or another semiconductor material which may be well suited for making transducers.

6.7 PHYSICAL EFFECTS IN SILICON

In Table 6.12, a summary is presented of known physical effects in silicon. In this table the main emphasis is on silicon, to be used for input and output transducers respectively for energy conversion into and from the electrical energy domain. For the manufacturing of LEDs, for instance, GaAs is the required material.

In the same fashion Table 6.13 depicts known physical effects in compatible technologies, where 'compatible' has to be understood as referring to technologies which are about the same as applied for silicon technology. For instance, the vacuum deposition of a nickel layer on silicon requires the same technique and the same equipment as for the aluminum interconnection pattern on silicon wafers. The same holds for gallium arsenide, of which it is known that the material can perform the same functions with higher speed and at higher temperatures than can be realized in silicon. Besides this effect, GaAs posesses some additional characteristics related to direct bandgap materials.

As can be seen from this table GaAs is applied in LEDs and at the beginning of integrated circuit technology. Very fast multipliers are already manufactured in GaAs and are commercially available. Numerous research centers are investigating GaAs and its potentials.

Concerning the different possibilities of silicon and other compatible tech-

Table 6.12 Summary of known physical effects in silicon

	Electrical			*In/Out*	
	Self-generating	*Modulating resistance, inductance capacitive*	*Modulating diode*	*Modulating transistor*	*IC transducer*
rad	Solar cell	Photo conductor	Photo diode	Photo transistor	Photo IC
me	?	Piezo resistor	Piezo junction	Piezo transistor	Piezo IC
th	Seebeck effect	$R = f(T)$	$I_{rev} = f(T)$	$U_{BE} = f(T)$	Temperature IC
el	Diffused resistance	MOSFET	Uni-Junction transistor	Dual gate MOSFET	IC
ma	Diffused coil	Magneto resistor	Magneto diode	Hall MOSFET	Hall IC
ch	Galvanic effect	$C = f(\text{conc})$	?	ISFET	?

Table 6.13 Summary of compatible physical effects in electrical input or output transducers

	Compatible techniques	*GaAs*
Radiant	CdS	LED
Mechanical	CdS/piezo-FET	?
Thermal	metal layer R(T)	?
Electrical	vacuum deposition Rs	transistors/ICs
Magnetic	vacuum deposition Ni layer	?
Chemical	aluminium oxide layers	?

Table 6.14 Review of the four different classes of transducers

Type		*Notation*	*Abbreviation*
Modulating (30)	In	$[x, \text{el}, z]$	MI
Modulating (30)	Out	$[x, y, \text{el}]$	MO
Self-generating (5)	In	$[x, \text{el}, 00]$	SI
Self-generating (5)	Out	$[\text{el}, y, 00]$	SO

nologies for the production of transducers, we have to consider four classes of transducers. In each case an energy conversion is involved in which the electrical energy is implied for input or output. The four classes are summarized in Table 6.14, in which for every class the electrical signal information is denoted, and x, y and z represent the six types of energy.

The numbers in parentheses denote the (theoretical) number of sensors or actuators. In the last column the abbreviations applied to the four classes are given. The five different energy conversions in semiconductors are discussed in more detail in the following sections. You can see here that six different energy domains imply five energy conversions only, because each time two energy domains are involved for one energy conversion.

6.8 RADIANT ENERGY IN SILICON MICROTRANSDUCERS

MI [el, el, ra]

PHOTOCONDUCTOR A photoconductor in Si consists of a p-type diffused strip into an n-type substrate with ohmic contacts at both ends of the strip. An auxiliary electrical energy source is required to detect a change in conductivity due to radiation.

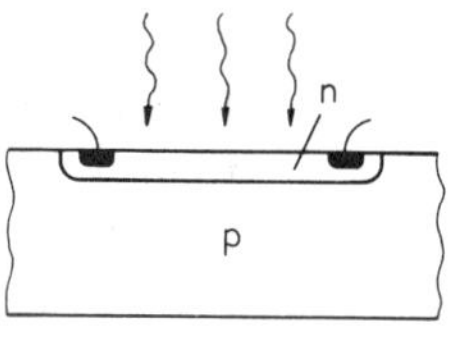

PHOTODIODE This device is a reverse-biased pn-junction in which electron-hole pairs are generated due to incident radiation in or close to the depletion layer. The reverse current is a measure of the incident light intensity. A large variety of photodiodes is manufactured.

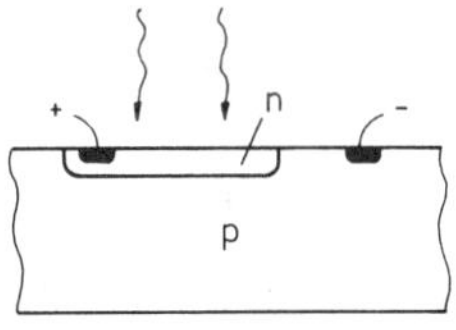

PHOTOTRANSISTOR This device can be considered as a photodiode connected in parallel with the base collector junction of a bipolar transistor. Electron–hole pairs are generated by incident light in or near the depletion layer. The reverse current in the base collector current is amplified by the transistor, resulting in a large change of the collector current due to incident light variations.

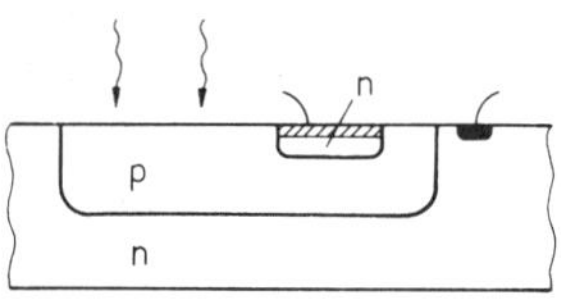

PHOTO-INTEGRATED CIRCUIT Complete light-sensitive integrated circuits with electronics are manufactured in a great variety.

PHOTODIODE ARRAYS An arrangement of photodiode arrays in two dimensions. Applying an x- and y-line geometry in combination with MOSFET switches, such arrays are manufactured for optical character recognition.

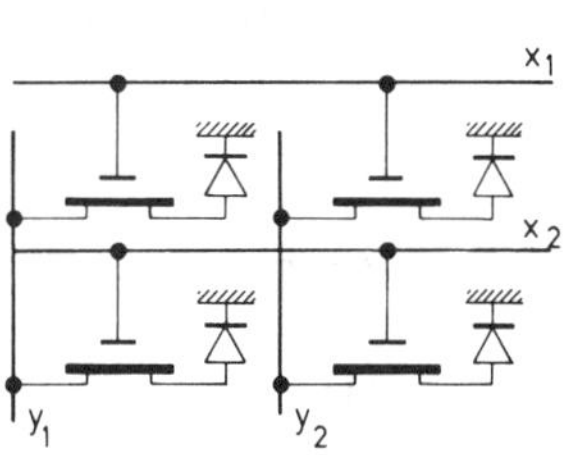

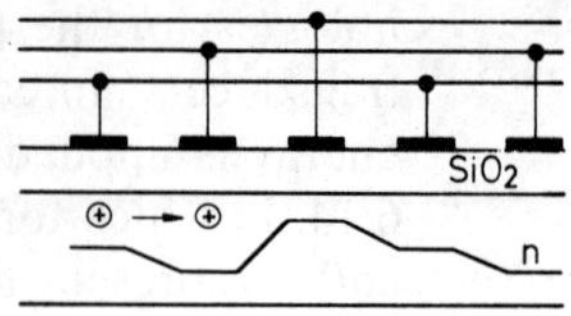

CHARGE COUPLED DEVICES (CCD) The devices consist of arrays of closely spaced metal electrodes on top of an oxidized silicon substrate. Minority carriers can be stored in potential wells beneath the electrodes or can be freely moved from one electrode to another electrode by application of suitable voltages to the electrodes.

MO [ra, ra, el]

So far no effects are known in silicon modulating transparency or reflectivity by an electrical field.

SI [ra, el, 00]

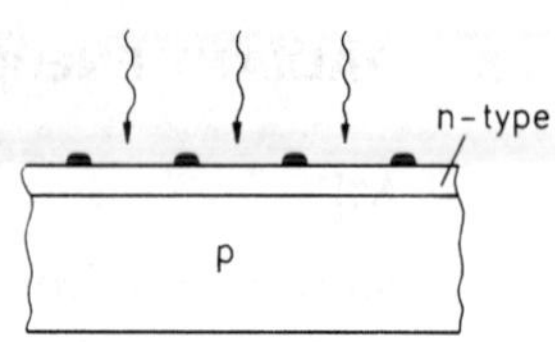

PHOTOVOLTAIC EFFECT Radiant energy is converted into electrical energy without the need of an auxiliary energy source. Incident light falls on the depletion layer of a pn-junction and generates electron–hole pairs, creating a potential difference across the pn-junction. Every type of radiant energy with a wavelength shorter than 1.1 μm provides enough energy to bridge the bandgap energy of 1.1 eV. For instance, photons, γ-radiation, X-rays, neutron radiation, and so on, produce the required energy. The theoretical maximum efficiency is about 18%.

SO [el, ra, 00]

In direct bandgap materials, with conservation of momentum, photons can be emitted with high efficiency when recombination of electrical generated electron–hole pairs occurs. The electrical generation of electron-hole pairs can be realized by forward biasing of a pn-junction. Silicon is not a direct bandgap material and therefore the emission of radiant recombination energy is most inefficient. Besides this, the emission occurs in the infrared spectrum. Only at very high reverse-biased voltages across a silicon pn-junction avalanche breakdown is realized, producing white light with low efficiency.

6.8.1 Radiant energy in compatible technology

MI [el, el, ra]

PHOTOCONDUCTIVITY For certain applications the spectral response of silicon is not the desired one. A customized

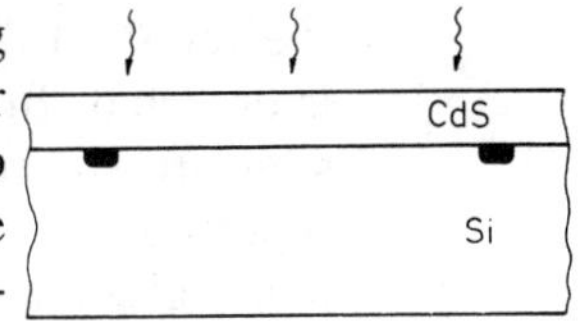

spectral response can be realized by evaporating suitable materials on top of an oxide layer of SiO_2. For instance, CdS with a wavelength of 0.7 μm or InSb with a wavelength of 7 μm are suitable materials. The change in conductivity can be detected with the underlying electronic circuitry.

MO [ra, ra, el]

LIQUID CRYSTAL DISPLAYS (LCD) LCDs consist of two flat electrodes of indium oxide between two glass plates between which slightly ordered long molecules of, for instance, methoxybenzolbutylaniline (MBBA) are placed. One electrode is made transparent and by means of rather small voltages the optical properties of the MBBA material can be changed. By applying suitable materials and doping, LCDs are made in several colors.

SI [ra, el, 00]

So far this effect is not reported in compatible technology.

SO [el, ra, 00]

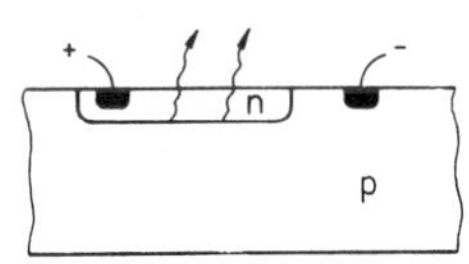

LIGHT EMITTING DIODES (LED) Light emitting diodes are made of direct bandgap materials like GaAs with suitable dopes, such as phosphorus, indium, cesium, etc. The complete visible spectrum can be covered including blue. This is illustrated in Fig. 6.12, in which

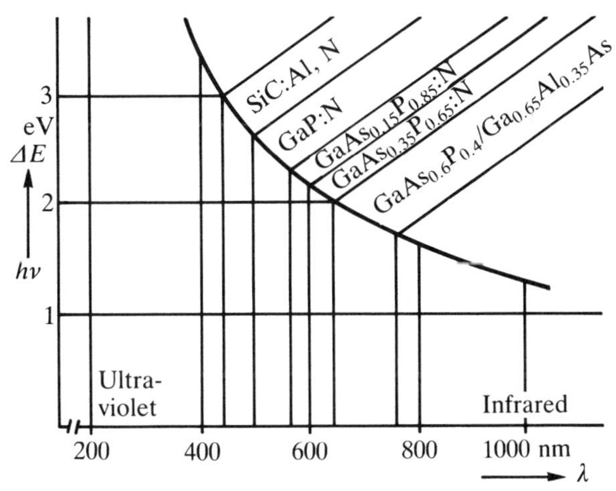

Figure 6.12 Recombination or direct bandgap energy versus wavelength. The required composite material and epitaxial layer are also shown (courtesy Delft University Press)

the required recombination energy versus wavelength is depicted. In the same figure, the respective composites are shown. The efficiency is high: up to 50% is reported. It is feasible that the above mentioned materials can be combined with silicon technology by applying them directly on top of the silicon wafer.

INJECTION LASER In a pn-junction of GaAs with polished surfaces perpendicular to the junction, an optical resonance cavity can be created. Due to the bandgap of GaAs of 1.35 eV the generated wavelength is about 9000 Å. Other suitable materials are GaP, GaSb, ZnS, PbS, etc. The efficiency is about 75%.

6.9 MECHANICAL ENERGY IN SILICON MICROTRANSDUCERS

MI [el, el, me]

PIEZORESISTOR In response to strain, a diffused resistor in silicon changes its value. Silicon is therefore used for a large variety of strain gauges. The temperature dependency of silicon strain gauges is rather high. For the measurement of pressures a very thin n-type silicon layer in which four p-type doped resistors are diffused in a Wheatstone bridge configuration is applied as a diaphragm.

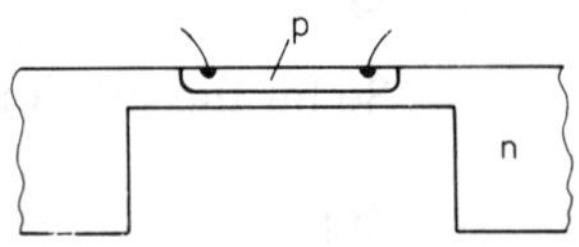

PIEZOJUNCTION When a silicon pn-junction is stressed the current-to-voltage characteristics change. This is caused by the piezoresistive effect and bandgap changes if a mechanical load is applied.

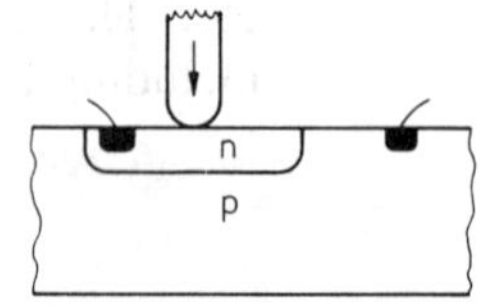

PIEZOTRANSISTOR A transistor shows the same strain sensitivity as a piezojunction, but in addition it shows a current gain factor which is also strain sensitive. For MOSFETs the same piezoresistivity is reported.

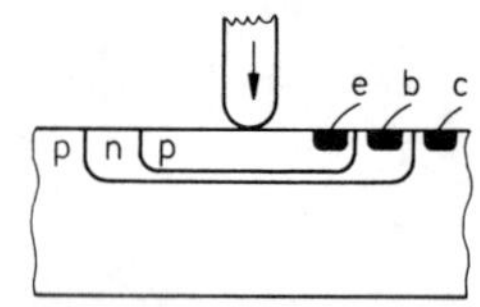

LATERAL PHOTO-EFFECT When a light spot illuminates a silicon solar cell a voltage is generated across the ohmic contacts as a function of the position of the light spot. With this effect large two-dimensional light spot sensitive photodiodes can be made.

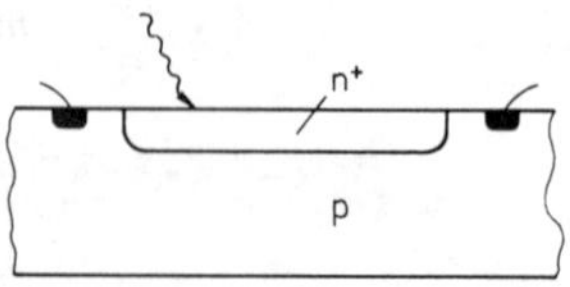

INTEGRATED RESISTIVE PRESSURE TRANSDUCER A complete Wheatstone bridge combined with temperature corrective circuitry and signal conditioning circuitry is integrated on a silicon chip. The piezoresistive effect is then applied. Very sophisticated etching techniques making strain bars in silicon are applied.

INTEGRATED TRANSISTOR PRESSURE TRANSDUCER A couple of transistors can be used in which one is subjected to stress by a diamond stylus on the emitter surface. Applying the required electronic circuitry on the same chip, as for instance current sources, operational amplifiers, and so on, fully integrated pressure transducers are manufactured.

MO [el, me, el]

So far the electrical modulation of a mechanical energy in silicon is not known.

SI [me, el, 00]

Silicon does not show the piezoelectric effect due to the symmetry in the lattic of silicon. So in silicon no self-generating transducer is available. The piezoelectric effect requires an asymmetric centre of inversion symmetry.

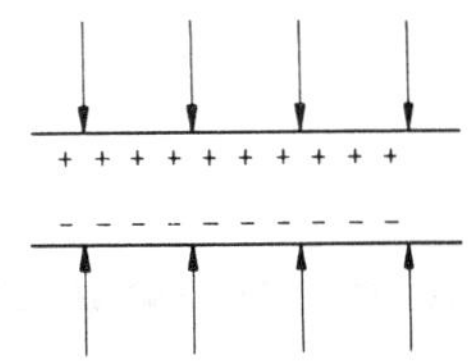

SO [el, me, 00]

So far this effect is not reported in silicon technology.

6.9.1 Mechanical energy in compatible technology

MI [el, el, me]

PIEZORESISTOR A great variety of materials is used for manufacturing the pressure-sensitive diaphragm or wires. Composites of platina-tungsten are reported.

PIEZO-FIELD EFFECT TRANSISTOR (PI-FET) The PI-FET is made by depositing a piezoelectric material such as ZnO or CdS, on top of a silicon wafer. Very sensitive pressure transducers can be made when the piezoelectric material is deposited between the channel of a MOSFET and the gate electrode. The piezoelectric layer is polarized applying a mechanical strain and the resulting electric field modulates the channel current. By integration techniques more electronic circuitry can be added. The frequency behavior is excellent and ranges from 0 to 20 MHz. Basically a tandem transducer is involved, because the first effect is piezoelectric; this effect is applied to modulate a MOSFET.

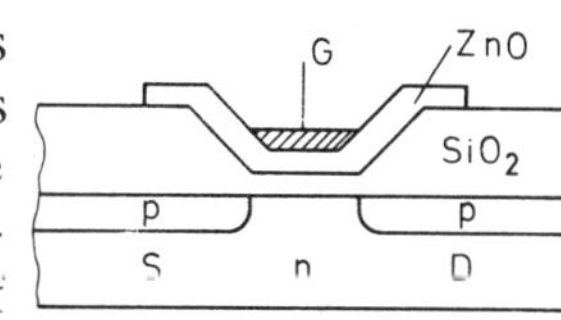

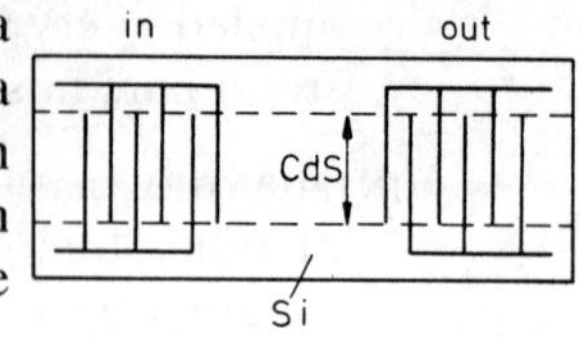

SURFACE ACOUSTIC WAVE FILTER (SAW-FILTER) When a piezoelectric material as CdS or ZnO is deposited on a silicon wafer or chip a surface acoustic wave filter can be manufactured. These filters can be combined with other electronic circuitry resulting in very versatile filters for microwave frequencies up to the GHz region.

MO [el, me, el]

So far this effect is not reported in a compatible technology.

SI [me, el, 00]

This effect is combined in a compatible technology in the piezo-field effect transistor.

SO [el, me, 00]

So far this effect is not known in a compatible technology.

6.10 THERMAL ENERGY IN SILICON MICROTRANSDUCERS

MI [el, el, th]

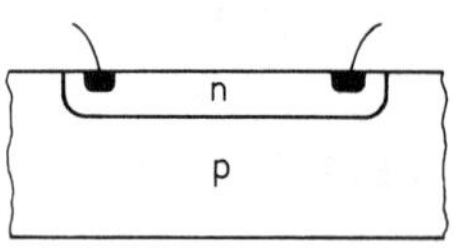

THERMORESISTOR In intrinsic (not doped) semiconductor material the temperature coefficient for conductivity is considerably large and shows an exponential behavior due to an exponential increase in charge carrier density of the electron–hole pairs. This material can be used in temperature measurement applications. Extrinsic (doped) semiconductor material shows a much less pronounced temperature coefficient of conductivity.

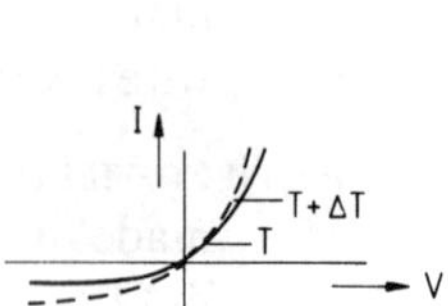

THERMODIODE Reverse- as well as forward-biased diodes display temperature-dependent behavior. The change is rather small and appears to be about −2 mV/°C for a forward-biased diode. (See also Chapter 7.)

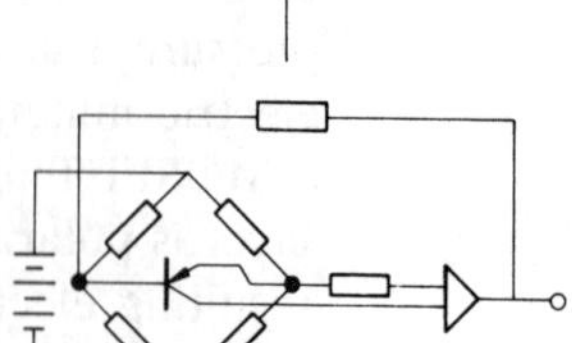

THERMOTRANSISTOR The base-emitter voltage of a transistor is a linear function of temperature if the emitter current is maintained constant. Compared with the diode, a considerable increase in sensitivity is obtained, ranging from 10 mV/°C to 200 mV/°C.

TEMPERATURE INTEGRATED SENSOR Very complete sophisticated integrated temperature sensors are manufactured applying for instance a type of current mirror

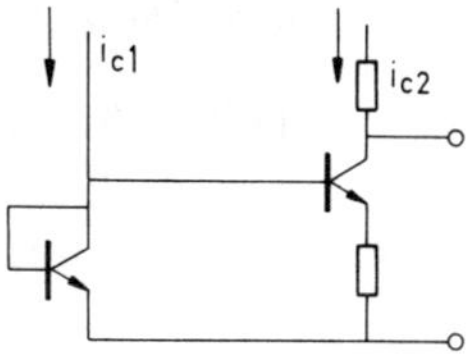

configuration of two matched transistors. The difference between the base-emitter voltages at different collector currents is proportional to the absolute temperature. An integrated Zener voltage reference and built-in operational amplifiers provide a transducer applicable for a temperature range between −25°C and +85°C offering a linear output of 10 mV/°C.

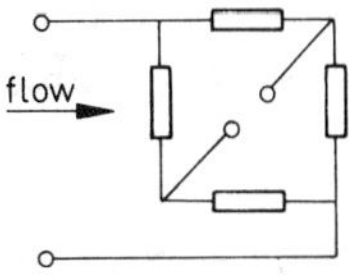

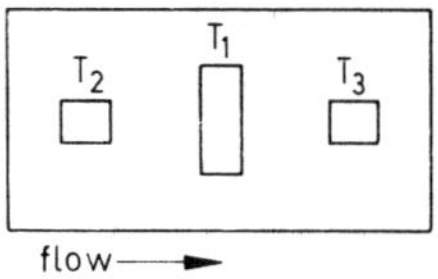

INTEGRATED SILICON FLOWMETER A flowmeter can be manufactured by applying a Wheatstone bridge configuration composed of four p-type ion-implanted resistors in an n-type silicon substrate. The four resistors have a large length-to-width ratio of about 85 and are integrated in a square configuration. The bridge current heats the chip well above ambient temperature. When the chip is placed plan-parallel in the airflow, the cooling effect of the four resistors is orientation – dependent with respect to the direction of the airflow. The signal due to a developed thermal imbalance is a measure of the flow velocity. The same principle can be applied with a transistor configuration showing more sensitivity but less linearity and stability. An improved version of the resistor configuration applying a double bridge configuration with thermal feedback offers excellent linearity and stability. This principle is discussed in more detail in Section 3.6.

MO [el, th, el]

The conversion of electrical energy into thermal energy is very simple indeed. The effect is caused by the interactions of passing current carriers with the lattice. The same effects can be found in diodes and transistors. Applying a matrix of such components combined with thermal paper offers the possibility of manufacturing a thermal matrix dot printer.

SI [th, el, 00]

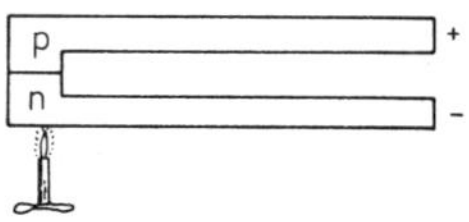

A voltage difference can be generated across differently doped silicon structures joined together when a temperature difference between the two regions is present. This is called the Seebeck effect. When a closed loop is made, current will flow. When a series connection of these different doped regions is made, very sensitive thermo-piles can be made suitable for long-range thermal measurements.

SO [el, th, 00]

This effect is well known and described under MO transducers, where a simple resistor can act as a self-generating output transducer, and a transistor in principle can be used as a modulating output transducer. A matrix of such components in combination with heat-sensitive paper can be used to make a thermal printer. The addressing electronics can be integrated on the same chip.

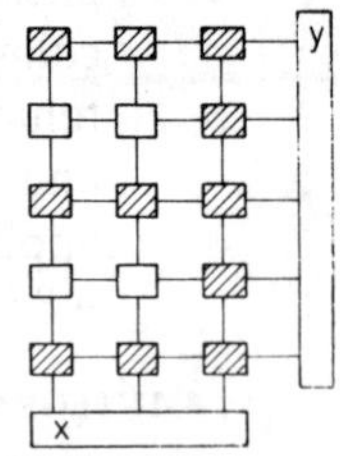

6.10.1 Thermal energy in compatible technology

A considerable variety of thermal energy transducers can be made in compatible technology because it is very easy to deposit a metal layer onto a silicon wafer. With suitable etching techniques, very fine, accurate and stable resistance networks can be fabricated. For this purpose an Ni–Fe alloy is often used. The limiting factor is of course the maximum allowed temperature for silicon before degradation occurs, which will happen above 175°C.

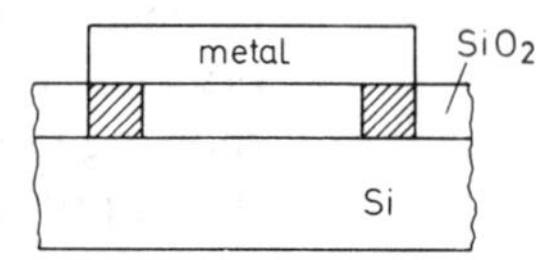

6.11 MAGNETIC ENERGY IN SILICON MICROTRANSDUCERS

MI [el, el, ma]

MAGNETORESISTOR The resistance of a carrying conductor will change due to a magnetic field. The effect is known as the Gauss effect and is rather small, depending strongly on the applied geometry of the diffused resistors. The largest effect can be obtained with a so-called Corbino-disk.

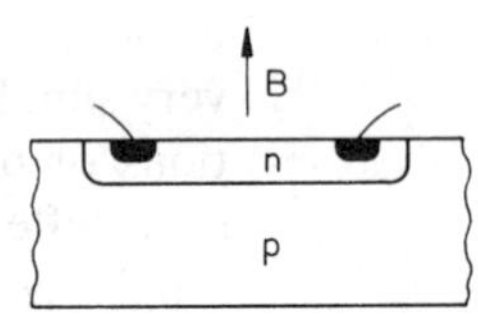

HALL EFFECT When a current-carrying semiconductor is subjected to a magnetic field a so-called Hall voltage will be generated across the semiconductor at right angles to both current and magnetic field direction. The polarity of the Hall voltage is dependent on the type of semiconductor material (p-type or n-type), the direction of the current and the direction of the magnetic field.

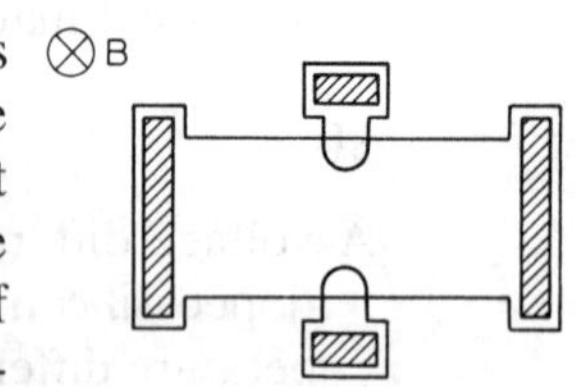

MAGNETODIODE In semiconductors, the charge carriers can be deflected from their straight path by applying a magnetic field, caused by the Lorentz force. Magneto-diodes show a high sensitivity to magnetic fields when they are manufactured with large differences in the recombination rates between the two regions. The diode characteristics change when a magnetic field is applied. Diodes made using SOS (silicon on sapphire) technology have a high sensitivity but are less linear than a Hall plate.

MAGNETOTRANSISTOR The magnetotransistor consists of a bipolar transistor structure and is composed of one emitter and two collectors. Without a magnetic field the collector current divides equally between both collectors. When a magnetic field is applied in the plane of the silicon chip perpendicular to the line connecting both collectors, an unequal current division will occur. Connecting two load resistors in the collectors, the difference in current is converted into a voltage difference and is a linear function of the magnetic field.

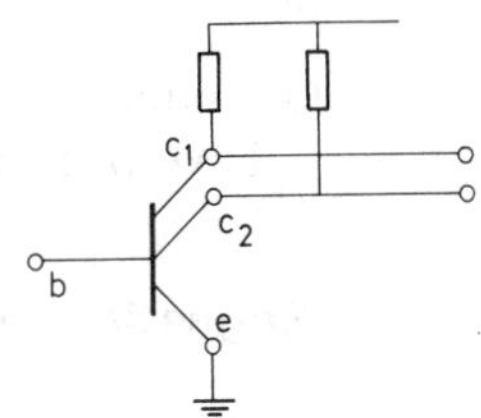

MAGFET The MAGFET consists of a MOSFET with two drain contacts analogous to that of the magnetotransistor and proves to have a comparable sensitivity.

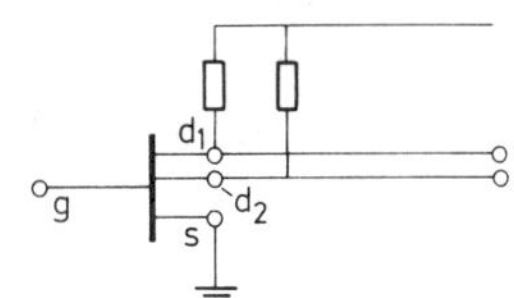

HALL MOSFET A Hall MOSFET consists of a MOSFET with a rather wide channel area and additional Hall contacts. The Hall voltage is inversely proportional to the charge carrier density and the thickness of the plate, but proportional to the current through the plate and the magnetic field. Sensitivities of 100 mV/mT are obtained.

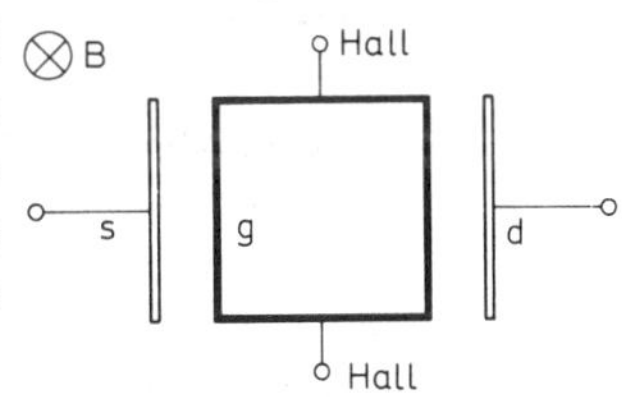

INTEGRATED MAGNETIC TRANSDUCERS Hall plates can be integrated together with signal-processing circuitry on the same chip. Applications are magnetic recording heads, brushless motors, displacement transducers, and so on.

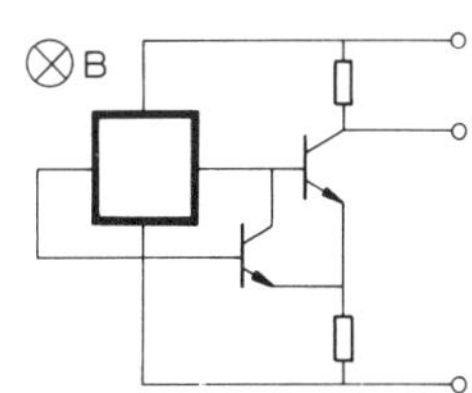

MAGNETOBRIDGE Four sensitive magnetoresistors are deposited on a single silicon chip in a Wheatstone bridge configuration. The bridge has a sensitivity of 1 mV at a magnetic field variation of 100 A/m and a supply voltage of 5 V. It can be used for speed and direction control of electric motors.

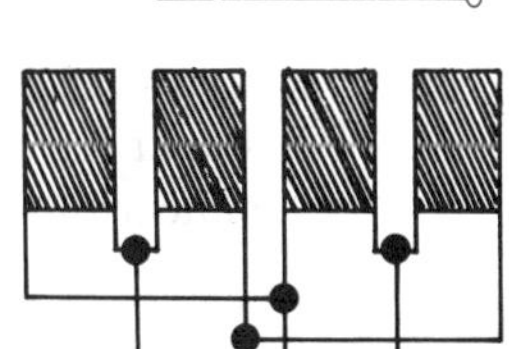

MO [el, ma, el]

So far, no applications of this effect are known.

SI [ma, el, 00]

This effect can be realized when a silicon coil made by diffusing a p-type winding in an n-type substrate is subjected to an alternating magnetic field. By Faraday's induction law, a relatively small voltage will be generated. Due to technological limits only a small number of windings can be integrated.

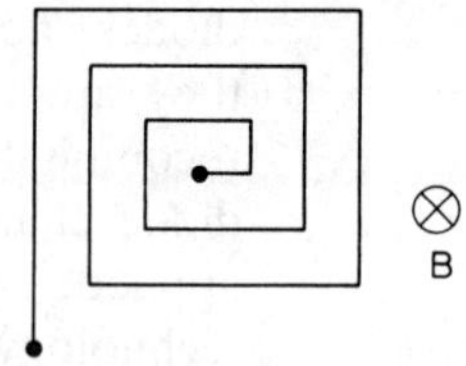

SO [el, ma, 00]

Of course this effect can be realized but no applications have been reported.

6.11.1 Magnetic energy in compatible technology

In some cases, the temperature sensitivity of magnetic transducers is unacceptably large. In that case the magnetic effect of other magnetic-sensitive materials can be used, depositing, for instance, layers of Ni-Co or Ni-Fe on top of a silicon dioxide layer with the underlying silicon substrate containing the signal-processing circuitry.

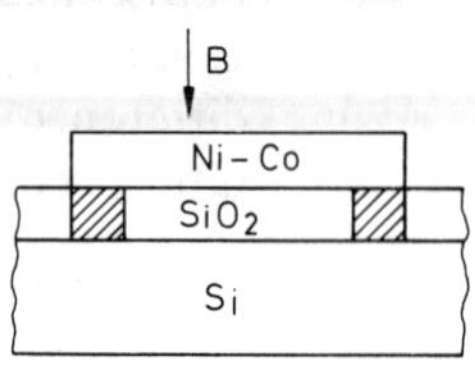

6.12 CHEMICAL ENERGY IN SILICON MICROTRANSDUCERS

The discussion of this type of energy conversion in semiconductors is the most difficult to classify, because it includes effects which are less easy to classify in one of the above classes and which are not purely chemical in their effects. A very serious problem concerning packaging of this class of sensors is also a major complicating factor in the manufacturing and lifetime of chemical sensors. Some types have a MTTF (non-repairable) of a couple of hours as, for instance, for the oxygen sensor.

MI [el, el, ch]

ION ELECTROLYTIC CONDUCTION A simple monolithic structure with two electrodes can be made such that the conductivity of an ionic solution on top of the silicon substrate can be determined. Here a major problem concerning packaging and corrosion arises.

HUMIDITY SENSORS A humidity sensor can be made by applying a silicon chip on top of a Peltier cooling element. On top of the chip a polycrystalline cam-structured geometry is implemented whose capacitance will change abruptly when water begins to condense on top of the silicon chip. A thermal energy classification might be justified here.

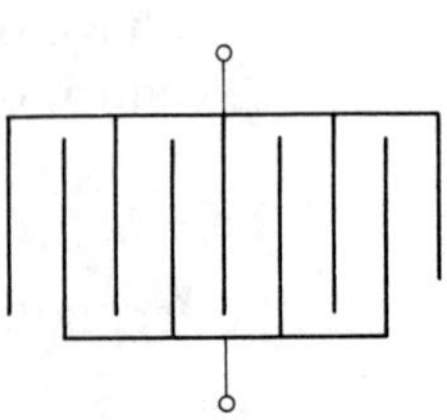

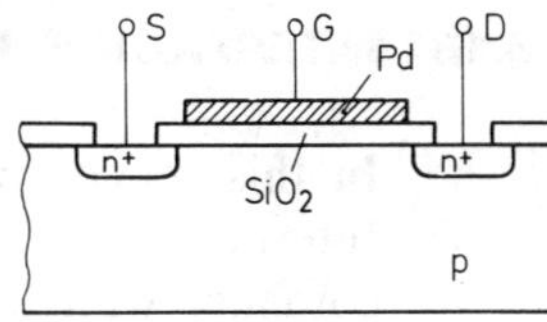

HYDROGEN SENSITIVE MOSFET A MOSFET with a palladium gate appears to be sensitive to hydrogen, showing a threshold voltage as a function of the hydrogen concentration. The sensitivity proves to be 10 ppm H_2. It has also been shown that this device is sensitive to H_2S and NH_3.

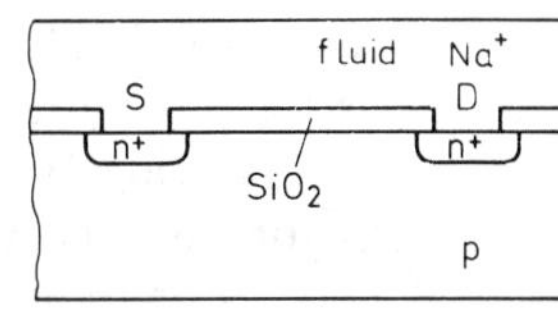

ION-SENSITIVE FIELD EFFECT TRANSISTOR (ISFET) When the gate electrode of a MOSFET is not metallized but covered with an ion-selective layer of Si_3N_4, Al_2O_3 polymer the device appears to be sensitive to ion concentrations in an electrolyte, providing the possibility of manufacturing pH sensor elements. Here again the packaging problem considerably limits the MTTF.

OXYGEN SENSOR Here the silicon substrate is used as carrier only, in which a spiral channel is etched, providing a chamber for an electrolytic solution of AgCl. The spiral-shaped chamber is covered with a membrane suitable for oxygen diffusion. Two silver electrodes are deposited on the substrate for use as anode and cathode and when a voltage is applied the charge carrier flow is a measure of the oxygen pressure in the chamber. The MTTF is a serious problem and is a mere couple of hours due to packaging problems.

MO [el, ch, el]

So far no chemical output transducers in silicon are known.

SI [ch, el, 00]

Galvanic cells can be realized by bringing together suitable electrodes and solutions of electrolytes, generating a voltage. The same effect can be obtained when a drop of an electrolyte is put on top of a silicon chip with appropriate electrodes. The generated voltage is a measure of the ionic concentration. A limited lifetime may be expected due to corrosion problems. No integrated chemical sensors are available so far.

This finishes the survey of energy conversions in silicon and compatible technologies. In Section 6.14 you will meet some other aspects concerning the most recent developments in technology and theory. In the next section we will discuss some possible bridge configurations and interfacing to sensors, for which operational amplifiers come into scope. Some basic knowledge of the operational amplifier is assumed.

6.13 INTERFACING TO SENSORS

In this section we will discuss some aspects of interfacing related to sensors. Interfacing can be defined as the physical connection of different systems in accordance with well defined protocols concerning the physical voltage levels, frequency differences, and so on. It will be clear that most sensors deliver a very weak signal level, ranging from some microvolts to a couple of hundreds of millivolts. In all cases an amplification is required for raising the signal level as well as for powering reasons.

A bridge configuration often can be applied but not in all kinds of sensors, as for instance with a microphone, solar cells, thermocouples, and so on. Self-generating transducers do not require an auxiliary energy source and generate their own voltage. They can be connected directly to an appropriate amplifier.

To become operational, modulating input transducers require a kind of biasing, which is always executed with an auxiliary energy source. To get rid of the common mode (supply) voltage a bridge configuration is preferred.

We can pose the following important questions:

1 If a bridge configuration can be applied, what is the bridge output voltage as a function of an induced arbitrary physical variation?
2 If no bridge configuration can be applied, which other configuration can be used?
3 What is the type of applied amplifier configuration?
4 Is any kind of filtering required?
5 What is the sensor's noise level?

Due to other complications, in the discussion we will lay emphasis on bridge configurations.

6.13.1 Sensors in bridge configurations

We shall consider the most common possibilities for implementing sensor elements in a bridge configuration. Bridge configurations are used with 1, 2 and 4 sensor elements, and in particular the last of these offers great advantages in integrated circuits. Besides this, in some measuring systems more than one bridge configuration is encountered, which can offer improved characteristics as we have seen in Section 3.6. To make comparisons between the different configurations the easiest way is to calculate the bridge output voltages in the respective cases.

1 Bridge configuration with one sensor element

The applied configuration is illustrated in Fig. 6.13, in which the supply voltage source V_{CC} and one sensor element R_S is depicted. Each time it is assumed that the change in physical parameter to be measured results in a change in the resistive sensor element of $R(1 + \delta)$. Then solving the bridge equation for V_{AB} and presuming all bridge resistances are equal at room temperature, we find

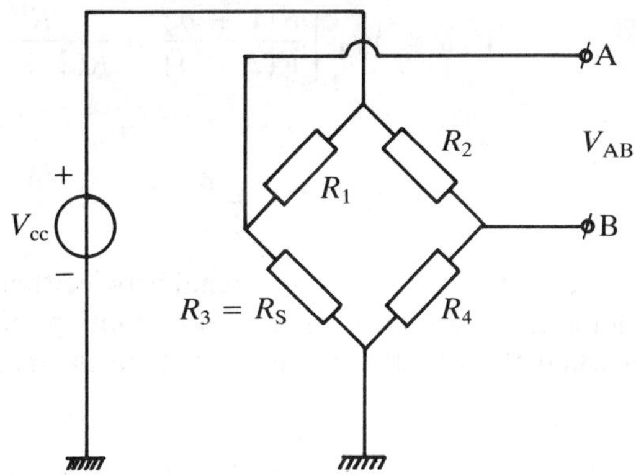

Figure 6.13 Bridge configuration with one sensor element

$$V_{AB} = V_{CC}\left[\frac{R(1+\delta)}{R + R(1+\delta)} - \frac{R}{2R}\right] \tag{6.14}$$

or

$$V_{AB} = V_{CC}\frac{\delta}{4 + 2\delta} \approx V_{CC}\frac{\delta}{4} \tag{6.15}$$

when δ is positive and $R_S = R_1 = R_2 = R_3 = R_4 = R$ at room temperature, all showing the same temperature coefficient.

*2 **Bridge configuration with two sensor elements***

The bridge configuration with two sensor elements is depicted in Fig. 6.14.

When the same assumptions are made as before, we find for the output signal V_{AB}

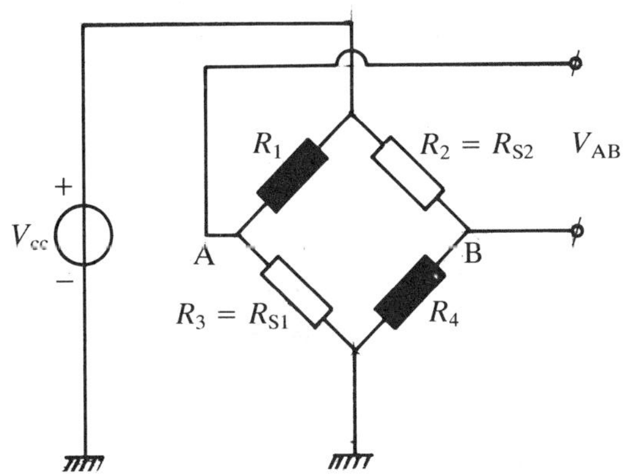

Figure 6.14 Bridge configuration with two sensor elements

$$V_{AB} = V_{CC}\left[\frac{R(1+\delta)}{R(2+\delta)} - \frac{R}{R(2+\delta)}\right] \qquad (6.16)$$

or

$$V_{AB} = V_{CC}\frac{\delta}{2+\delta} \approx V_{CC}\frac{\delta}{2} \qquad (6.17)$$

As can be seen, the bridge output signal is twice that of the configuration with one sensor element. Besides this aspect, in certain applications the output voltage changes sign when the physical parameter changes sign.

* * *

Example 6.3

Suppose the sensor elements are strain-sensitive elements and one strain gauge element is subjected to strain and the other to stress; then the output voltage will change sign when strain and stress are reversed for the respective strain gauges. This is illustrated in Fig. 6.15, where a metal bar provided with two strain gauges is sequentially subjected to a force F_1 and a force F_2. So the force F_1 causes a strain variation converted into a change in resistance of $R(1 + \delta)$ in R_{S2} and a stress variation converted into a change in resistance of $R(1 - \delta)$ in R_{S1}.

* * *

3 Bridge configuration with four sensor elements

When four sensor elements are applied in one bridge configuration it is required that the four sensor elements pairwise with opposite sign are subjected to the physical parameter. If this condition is not fulfilled a zero output signal will be the result. In most cases this condition can easily be fulfilled even when fully integrated bridge sensors are applied. This four-sensor configuration offers another great advantage, which is more sophisticated than in the case with two sensor elements. Besides the aspect of changing sign when the physical parameter changes sign,

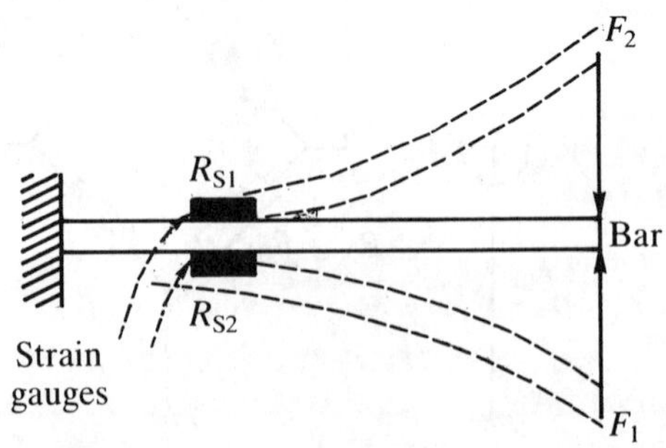

Figure 6.15 Metal bar sequentially subjected to forces F_1 and F_2 generating output voltages with opposite sign

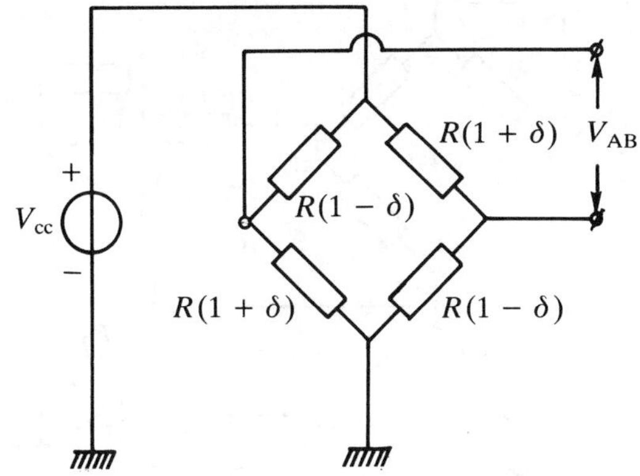

Figure 6.16 A four-sensor bridge configuration

when the *direction* of the physical parameter changes the bridge output voltage changes, offering a directional sensitivity. So in this case, a *vector* measurement is obtained, which means the *magnitude* and the *direction* of the physical parameter can be measured. This configuration is depicted in Fig. 6.16.

In this case the output voltage V_{AB} equals

$$V_{AB} = V_{CC}\left[\frac{R(1+\delta)}{R(1+\delta)+R(1-\delta)} - \frac{R(1-\delta)}{R(1+\delta)+R(1-\delta)}\right] \quad (6.18)$$

or

$$V_{AB} = V_{CC}\delta \quad (6.19)$$

which is twice that of the configuration with two sensor elements and four times that of the configuration with one sensor element. Note that in all three configurations the supply voltage is involved in the expression and thus can influence to a large extent the accuracy of the output. A well-controlled reference supply voltage is a prerequisite.

* * *

Example 6.4

Suppose the four-sensor configuration consists of an integrated circuit with four p-type doped resistive elements with a positive temperature coefficient δ, as depicted in Fig. 3.28. Suppose further that the bridge is thermally heated by its own dissipation well above ambient temperature, then an airflow will cool down the sensor configuration, showing a directional sensitivity. The output signal changes sign when the airflow changes its direction by 90°. This application is called an anemometer (*anemos* is Greek for 'wind').

* * *

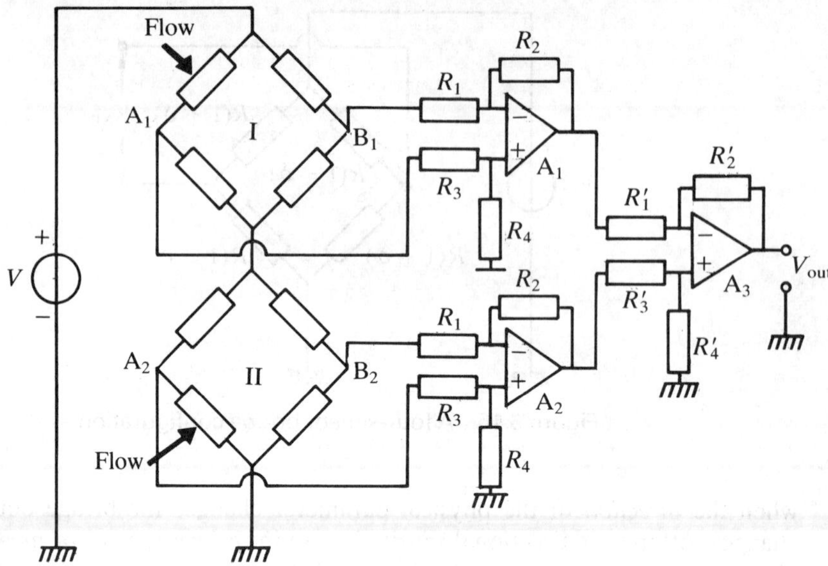

Figure 6.17 An integrated silicon double bridge anemometer

4 A double bridge arrangement
One application of an arrangement with two bridges is discussed in Section 3.6, but the use of this double bridge configuration was devoted to biasing, resulting in a constant voltage and current behavior in one bridge only. Another application of this configuration is illustrated in Fig. 6.17.

Referring to the same integrated sensor of Fig. 3.28, by careful inspection of the interconnection pattern it can be seen that both bridges are connected in opposition with each other. When both bridges are series connected to a voltage source and the sensor is again applied as an anemometer, the same airflow will generate two output voltages V_{A1B1} and V_{A2B2} with opposite sign. If a three-amplifier configuration is applied as depicted in the same figure, the output signals will be summed and the common-mode signals will be suppressed. Of course, high-quality chopper amplifiers are preferred for this configuration.

6.13.2 Biasing circuits

Application of a split power supply for a bridge sensor configuration can avoid common-mode input voltages at the input of the amplifier, hence reducing to a large extent the CMRR requirements. For instance, an amplifier with 80 dB CMRR adds 100 μV error to the output signal per volt common-mode input voltage. In some cases, the bridge is supplied via a switched power supply to reduce power dissipation. A perfectly symmetrical power supply is then necessary, providing a strongly reduced offset voltage. A schematic layout of such a configuration is depicted in Fig. 6.18.

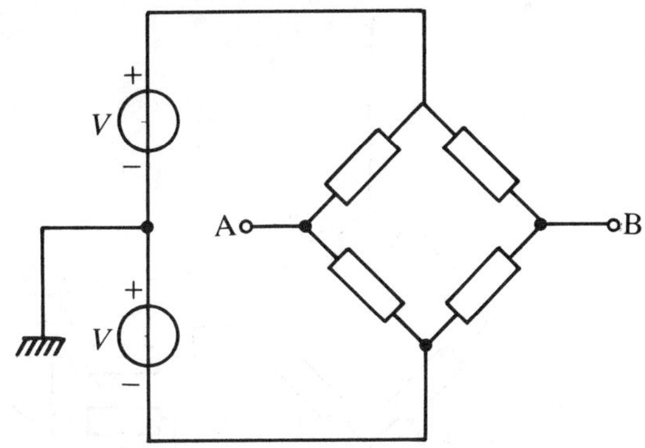

Figure 6.18 A symmetric power supply to reduce CMRR demands

Another current-biasing circuit is depicted in Fig. 6.23 and will be discussed in the next section.

6.13.3 Bridge amplifier configurations

In this section we shall discuss several interfacing connections between the bridge and the applied amplifier configuration. In most cases operational amplifiers are used due to their flexibility as a general building block. In all derivations we assume idealized operational amplifiers, i.e. with infinite open-loop gain, no input currents and hence infinitely large input impedances.

Circuit 1

In the circuits in Fig. 6.19 a standard differential amplifier is connected to the output of the respective bridge. The circuit of Fig. 6.19a has one sensor element, and that of Fig. 6.19b has two.

The output voltages can be derived from (6.15) and (6.17). For Fig. 6.19a,

$$V_{\text{out}} = AV\delta/4 \tag{6.20}$$

in which $A = R_2/R_1$ and for Fig. 6.19b with the same expression for A the value is twice of that of Fig. 6.19a, or

$$V_{\text{out}} = AV\delta/2 \tag{6.21}$$

Circuit 2

As can be seen from the expression for the configuration with one sensor element in Fig. 6.13 and equation (6.15), the exact result is a non-linear expression in δ. For a bridge configuration with one sensor the behavior can be linearized if we apply the circuit depicted in Fig. 6.20.

For this circuit with ideal amplifiers it can be said that node A and node B behave as a virtual common. With this assumption the following current equations for node A and node B will be valid:

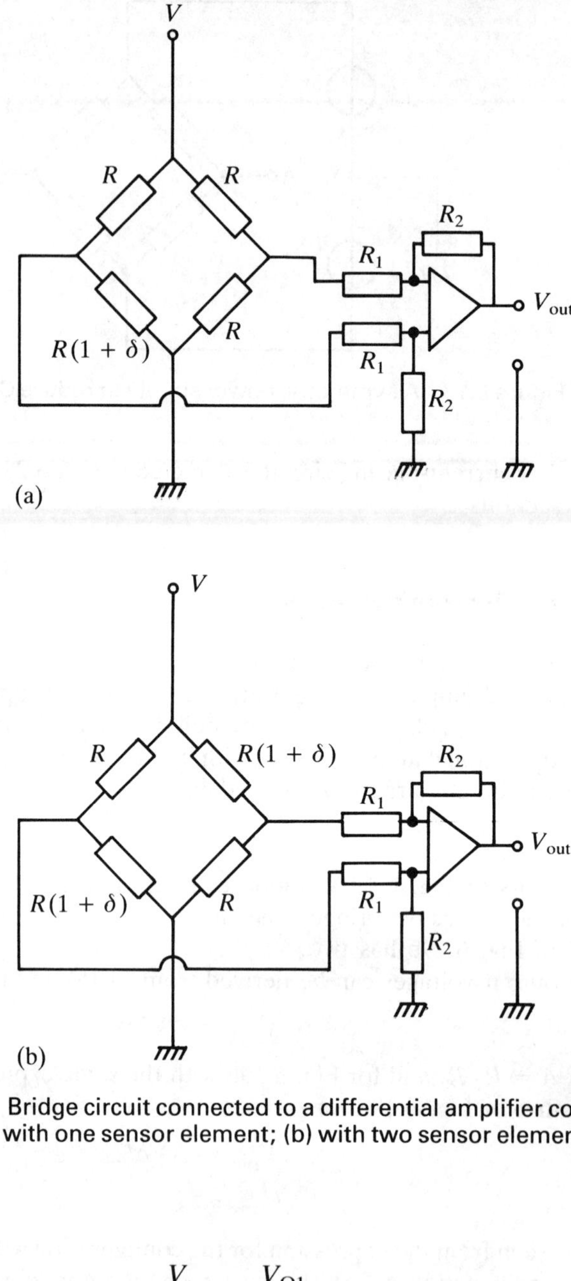

Figure 6.19 Bridge circuit connected to a differential amplifier configuration: (a) with one sensor element; (b) with two sensor elements

For node A,

$$\frac{V}{R} + \frac{V_{O1}}{R(1 + \delta)} = 0 \tag{6.22}$$

and for node B,

$$\frac{V}{R} + \frac{V_{O1}}{R} + \frac{V_{out}}{R_F} = 0 \tag{6.23}$$

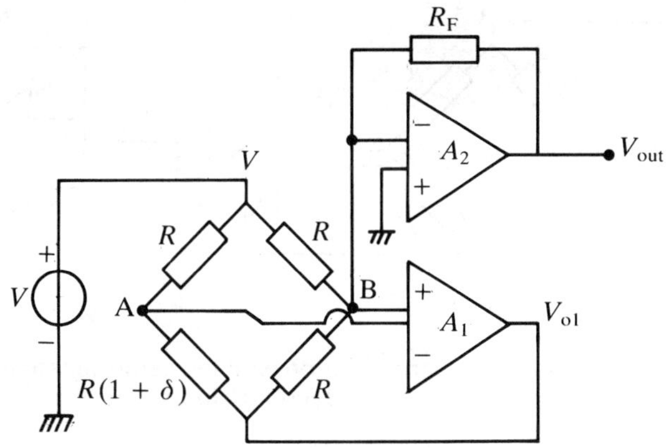

Figure 6.20 Linearization configuration of a bridge circuit with one sensor element applying feedback

Solving both equations for V_{out} yields a linear equation in δ.

$$V_{out} = \frac{R_F}{R}\delta V \tag{6.24}$$

A more accurate and general solution provides the following expression for V_{out} (see Problem 6.12):

$$V_{out} = A\delta V - \delta V + \tfrac{1}{2}A\delta^2 V \tag{6.25}$$

in which $A = R_F/R$.

The amplifier A_1 acts as a comparator, maintaining $V_A = V_B$ under all circumstances. Amplifier A_2 will deliver the required current via R_F to maintain this equality.

Circuit 3

An alternative feedback circuit is depicted in Fig. 6.21, in which the sensor element is implemented in the feedback loop.

An almost exact solution for the output voltage V_{out} can easily be obtained considering the amplifiers separately. The result is

$$V_{O1} = -\tfrac{1}{2}V\delta \tag{6.26}$$

and

$$V_{out} = -\frac{R_F}{R_1}V_{O1} = \tfrac{1}{2}V\delta\frac{R_F}{R_1} \tag{6.27}$$

Circuit 4

In the circuit of Fig. 6.22 two single-ended amplifiers are used. To reduce the first offset term a matched pair of amplifiers is preferred. Chopped amplifiers can offer the desired specifications, reducing strongly offset and drift behavior. The result is

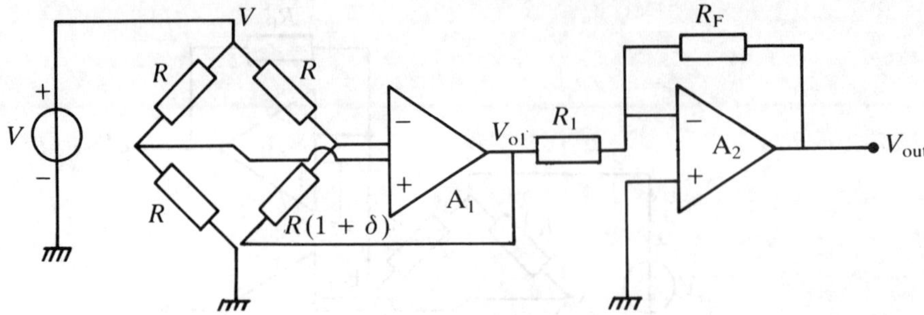

Figure 6.21 A bridge configuration with the same sensor implemented in the feedback loop

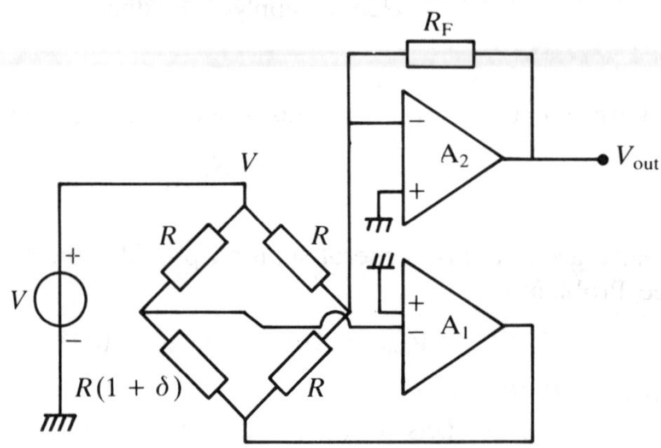

Figure 6.22 Configuration with two single-ended amplifiers

$$V_{out} = V\delta\frac{R_F}{R} \tag{6.28}$$

Circuit 5

In the circuit of Fig. 6.23 a current-biasing amplifier configuration is implemented. This also reduces non-linearity but instability might easily occur due to stray capacitances. Note that two sensors are implied in the bridge configuration. The expression for the output voltage can be obtained applying the current node equations for A and B. The result is (see Problem 6.15):

$$V_{out} = A\tfrac{1}{2}(V - V_{ref})\delta \tag{6.29}$$

or

$$V_{out} = A\tfrac{1}{2}(V - IR_5)\delta \tag{6.30}$$

in which $A = R_2/R_1$ is the gain factor of A_1.

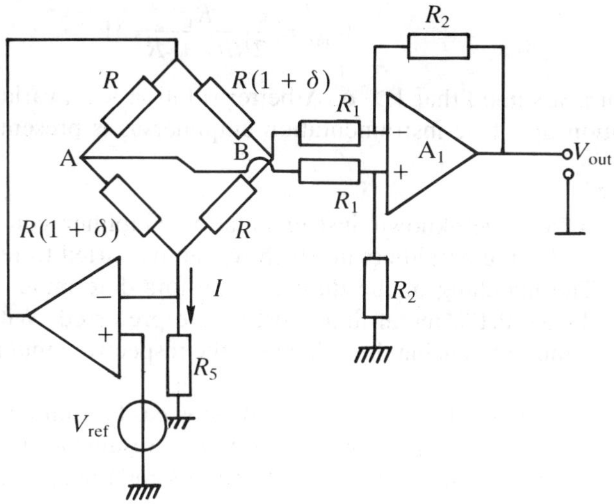

Figure 6.23 A current-biasing bridge amplifier configuration

Besides the stability problem the stability of the reference voltage strongly influences the accuracy of this circuit.

Circuit 6

In Fig. 6.24 a single differential amplifier is applied with which the gain factor can be varied. The disadvantage of this circuit is that any variation in R_1 disturbs the balance of the amplifier's input impedance, resulting in a change in the offset.

Note that the plus and minus sign of the supply voltage V is reversed compared with a usually connected supply circuit. This is performed to obtain a more simplified expression for V_{out}. The expression for V_{out} is not strictly linear but may be approximated with

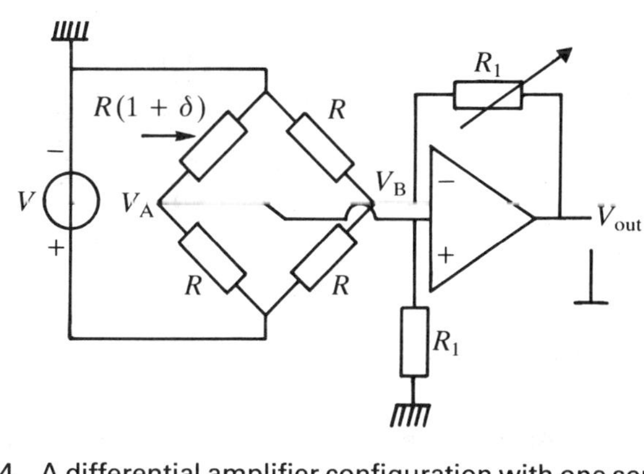

Figure 6.24 A differential amplifier configuration with one sensor bridge element

$$V_{out} = \frac{R_1^2}{2RR_1 + R^2}\delta V \tag{6.31}$$

where it is assumed that $1 \gg \delta$. A better solution for a variable gain setting is the application of a true instrumentation amplifier as is presented in circuit 7.

Circuit 7

In Fig. 6.25 a well-known instrumentation amplifier configuration is depicted consisting of three amplifiers in which R_1 can be varied to realized a variable gain setting. The matching of the different resistors determines to a large extent the CMRR, hence 0.1% metal film resistors are preferred. With R_1 the gain can be varied without influencing the balance of the respective input impedances of A_1 and A_2.

The expression for the output voltage is easily found and may be derived by calculating the current in R_1 and R_2 twice. We presume again that the amplifiers are ideal, that is to say, there are no input currents and the open-loop gain is infinite for all three amplifiers.

$$I = \frac{V_{out1} - V_{out2}}{2R_2 + R_1} \tag{6.32}$$

and

$$I = \frac{V_A - V_B}{R_1} \tag{6.33}$$

because there is one current only in R_1 and R_2, and V_A and V_B can be found to exist across R_1. Combining both equations yields

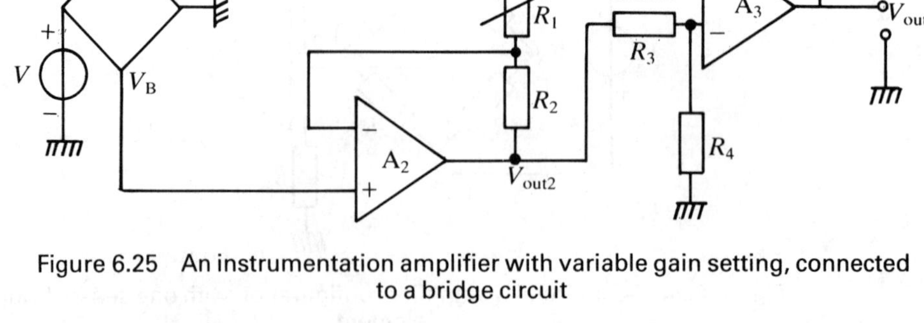

Figure 6.25 An instrumentation amplifier with variable gain setting, connected to a bridge circuit

$$\frac{V_{\text{out1}} - V_{\text{out2}}}{2R_2 + R_1} = \frac{V_A - V_B}{R_1} \tag{6.34}$$

or

$$V_{\text{out1}} - V_{\text{out2}} = \frac{R_1 + 2R_2}{R_1}(V_A - V_B) \tag{6.35}$$

and for the third stage is found

$$V_{\text{out}} = \frac{R_4}{R_3}(V_{\text{out1}} - V_{\text{out2}}) \tag{6.36}$$

Substituting equation (6.35) in (6.36) we find

$$V_{\text{out}} = \left(1 + \frac{2R_2}{R_1}\right)\frac{R_4}{R_3}(V_A - V_B) \tag{6.37}$$

The configuration appears to have a very high input impedance and is obtainable as one building block. As will be discussed in Chapter 9, low source impedances produce low noise, but designing the aluminum interconnection pattern on the chip parallel to the sensor area as shown in Fig. 3.28 provides a considerable reduction in the source resistance while maintaining a good sensitivity. For instance, the source resistance of the depicted chip is 50 Ω while the sheet resistance is about 5000 Ω.

Fig. 6.26 shows a complete schematic suitable for n transducers and n actuators

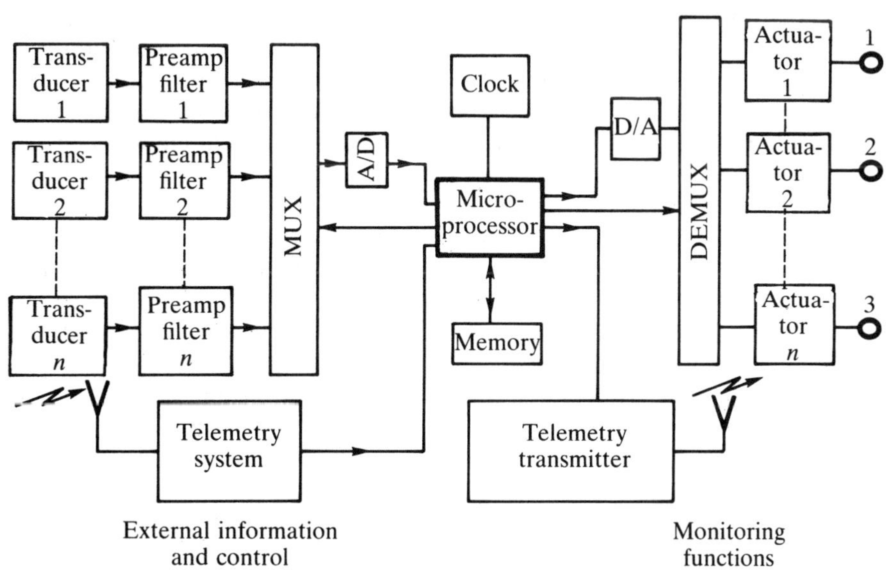

Figure 6.26 Complete interface schematic with telemetry receiver and transmitter for n transducers and n actuators

with remote control provisions for transmitting and monitoring information. Especially in the medical field, applications for mobile patients can be found. There is an increasing need for wireless transmitting and monitoring the patient's condition, offering a very feasible type of interfacing. Also, we can find this type of application in agriculture. Farmers can monitor their cattle for food consumption and milk production individually by remote control.

6.14 FUTURE TRENDS

In discussing future trends for transducers and especially for silicon microtransducers, two different directions can be distinguished. The first is a purely technological one whereas the second appears to have a more fundamental character.

6.14.1 Technological trends

In the development of transducers an increasing trend is present to apply silicon technology due to the advantages discussed in Section 6.6. Of course, some problems and disadvantages exist and should be mentioned to put silicon technology in the right context. From this consideration some trends can be recognized of what may be expected in the future.

1 Starting the development of an integrated sensor requires considerable effort and time before a marketable sensor is available. For numerous sensors the time of development is ten years or more. At a first glance this seems an unbelievable amount of time to invest, but the development of a sensor is really a complicated affair. Many problems have to be solved in a multidisciplinary environment. A sensor is not pure electronics and although proving a physical effect is not too difficult, manufacturing a reliable sensor with linear characteristics and good response time is a formidable task. Besides this, often for every application an appropriate packaging must be developed. All these factors must be translated into a price performance ratio and it is not difficult to understand that this will only start to be worthwhile when large series can be fabricated. It is a lucky circumstance that the need for sophisticated transducers is increasing because we are becoming more and more dependent on electronic systems. Transducers are the sense organs for every system communicating with the outside world. It will be clear that when large series are manufactured the result is a considerable cost reduction.

2 The application of silicon is limited to the allowed applicable temperature range for silicon. The applicable temperature range lies between −100° and 200°C. Below −100°C, the applied doped materials in silicon are hardly activated, and above 200°C, degradation of silicon will occur. Here GaAs can provide new possibilities because the bandgap of GaAs is 1.43 eV which is 0.31 eV larger than

that of silicon. The consequence is a higher allowed operational temperature. It is not improbable that GaAs may also be applied for a great variety of transducers, as will other III–V semiconductor components as long as they are compatible with planar silicon technology concerning cost aspects.

3 The bonding technique used for connecting the chips with their outer environment is very sophisticated and is performed fully automatically, and can be executed, for instance, with a speed of 8 bonds per second. But the gold or aluminum wire bonds have a thickness of 1 mil or 25 μm which is thinner than a hair, and so they are highly vulnerable when they are exposed to open air which is required for some transducers. Smart packaging can solve this problem.

It may be expected that in the future, increasingly fully integrated sophisticated transducers will be fabricated for several physical parameters combined with computing power on one single chip. However, this development may be considered as a pure technological development.

6.14.2 Fundamental developments

The discussed energy conversions are well known from the literature and can be described in detail in a macroscopic way. But in the last decades less attention has been paid to more fundamental questions such as what is really happening in an energy conversion process and to what fundamental processes are energy conversions bound. This interesting area of investigation is born from a more *fundamental* approach in measuring techniques.

As we have seen in the introduction, information is a kind of order to be recognized from its environment and is always bound to an energy carrier or mass. Recently Middelhoek and coworkers of Delft University of Technology started a new investigation concerning the underlying physical processes in transducers and measuring techniques. Here a microscopic approach of the energy conversions comes into scope. For this purpose a slightly different division of the energy domains is defined. This division can provide a better insight into the underlying process of energy conversion which will be shown by an example. The classification is shown in Table 6.15.

Table 6.15 Review of an alternative energy classification

Mechanical energy	is assigned to the kinetic energy of uncharged particles
Charge energy current	is assigned to the kinetic energy of charged particles
Gravitational energy	is assigned to mass
Electrical energy	is assigned to charge
Magnetic energy	is assigned to moving charge
Radiation energy	is assigned to periodical movement of charge
Thermal energy	is assigned to kinetic energy *in* matter
Chemical energy	is assigned to electromagnetic forces *in* matter

Considering the energy conversions in a microscopic fashion, all possible conversions can be assigned to one of the following three classes.

1 Collisions
If the charge of the particles differ, a conversion between mechanical energy and charge currents can occur caused by a transfer of kinetic energy through collision.

2 Changes in position
If a particle with a certain kinetic energy moves in a space where a gradient in potential energy is present, the potential energy of the particle can change. The change of energy is converted into a change in the kinetic energy of the particle.

3 Interactions
Energy conversions generated by interactions between internal and external potential energy.

Of course in all cases the law of energy conservation is applicable, which means that a decrease in one type of energy will give rise to the same energy increase in another type of energy at the same time and place.

* * *

Example 6.5

Suppose a temperature gradient is generated in a piece of material, a heat current will flow, generally consisting of two compensating charge currents. But if this electrical equilibrium is distorted, electrical effects will be generated like the Seebeck effect or the Ettinghausen–Nernst effect.

* * *

6.15 PROBLEMS

6.1 Describe a general layout of a data-collecting system.

6.2 Describe the main reasons why integrated silicon transducers can be important.

6.3 What are the definitions of an intensive and an extensive quantity?

6.4 What is the difference between a cross-effect and a direct effect?

6.5 What is the difference between the internal and the external photoelectric effect?

6.6 What is understood by a direct and indirect bandgap material?

6.7 What is the characteristic difference between a GaAs and a GaP diode?

6.8 Two physical effects can be applied for pressure transducers. Which are they?

6.9 Why is the temperature coefficient of the piezo-resistance effect in intrinsic silicon larger than in extrinsic silicon?

6.10 What are thermoelectric effects?

6.11 Which effect(s) can be used for manufacturing an output transducer denoted as [el, ma, 00]?

6.12 Derive expression (6.25). (*Hint:* apply the current node equations for node A and B.)

6.13 Prove the expression for the output voltage for Fig. 6.21. Apply the same hint as before for a general solution.

6.14 Derive the expression for the output voltage for Fig. 6.22.

6.15 Answer the same question for Fig. 6.23.

6.16 Answer the same question for Fig. 6.24.

Seven

OFFSET AND DRIFT

7.1 INTRODUCTION

For two reasons it can be said that linear amplification is one of the most important operations in the field of electronics. First, a correct matching to the signal source can be achieved, and second, for reasons of sensitivity linear amplification will always be necessary. The importance of this sensitivity aspect can easily be illustrated by the very low signal source levels which are always encountered in the medical and biological field. Signal sources with mV and μV voltage level are usual.

In the problem of linear amplification four different aspects can be distinguished which mainly determine the final accuracy. These are:

1 Offset and drift
2 Guarding
3 Noise
4 The application of universal active elements

In the following chapters the first three items will be emphasized and discussed in detail, and in less detail the application of active elements, because they can be found in every textbook about electronics. Generally it can be said that ultimately the overall accuracy of a measurement set-up is determined by the offset, drift, guarding and noise characteristics of all components involved. Guarding and noise will be discussed in Chapters 8 and 9 respectively; this chapter considers offset and drift. First we will discuss the principles of offset voltage and offset current in active components. Secondly, the drift behavior of active components is treated.

7.2 PRINCIPLES OF OFFSET VOLTAGE AND CURRENT

It is well known that every active element requires a certain biasing current and voltage in order to make this element operative in the desired region. With the input signal source put to zero, the effect of this biasing will result in a certain

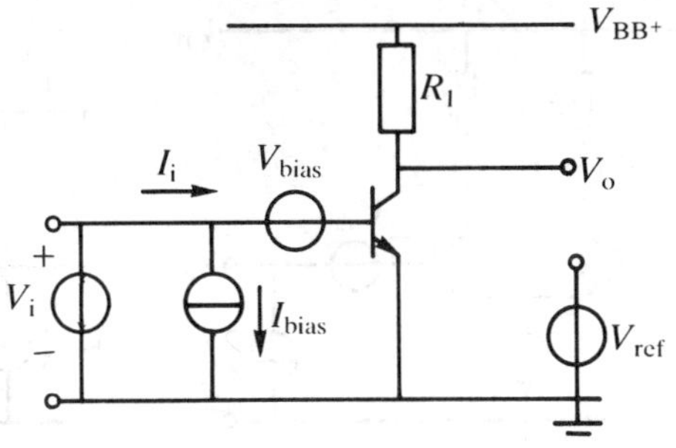

Figure 7.1 Bipolar transistor with biasing sources

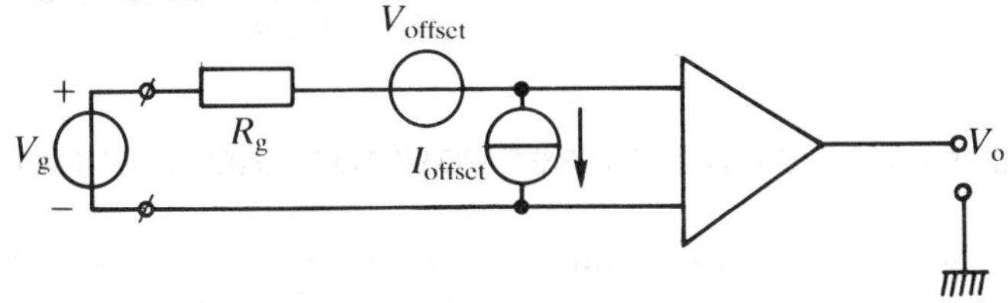

Figure 7.2 Differential amplifier with respective biasing sources I_{bias} and V_{offset}

output voltage level. This is illustrated in Fig. 7.1 for a single bipolar transistor and in Fig. 7.2 for an amplifier with a differential input stage.

In Fig. 7.1 a bias current I_{bias} and a bias voltage source V_{bias} are depicted.

In the case of the differential amplifier it makes sense to talk about differences in biasing voltages and differences in biasing currents at both input terminals because these are the resulting parameters. We will return to that later on.

By definition, the required input voltage to be applied to make the output voltage exactly zero, is called the input offset voltage of the active element. In the same way, by definition the required input current to be applied to make the output voltage exactly zero, is called the input offset current of that active element.

In both definitions it is presumed that the signal source is set to zero. Both conditions must be fulfilled to make the output zero. The required voltage and current to make the output voltage zero can be considered to be delivered by two independent sources V_{bias} and I_{bias} in the single-stage case and V_{offset} and I_{offset} for the differential stage. These sources are often interpreted as interfering sources which can be drawn outside the amplifier at the input side. This concept allows one to talk about a bias-free amplifier, which is, of course, an imaginary, idealized amplifier. Usually, for operational amplifiers, the (input) offset voltage and the (input) bias current are specified by the manufacturer.

Fig. 7.3 illustrates the differential amplifier in more detail. In this figure for symmetry reasons two bias voltage sources and two bias current sources can be recognized, from which Fig. 7.2 is derived. In Fig. 7.3b the bias- and offset-free amplifier is drawn with the respective offset and bias sources. This concept will be met again in Chapter 9.

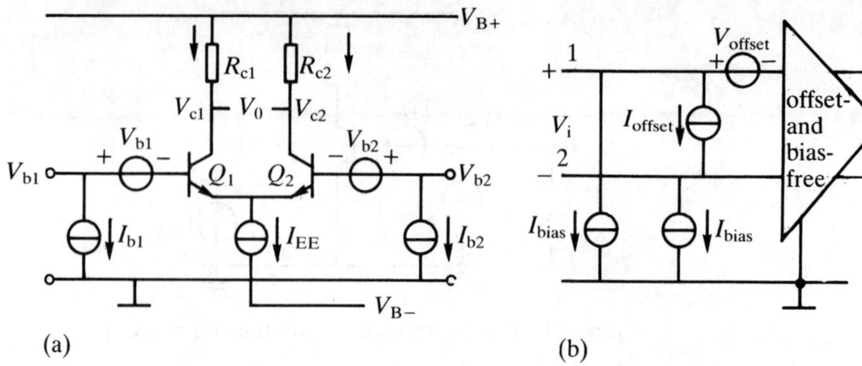

Figure 7.3 (a) Differential amplifier with biasing sources; (b) equivalent circuit with offset sources

7.3 CALCULATION OF THE OFFSET VOLTAGE AND CURRENT

In this section the expressions for the offset voltage and current of the bipolar transistor and the field-effect transistor will be derived.

7.3.1 The bipolar transistor

To calculate the offset voltage and current for the bipolar transistor we use the following expression:

$$I_C = K\frac{A_E}{W_B}\exp(-qV_{GAP}/kT)[\exp(qV_{BE}/kT) - 1] \tag{7.1}$$

where

I_C = the DC collector current
K = a temperature-dependent quantity proportional to T^3
A_E = the emitter surface
q = specific electronic charge, 1.6×10^{-19}
V_{GAP} = bandgap between valence band and conduction band
T = absolute temperature in Kelvin
V_{BE} = base-emitter voltage
W_B = base thickness
k = Boltzmann's constant

Neglecting the −1 term, taking the natural logarithm of equation (7.1) and assuming both $V_i = 0$ and $V_o = 0$, an expression for the base-emitter voltage is found:

$$V_{BE} = \frac{kT}{q}\ln\left(\frac{I_C W_B}{KA_E}\right) + V_{GAP} \tag{7.2}$$

Due to the large value of K the first term will be negative, and to find the magnitude of V_{BE} with $V_{output} = 0$ the actual value for I_C must be substituted. Then the bias current with $I_I = 0$ and $V_o = 0$ equals

$$I_{bias} = I_C/\beta \tag{7.3}$$

where β is the current gain factor. The order of magnitude for V_{BE} is 0.6 to 0.7 V and for the current appears to be 1 to 5 μA. Note that for a single stage V_{BE} equals V_{offset} and I_B equals I_{bias}. Of course in a single stage these values are always present and no precautions are available to compensate them effectively other than by external sources.

7.3.2 The field-effect transistor

The offset behavior of a field-effect transistor can be subdivided into *four* regions and is determined by the region in which the transistor is operative. Apart from this we have to distinguish between two types: the junction and the isolated gate field-effect transistor. This last type is usually called a MOS-type transistor. In Fig. 7.4 the characteristic of a FET is drawn with the four regions to be distinguished.

The first operative region is called the *linear region* and is characterized by a strong linear relationship between I_D and V_{DS}. The second region is called the *weak inversion* region and shows a strong exponential behavior. The third region is called the *boundary region* between weak and strong inversion and is characterized by quadratic behavior. The fourth region is known as the region for *strong inversion* and shows a quadratic behavior. This makes the discussion of the offset behavior a little more complicated. We shall confine the discussion to the most frequently applied region of strong inversion.

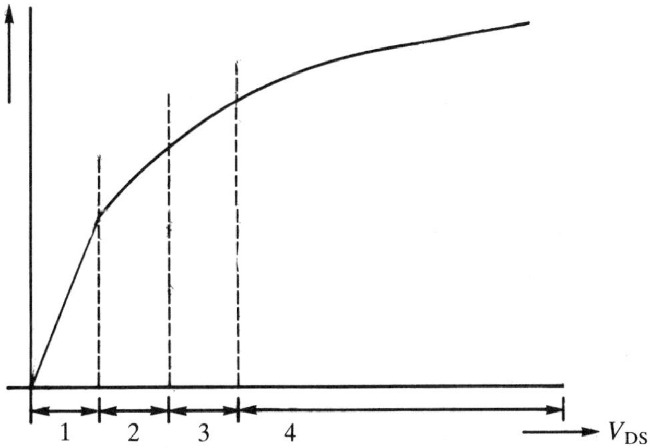

Figure 7.4 Typical characteristic for a field-effect transistor

1 The junction field-effect transistor
For the junction field-effect transistor, the offset behavior can be derived from the equation

$$I_D = \Gamma \frac{W}{L} V_p \left[1 - 3\frac{V_{GS}}{V_p} + 2\left(\frac{V_{GS}}{V_p}\right)^{3/2} \right] \tag{7.4}$$

where

I_D = the drain currrent
Γ = a geometrical and physical constant of the junction FET
V_p = the pinch-off voltage
V_{GS} = the gate source voltage
W = the channel width
L = the channel length

This expression can be approximated by

$$I_D = \Gamma \frac{W}{L} V_p \left[1 - 3\frac{V_{GS}}{V_p} \right]^2 \tag{7.5}$$

If V_{GS} is made explicit we find the expression for the offset voltage for a junction FET:

$$V_{offset} = V_{GS} = V_p \left\{ 1 - \sqrt{\frac{I_D}{\Gamma V_p W/L}} \right\} \tag{7.6}$$

2 The MOS field-effect transistor
For a MOS field-effect transistor in the same region, the expression for strong inversion is

$$I_D = \frac{K_p W}{2nL}[V_{GS} - V_p]^2 \tag{7.7}$$

where

I_D = the drain current
K_p = a geometrical and physical constant of the MOSFET
W = the channel width
n = a slope factor related to the actual inversion region
L = the channel length
V_{GS} = the gate source voltage
V_p = the pinch-off voltage

The pinch-off voltage, V_p, is defined as the gate source voltage, V_{GS} required to cause pinch-off of the channel to zero width. Note the difference with the threshold voltage V_t, which is defined as the gate-to-source voltage required to cause an inversion layer.

Solving equation (7.7) for V_{GS} delivers the desired expression for the offset voltage of the MOS field-effect transistor:

$$V_{\text{offset}} = V_{\text{GS}} = V_{\text{p}} + \sqrt{\frac{I_{\text{D}}2nL}{K_{\text{p}}W}} \tag{7.8}$$

Characteristic values for the offset voltages for both the junction FET and the MOSFET are several volts. For the current the values are in the order of 10–100 pA.

7.4 OFFSET BEHAVIOR OF DIFFERENTIAL AMPLIFIER STAGES

Here again we have to make a distinction between the bipolar and the field-effect transistor configuration. We refer to the layout of Fig. 7.3 where by definition the offset voltage of the differential stage can be written

$$V_{\text{offset}} = |V_{\text{off1}} - V_{\text{off2}}| \tag{7.9}$$

or

$$V_{\text{offset}} = |V_{\text{bias1}} - V_{\text{bias2}}|_{V_{\text{out}}=0} \tag{7.10}$$

and for the offset current by definition

$$I_{\text{offset}} = |I_{\text{b1}} - I_{\text{b2}}|_{V_{\text{out}}} \tag{7.11}$$

The referred configurations are so-called coupled pairs.

7.4.1 The bipolar differential input stage

Applying equation (7.2) for the bipolar differential input stage we find

$$V_{\text{offset}} = \frac{kT}{q}\left[\ln\frac{I_{\text{C1}}W_{\text{B1}}}{KA_{\text{E1}}} - \ln\frac{I_{\text{C2}}W_{\text{B2}}}{KA_{\text{E2}}}\right] \tag{7.12}$$

in which it is presumed that both transistors have the same temperature and equal Ks. The condition $V_{\text{out}} = 0$ delivers the expression

$$I_{\text{C1}}R_{\text{C1}} = I_{\text{C2}}R_{\text{C2}} \tag{7.13}$$

Combining equations (7.12) and (7.13) yields

$$V_{\text{offset}} = \frac{kT}{q}\ln\frac{R_{\text{C2}}W_{\text{B1}}A_{\text{E2}}}{R_{\text{C1}}W_{\text{B2}}A_{\text{E1}}} \tag{7.14}$$

Typical values for the offset voltage of the bipolar differential stage are 0.5–5 mV. This offset is easily nulled via R_{C1} or R_{C2}, but as can be seen from the expression for the offset, *symmetry* for all components involved plays a key role in the offset behavior of the differential stage. This is true for the geometrical dimensions as well as for temperature and doping concentration levels for the transistors and resistors involved.

The offset current can be derived from the expressions (7.10) and

$$V_{out} = R_{C1}I_{C1} - R_{C2}I_{C2} = 0 \tag{7.15}$$

Equation (7.15) with (7.3) yields

$$\beta_1 R_{C1} I_{b1} - \beta_2 R_{C2} I_{b2} = 0 \tag{7.16}$$

or

$$I_{b1}/I_{b2} = \beta_2 R_{C2}/\beta_1 R_{C1} \tag{7.17}$$

Rearranging (7.11) yields for the offset current

$$I_{offset} = I_{b2}(I_{b1}/I_{b2} - 1) \tag{7.18}$$

Substituting (7.17) in (7.18) gives

$$I_{offset} = I_{b2}\left(\frac{\beta_2 R_{C2}}{\beta_1 R_{C1}} - 1\right) \tag{7.19}$$

Here again, the symmetry of βs and R_Cs stipulates the offset current.

7.4.2 The junction-FET differential input stage

This configuration, often called a source coupled pair, is depicted in Fig. 7.5. If the same definitions, (7.10) and (7.11), are applied for the offset voltage of a junction-FET differential input stage, we find

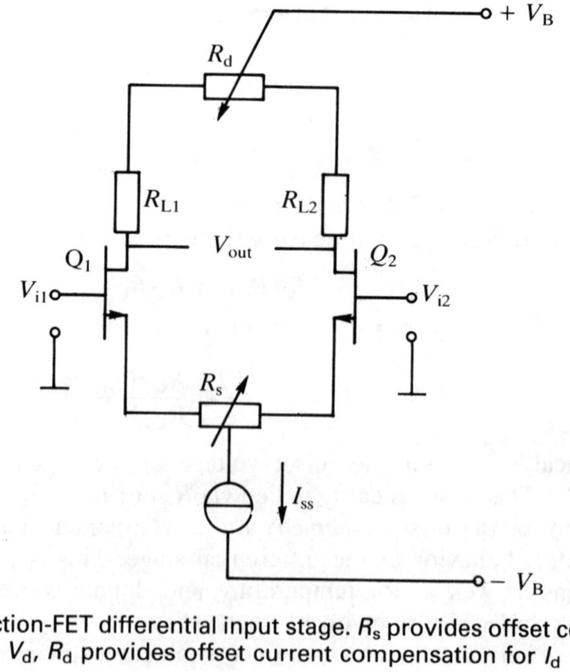

Figure 7.5 Junction-FET differential input stage. R_s provides offset compensation for V_d, R_d provides offset current compensation for I_d

$$V_{offset} = V_{GS1} - V_{GS2}$$

or

$$V_{offset} = V_{p2} - V_{p2} + \sqrt{\frac{I_{D2}}{\Gamma V_{p2}(W/L)_1}} - \sqrt{\frac{I_{D1}}{\Gamma V_{p1}(W/L)_2}} \tag{7.20}$$

The offset current of a FET is generally neglected because of the very low value of the gate current. The gate currents for junction-FETs are in the order of nanoamps to picoamps, and for the offset current some tens of picoamps are reported. This will be discussed again in a later section. If desired, the expression for the offset voltage can be rearranged in relative ratios in W/L and the load resistances R_L.

7.4.3 The MOSFET differential input stage

Applying again the definitions for the offset voltage, the following expression can be derived:

$$\begin{aligned} V_{offset} &= V_{GS1} - V_{GS2} \\ &= V_{p1} - V_{p2} + \sqrt{\frac{I_{D1} 2nL_1}{K_p W_1}} - \sqrt{\frac{I_{D2} 2nL_2}{K_p W_2}} \end{aligned} \tag{7.21}$$

This is equivalent to

$$V_{offset} = \Delta V_p + \frac{(V_{GS} - V_p)}{2}\left[\frac{-\Delta R_L}{R_L} - \left(\frac{\Delta(W/L)}{(W/L)}\right)\right] \tag{7.22}$$

where

$$\Delta V_p = V_{p1} - V_{p2} \tag{7.23}$$

$$V_{GS} = \tfrac{1}{2}(V_{GS1} + V_{GS2}) \tag{7.24}$$

$$\Delta R_L = R_{L1} - R_{L2} \tag{7.25}$$

$$R_L = \tfrac{1}{2}(R_{L1} + R_{L2}) \tag{7.26}$$

$$W/L = \tfrac{1}{2}([(W/L)_1 + (W/L)_2] \tag{7.27}$$

$$\Delta(W/L) = (W/L)_1 - (W/L)_2 \tag{7.28}$$

The same remark is applicable to the offset current as to the case of the junction-FET. In the next section we will come to that again. In most cases, it is easy to compensate for any offset in order to get a nulled output when no signal is present. An example of an offset nulling procedure for a differential input stage for current and voltage is depicted in Fig. 7.5. This nulling can be performed also by trimming techniques (by laser) on the integrated circuit itself. But much more important are drift phenomena, which will be discussed in the next section.

7.5 DRIFT BEHAVIOR OF OFFSET CURRENT AND VOLTAGE

Drift behavior of the offset current and voltage is much more crucial and complex than offset itself. In most cases, good measurements are available to prevent or compensate for any offset of either current or voltage. Drift phenomena are characterized by slow variations of the offset sources due to several causes and finally, in connection with noise, they determine the lowest limit which can be measured with specified accuracy. Drift is more difficult to overcome and is still one of the most discussed subjects in analog electronics. Four main causes generating drift can be mentioned:

1 stochastic variations in electrical conductivity
2 aging, humidity, mechanical vibrations, etc.
3 power supply variations
4 temperature changes affecting parameters.

7.5.1 Stochastic variations in electrical conductivity

Especially at low frequencies stochastic variations may influence the drift behavior. This is especially the case where $1/f$ noise comes into scope, which may influence the drift behavior considerably. Examples of components which have to be considered as large $1/f$ drift sources are carbon composite resistors, electrolytic capacitors, and zener diodes. Possible precautions are frequency shifting, filtering and the choice of components with little $1/f$ noise. More details can be found in Section 10.4.

7.5.2 Aging, humidity and vibrations

Electrical conductivity can be influenced considerably by aging and humidity. Aging can be considered as the changing of electrical properties of components with passing time. Chemical and physical changes in the material from which the components are made are likely to occur as time passes. This can be considered as a wear-out process. Possible precautions are, for instance, derating in load conditions. More details about aging can be found in Chapter 5.

Humidity will also have an influence on conductivity. To prevent this, effective measures are available, such as hermetic sealing and slightly raising the temperature of humidity-sensitive components.

Vibrations are mainly caused by hazardous environmental conditions as in missiles, aircraft, ships and mobile electronic equipment. These conditions can easily influence the electronic parameters and thus the working performance of the electronic components. Suitable stable and rigid constructions, especially of connectors and plugs, can eliminate these problems.

7.5.3 Power supply variations

In general, power supply variations will affect either the output voltage or the output current, for which two factors can be mentioned.

By definition the voltage *power supply rejection ratio* is the equivalent input voltage which must be connected to the input to keep the output voltage constant when the supply voltage changes by an amount V_B because the relationship between the input and output voltages is the gain factor. For a single bipolar amplifier stage as shown in Fig. 7.1 this equals

$$\left|\frac{V_i}{V_B}\right|_{V_{out}=\text{const}} = \frac{r_c}{R_c} \tag{7.29}$$

where

$r_c = \dfrac{kT}{qI_C}$
k = Boltzmann's constant, 1.38×10^{-23} J/K
T = absolute temperature
q = specific charge, 1.6×10^{-19} C
I_C = collector current
R_c = collector resistance

In the case of a single stage, the supply voltage will directly influence the output voltage in a one-to-one relationship. Usually this figure of merit is expressed in dB. For instance, if $r_c = 25\ \Omega$ and $R_c = 2.5\ \text{k}\Omega$ then the power supply rejection ratio, usually abbreviated to PSRR, is 10 mV/V or 40 dB.

For the power supply rejection ratio concerning the current we find

$$V_{out} = V_B - I_C R_c \tag{7.30}$$

or

$$V_{out} = V_B - I_C R_c = 0 \tag{7.31}$$

and with $I_C = \beta I_i$ we find

$$\left|\frac{I_i}{V_B}\right|_{V_{out}=\text{const}} = \frac{1}{\beta R_c} \tag{7.32}$$

in which β is the current gain factor. A large value of β will deliver a large value for the rejection ratio in dB. This value is often not specified by the manufacturer but may be of importance when current sources are implemented.

It will be clear that when the same definitions are applied to balanced amplifier configurations as is a differential stage, the power supply rejection ratio for the voltage and the current will increase considerably and a value of 120 dB can easily be obtained. Here too, symmetry aspects determine the final rejection ratio and for the current the tail current source involved will be dominant. Assume that for a differential amplifier, a 10 μV/V is specified, then this equals 100 dB.

7.6 THE INFLUENCE OF TEMPERATURE

The influence of temperature is caused by the fact that most electronic parameters show a temperature-dependent behavior. This temperature drift is one of the main causes of offset drift and, as said already, is much more cumbersome to compensate for. It should be emphasized here that, in most cases, offset nulling does not mean drift nulling. In general this offset nulling is only valid for one temperature. Firstly we will demonstrate the causes of temperature drift with several examples.

* * *

Example 7.1

Thermocouple effects caused by temperature gradients across dissimilar metals can generate voltages proportional to the temperature difference between the junction and the measurement end of the metal. This voltage can range from zero to hundreds of microvolts per degree Kelvin depending on the materials involved. Small temperature differences can easily exist across small distances of several millimeters. It will be clear that this implies considerable problems when low signal levels are present. Temperature gradients can be found across resistors and package–to–printed circuit board interfaces. It has been shown that temperature differences can exist on silicon integrated circuits as well, despite their relatively good specific thermal conductivity.

* * *

Example 7.2

Suppose a low-drift amplifier is monitored over a 10-minute period and, during this period, it appeared to have an input referred offset variation of ±5.0 μV. When the circuit is shielded from draught the offset variation is reduced to ±0.5 μV, which results in a factor of improvement of 10.

* * *

Example 7.3

Carbon composite, oxide and some metal-film resistors can cause large thermocouple errors. Wire-wound resistors manufactured from manganese are best, since they only generate about 2.0 μV/K with copper as reference.

* * *

Example 7.4

Suppose the gain-setting resistors of an amplifier cannot track with temperature, then a gain error will result. For instance, if a gain-of-1000 amplifier with a constant input of 10 mV appears to have a temperature mistrack of 0.5% over the desired operating temperature range, the error at the output will be 50 mV, which referred to the input is 50 μV. It is important that the gain-setting resistors have the same temperature coefficient and have a close temperature tracking.

* * *

7.7 TEMPERATURE DRIFT IN BIPOLAR TRANSISTORS

The temperature behavior of a bipolar transistor can be calculated by differentiating equation (7.2) with respect to T. This results in

$$\left.\frac{\partial V_{BE}}{\partial T}\right|_{I_C} = \frac{k}{q}\ln\frac{I_C}{A} \tag{7.33}$$

Multiplying the right-hand term by T/T and substituting again equation (7.2), this yields

$$\left.\frac{\partial V_{BE}}{\partial T}\right|_{I_C} = \frac{1}{T}[V_{BE} - V_{GAP}] \tag{7.34}$$

For a single bipolar transistor $V_{BE} \approx 0.6$–0.7 V, the bandgap voltage is ≈ 1.2 V, and substituting $T = 300$ K, the temperature coefficient of the offset voltage of the bipolar transistor is approximately -2 mV/K.

For the temperature dependency of the offset current we find

$$\begin{aligned}\left.\frac{\partial I_B}{\partial T}\right|_{I_C} &= \left.\frac{\partial (I_C/\beta)}{\partial T}\right|_{I_C} \\ &= \frac{1}{\partial T}\left(-\frac{I_C\partial\beta}{\beta^2}\right)\end{aligned} \tag{7.35}$$

where I_C is considered to be a constant. Substituting again $I_C = \beta I_B$, the temperature coefficient for the offset current becomes

$$\left.\frac{\partial I_B}{\partial T}\right|_{I_C} = -I_B\frac{\delta\beta}{\beta\delta T} \tag{7.36}$$

This value appears to be smaller than 1% of I_B per Kelvin.
It is very easy to find an expression for the temperature dependency of the offset voltage of the *balanced* bipolar amplifier stage. Differentiating expression (7.14) delivers the desired result, which is

$$\frac{\partial V_{offset}}{\partial T} = \frac{k}{q}\ln\frac{R_{C2}W_{B1}A_{E2}}{R_{C1}W_{B2}A_{E1}} \tag{7.37}$$

and multiplying the right hand-term by T/T and substituting the original expression for the offset we find

$$\frac{\partial V_{\text{offset}}}{\partial T} = \frac{V_{\text{offset}}}{T} \tag{7.38}$$

Suppose the offset voltage for a balanced amplifier is 2 mV and $T = 300$ K, then the temperature coefficient of the offset voltage is

$$\frac{\partial V_{\text{offset}}}{\partial T} = \frac{2 \times 10^{-3}}{300} = 7.7\ \mu\text{V/K}$$

Theoretically it is possible to null this coefficient by controlling the ratio of R_{c2}/R_{c1} and making the ln term 1, but in practical circumstances, other temperature-dependent parameters remain, reducing the effect to several µV/K. One of the most important reasons for this is the small temperature differences which still can be generated on a single piece of silicon, despite a relatively good thermal conductivity of 84 W/mK. This can be caused by dissipating differences on the chip itself, so it might be important to design the input stage and the output stage calculating the thermal symmetry axis and placing the transistors around it.

The temperature coefficient of the offset current is found by differentiating expression (7.11)

$$\begin{aligned}\frac{\partial I_{\text{off}}}{\partial T} &= \frac{\partial}{\partial T}(I_{b1} - I_{b2}) \\ &= I_{B2}\frac{\partial \beta_2}{\beta_2 \partial T} - I_{B1}\frac{\partial \beta_1}{\beta_1 \partial T}\end{aligned} \tag{7.39}$$

and if

$$\frac{\partial \beta_2}{\beta_2 \partial T} = \frac{\partial \beta_1}{\beta_1 \partial T} \tag{7.40}$$

then

$$\frac{\partial I_{\text{off}}}{\partial T} = I_{\text{off}}\frac{\partial \beta}{\beta \partial T}$$

which is a factor of $I_{\text{off}}/I_{\text{bias}}$ smaller than in the case of the single-stage amplifier.

We have yet to consider the case when the two transistors composing the input stage do not have the same temperature. Then the offset becomes, with all other parameters equal,

$$V_{\text{off}} = V_{\text{BE1}} - V_{\text{BE2}} = (T_1 - T_2)\frac{k}{q}\ln\frac{I_C W_B}{K_{AE}} \tag{7.41}$$

Rearranging equation (7.2),

$$\frac{kT}{q}\ln\frac{I_C W_B}{KA_E} = V_{\text{BE}} - V_{\text{GAP}} \tag{7.42}$$

Multiplying equation (7.41) by T/T and substituting equation (7.42) in (7.41), the result for V_{off} as a function of temperature T yields

$$V_{off} = \Delta T(V_{BE} - V_{GAP})/T \tag{7.43}$$

where

$$\Delta T = T_1 - T_2$$

Suppose the chip temperature is 353 K and the resident temperature difference on the chip is 1°C, then the temperature coefficient of the offset voltage becomes

$$V_{off}|_T = (0.6 - 1.15)/353 = -1.56 \text{ mV/K}$$

which is a considerable and rather poor value. Reducing the temperature differences between components and transistors will improve this value to a large extent and can be realized by applying integrated technology only.

7.8 TEMPERATURE DRIFT IN FETS

Here again we have to deal with two types of transistors, which require a different description.

7.8.1 The junction field-effect transistor

Two main causes are responsible for the temperature drift in junction-FETs:

1 With increasing temperature the channel resistance will increase.
2 The thickness of the depletion layer will decrease when temperature increases. This results in a lower channel resistance.

As can be seen, both causes oppose each other, but a complete cancellation is obtained only in the region well above pinch-off voltage. Fig. 7.6 shows the

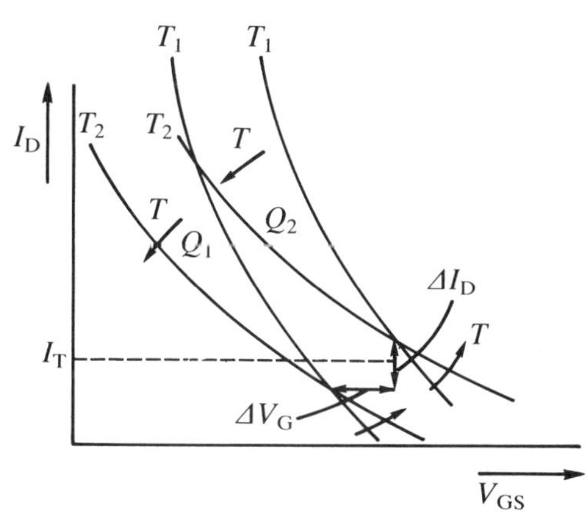

Figure 7.6 Behavior of the transconductance of two junction-FETs at two different temperatures

transconductance of two junction-FETs at two different temperatures T_1 and T_2.

To calculate the temperature coefficient the expression for a reverse-biased diode is applicable, thus from (7.1),

$$I_D = I_s = -K\frac{A_E}{W_B}\exp(-qV_{GAP}/kT) \tag{7.44}$$

where I_s is the saturation current, and all other parameters are as mentioned before. Differentiating (7.44) with respect to T delivers

$$dI_D = -K\frac{A_E}{W_B}(-qV_{GAP}/k)\exp(-qV_{GAP}/kT)d(1/T) \tag{7.45}$$

Substituting again, equation (7.44) in (7.45) gives

$$\frac{dI_D}{dT} = +I_D\frac{q}{kT^2}V_{GAP} \tag{7.46}$$

If we take $T = 300$ K and $V_{GAP} = 1.15$ V, then substituting these values in (7.46) the temperature coefficient equals

$$\frac{dI_D}{dT} = 0.147 \times I_D \qquad \text{(A/K)} \tag{7.47}$$

This is a rather high value (±15%), but owing to the very low absolute value of the gate current of a few nanoamperes the influence remains rather low.

Offset voltage drift in junction-FETs is not easy to calculate because several terms in the expression for the offset have different temperature coefficients and partly cancel each other. This is especially the case at one value for I_D. The value of the temperature-dependent offset voltage for a source coupled junction-FET pair appears to be some microvolts per Kelvin (see Table 7.1).

7.8.2 The MOS field-effect transistor

Determining the temperature coefficient of a MOSFET is much more complicated than is the case with the junction-FET, but here a very low value of the offset current is involved, picoamperes, and so it makes little sense to calculate this value which still has a negligible influence. As will be seen, the $1/f$ noise is more important and therefore, in general, applications of MOSFETs in DC input stages is not found. But due to recently obtained large improvements in CMOS technology, more and more linear DC CMOS amplifiers can be met in a great variety of integrated (sensor) applications.

It is interesting to compare several types of amplifiers with their state-of-the-art specifications. This is depicted in Table 7.1 for three different types of amplifiers in which the well-known 741 is mentioned for reference reasons.

In the last column, a so-called chopped amplifier is mentioned, in which the offset is automatically nulled at regular time intervals. This is realized with an internally generated frequency and two added capacitors by which the fault voltage

Table 7.1 Comparison of three types of amplifiers

	Bip. 741	*FET-ampl.*	*Chopped ampl.*
Input offset voltage (25°C)	1	±20	±0.001 mV
V_{off} temp. coeff.	15	±10	±0.01 μV/K
Input bias current (25°C)	80 000	−10	35 pA
Input offset current (25°C)	20 000	±2	0.5 pA
I_{off} temp. coeff.	0.5 nA/K	doubles every 10 K	
CMRR	90	80	130 dB
PSRR	96	86	130 dB
Offset voltage versus time	n.s.	±100	0.100 μV/m

Note: (n.s. = not specified; μV/m = μV/month)

is nulled. A possible principle of operation is illustrated in Fig. 7.7, in which an internal oscillator, a main amplifier and a so-called null amplifier are shown.

The oscillator drives two electronic switches, A and B. Both amplifiers can be nulled, whereas the main amplifier directly connects the input to the output. The null amplifier measures its own offset when the A switches are closed and B switches are open. When the B switches are closed and the A switches are open, the nulling amplifier measures the output of the main amplifier. The required A and B switches are analog MOSFET switches and are controlled by the local integrated oscillator for which only two capacitors should be added externally. The two capacitors store the required compensating voltages. The extremely low offset voltage of less than 1 μV is realized by comparing the output voltages of both amplifiers by an algebraic subtraction via the two capacitors.

Some other measures are available for eliminating offset voltage, such as

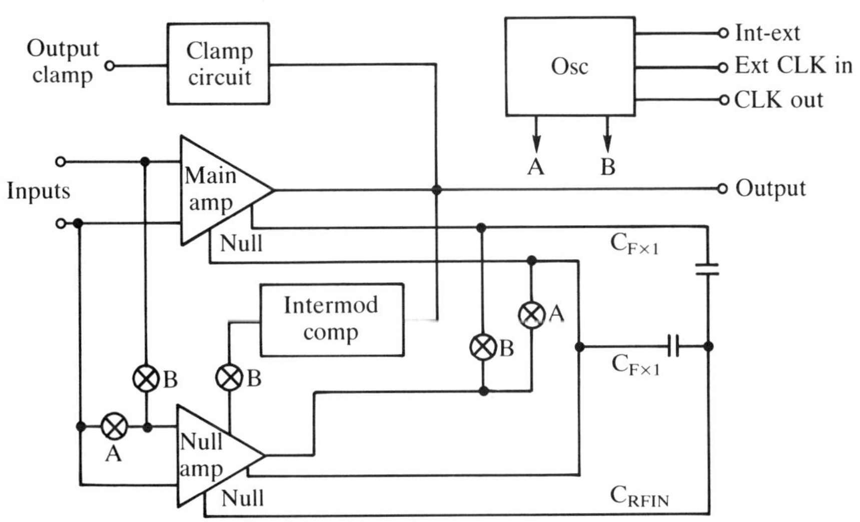

Figure 7.7 Principle of a chopped amplifier with extreme low offset (courtesy Intersil)

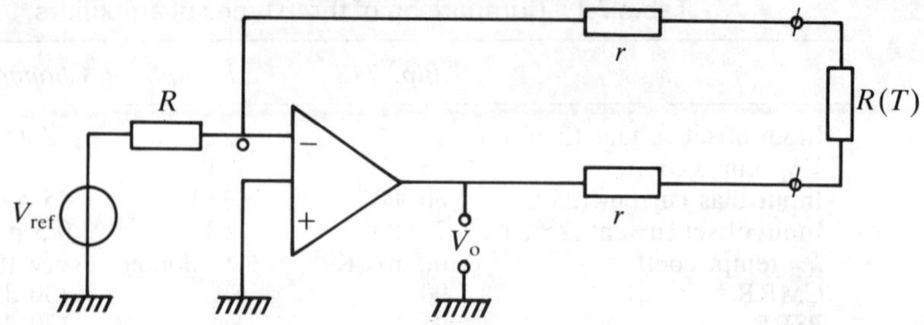

Figure 7.8 Circuit for Problem 7.3

isolating the interfering source. This can imply power supply stabilization, maintaining a constant temperature for all temperature-sensitive parts, and the choice of very low sensitive materials and components. Of course you will have to pay for it, but it can be worthwhile.

7.9 PROBLEMS

7.1 What is meant by offset and by drift? Explain why it is not always true that a realized nulling involves no drift.

7.2 The temperature coefficient of a differential amplifier is 3 μV/K. The gain setting factor is 1000 and the offset voltage is 1.5 mV at 25°C. By which value will the output voltage change when the ambient temperature changes by 15 K?

7.3 In Fig. 7.8 a temperature measurement set-up is drawn, in which $R(T)$ is a temperature-dependent platina resistance. The temperature behavior of $R(T)$ is described by $R(T) = R_0(1 + \alpha T)$, where $R_0 = 100\ \Omega$, $\alpha = 3.2 \times 10^{-3}$/K and T is the resistance temperature in °C. The resistance of the supply leads is represented by r, and the reference voltage source $V_{ref} = 7.25$ V. If the offset voltage is 100 μV and $I_{bias} = 100$ nA, calculate the output error in °C if $T = 0$°C.

7.4 Calculate the offset voltage change for a single junction FET when the drain current changes by 12%.

7.5 Derive an expression for the offset voltage for Fig. 7.5 when there is a mismatch of 10% in the W/L and all other parameters are supposed to be balanced.

7.6 Perform the same procedure as in Problem 7.5 for a source coupled pair composed of MOSFETs.

Eight

GUARDING AND SHIELDING

8.1 INTRODUCTION

This chapter includes a comprehensive discussion of guarding and shielding techniques which are available in electronic measuring systems in order to avoid any kind of interference. The causes and the nature of interference will be outlined and it will be seen how important guarding and shielding are with respect to the required common-mode rejection ratio (CMRR). The influence on the claimed accuracy of the applied measuring instrument will be investigated. We shall discuss pitfalls, what to do and what not to do, and will conclude with a collection of problems for practical training. Sometimes noise is also considered as interference, but because the nature of noise is a more fundamental physical process of the components and devices itself, we will treat noise in a separate chapter.

Guarding can be defined as a collection of measures and precautions taken to prevent stray currents entering electronic sensitive parts by connecting the appropriate parts of a circuit to earth or ground.

Shielding is the technique of placing electronic systems in a metal casing to prevent electrostatic and/or magnetic fields entering sensitive components. We will start our discussion with a few examples of how interference can occur and demonstrate some causes of interference.

8.2 THE NATURE AND CAUSES OF INTERFERENCE

Interference can be defined as any kind of physical influence on a given system which reduces the quality and performance of that system. This influence may cause either a reduction in accuracy or in CMRR, as will be seen later. If very low signal levels are present, the influence of noise will play a vital role but as said already, due to the character of noise, this requires a separate treatment and will be discussed in Chapter 9.

Two different types of sources of interference can be distinguished. Interference can be caused either by electrostatic fields or by magnetic fields. Both

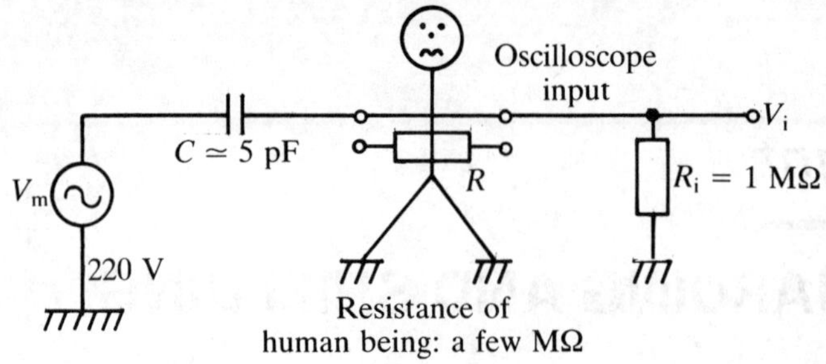

Figure 8.1 Direct mains interference of a human being touching the input terminals of an oscilloscope

causes will be treated in detail. Since most instruments obtain their operating power from the mains supply, one of the most general sources of interference is the mains (voltage) coming from the power supply with a mains frequency of 50 or 60 Hz depending on the country.

In most cases we will assume a fixed frequency of 50 Hz throughout this chapter. The mains can easily induce a voltage by parasitic capacitive coupling to the measuring system or to the signal source itself. This is shown in the following examples.

* * *

Example 8.1

Fig. 8.1 illustrates the situation where a person touches the input of an oscilloscope which has an input impedance of ≥1 MΩ. Assume that the transition impedance of the person is $R_{tr} = 500$ kΩ and the capacitor between the mains (50 Hz) and R_{tr} is $C \approx 5$ pF, which is a typical value for normal rooms, then an input signal V_i will appear on the CRT of which the magnitude can be calculated as follows:

$$V_i = V_m \frac{R_i}{R_{tr} + R_i + 1/j\omega C} \tag{8.1}$$

where

V_m is the mains voltage (assumed to be 220 V)
R_{tr} is the transistion impedance of the person
R_i is the oscilloscope's input impedance
C is the parasitic capacitor

As can be calculated, $|1/j\omega C| \gg R_{tr} + R_i$, hence neglecting these terms the denominator becomes

$$|1/j\omega C| = \frac{1}{2\pi 50 \times 5 \times 10^{-12}} \approx 600\ \text{M}\Omega \tag{8.2}$$

and the resulting interfering input voltage will be:

$$V_i = \left| \frac{R_i}{1/j\omega C} V_m \right| = |j\omega C R_i V_m| = 0.38 \ V_{cf} \tag{8.3}$$

So a considerable input voltage is induced at the input of the oscilloscope due to a capacitive coupling of the mains and a high input impedance, R_i. As will be seen, capacitive coupling will be one of the main causes of interference.

* * *

Example 8.2

Another example where interference has been a problem is the determination of an electroencephalogram (EEG) of a human being. This means monitoring the brain activities by the measurement of external voltages on the skin. For this purpose two electrodes must be attached to the patient's headskin and, via a small parasitic capacitor of, say, 2 pF, the patient is connected to the mains. This is illustrated in Fig. 8.2. In the same figure the electrical equivalent of the disturbing source is depicted.

To reduce the interference, the patient is connected to ground with a separate electrode and, via the parasitic capacitive coupling, a current can flow through the remaining resistance, R_c, of, for instance, 10 kΩ. Despite a carefully made connection to common, a rest-impedance will often remain. The magnitude of the interference current can be calculated with

$$I_i = \frac{V_m}{R_c + 1/j\omega C} \tag{8.4}$$

Substituting the given values and, again, assuming $|1/j\omega C| \gg R_c$, this yields, with $V_m = 220$ V for the current

$$I_i = |j\omega C| \, V_m = 0.14 \ \mu A \tag{8.5}$$

and the generated voltage difference over R_c can be calculated from

$$V_c = I_i R_c \tag{8.6}$$

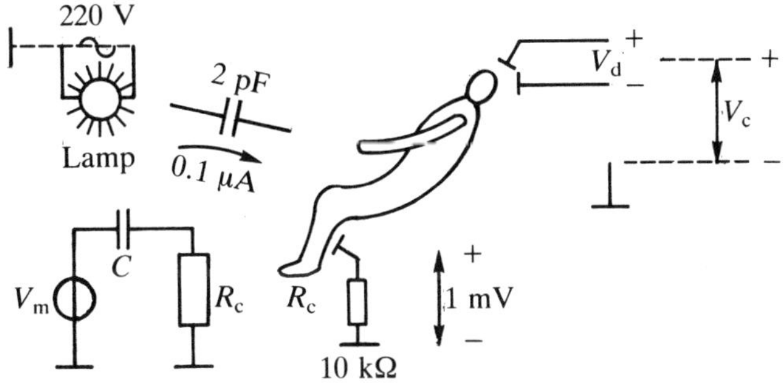

Figure 8.2 Recording problem of an electroencephalogram (EEG) and its parasitic small-signal equivalent circuit

Substituting the values results in $V_{cm} = 1.4$ mV, which is a factor of 1000 higher than the source voltage level to be measured, which is normally in the range 1–10 μV! Hence the interference completely obscures the required signal. Again, from this example, the conclusion can be drawn that, in circumstances where relatively high frequencies are present, stray capacitances play a vital role.

* * *

Apart from this capacitive coupling via the mains, there are still other tedious sources of interference present. The frequency of the mains (50 or 60 Hz) is in the middle of the frequency band of the brain's activities, which is in the range 1–100 Hz and, as a consequence, is sensitive to those frequencies. Besides this, the heart-muscle and other muscles generate common-mode voltages of about 1 mV. Both sources can influence the accuracy considerably.

These sensitive measurements require an amplifier with a balanced input configuration. In Fig. 8.3 (a) and (b) two alternative representations of a complete incremental model of such balanced amplifier input configurations are depicted. The signal source V_{cm} with internal resistance R_c represents the interfering source which is present between both electrodes and ground. Because this signal acts simultaneously on both electrodes 1 and 2, with equal sign, it is called a common-mode signal source.

The voltage source V_s with internal resistance R_s represents the measurand and is present between both electrodes 1 and 2. V_s is called the desired differential-mode signal V_d. R_{e1} and R_{e2} represent the two electrode impedances which very often fluctuate in magnitude and are of considerable level.

The input impedances of the amplifier's circuitry are denoted by R_{i1} and R_{i2} and should be much larger than R_{e1} and R_{e2} to avoid any attenuation of the small signal. If R_{e1} is not precisely equal to R_{e2} it is easily seen that V_{cm} induces a differential-mode signal, which will be added to the desired signal. We will denote both input signals of the amplifier with respect to ground by V_a and V_b.

Referring to Fig. 8.3, a definition for the differential mode signal and the common-mode signal can be found writing V_a and V_b as a sum and difference of two voltages as follows:

$$V_a = \tfrac{1}{2}(V_a - V_b) + \tfrac{1}{2}(V_a + V_b) \tag{8.7}$$

and

$$V_b = -\tfrac{1}{2}(V_a - V_b) + \tfrac{1}{2}(V_a + V_b) \tag{8.8}$$

In these equations, by definition the component $\tfrac{1}{2}(V_a - V_b)$ is called the differential mode signal because this signal appears with opposite sign at both inputs of the amplifier. In the same way, $\tfrac{1}{2}(V_a + V_b)$ is called the common-mode signal because this signal appears with the same sign at both inputs of the amplifier.

As you have seen already, it is important to recognize possible interferences due to small parasitic stray capacitances, especially when signal sources can supply low power only. In the past the only good protection against interference was placing the patient in a Faraday's cage which then prevents an external elec-

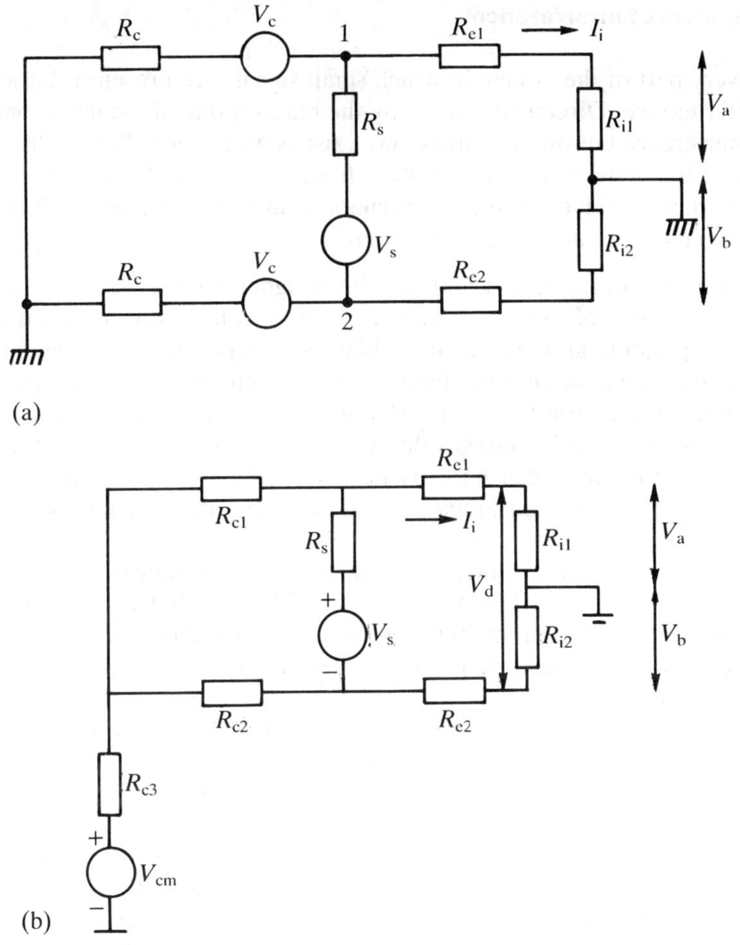

Figure 8.3 (a) Interfering signal source at the inputs of the applied balanced amplifier monitoring an EEG (b) alternative interference circuit

trical field passing through it. But, due to the availability of excellent differential amplifiers, this has become obsolete and is no longer required.

* * *

Example 8.3

When a measuring system is used where a digital voltmeter or a recorder is implemented and accuracies are claimed of about 0.1% or better for both instruments, great care must be taken in order to maintain this accuracy. Suppose a recorder has a guaranteed accuracy of 0.25% and is used with a full-scale deflection of 1 mV. Suppose further that a parasitic voltage of 20 μV is present. Then an error of 2% arises which completely destroys the claimed accuracy.

* * *

8.2.1 Sources of interference

Every part of the system in which small signals are present is basically sensitive to interference. Directly or indirectly the mains is one of the most common sources of interference but other sources may exist as well. The following list gives nine other potential sources of interference. Interference can be caused either by sources which can produce an electromagnetic field or by sources which have a thermal or mechanical influence on the system.

1 A temperature coefficient can be assigned to every part of a system, so the magnitude of certain parameters of a component is always a function of temperature and temperature changes. Temperature can influence the biasing of active components considerably. In very sensitive parts you have to reckon with temperature gradients caused by dissipation differences in components. This may cause thermopile voltages due to the Seebeck effect. Especially large temperature differences can be present due to dissipation differences of components mounted on printed circuit boards. Seebeck voltages are in the order of 40 μV/K.
2 Mechanical shocks can influence poorly installed and soldered connections.
3 Ignition equipment in cars produces a broad bandwidth of frequencies. The ground frequency lies between 50 and 200 Hz but higher frequencies are also present, ranging from 30 to 300 MHz and, for these frequencies, the spark-plug cables can act as a transmitting antenna.
4 In digital equipment, pulse-shaped signals are applied and the present frequency bandwidth is a function of the slope of the applied pulses and ultimately determines the magnitude of the influence.
5 Switches in power distribution systems can produce large pulses and then act as an interfering source. This is also the case in large production equipment systems in which a large number of switches must be controlled and activated. A very popular type of power controller producing a large amount of interference is the thyristor. The steeper the slope, the greater the interference. Switching at the precise moment the voltage crosses the zero level can reduce this type of interference.
6 The collector of an AC series electromotor can be interpreted as a fast switch, operating with a frequency between 1 and 10 kHz determined by the amount of lamellas and the number of revolutions per minute. This mechanical switch produces pulses with steep slopes and thus a broad frequency spectrum will result.
7 Very high voltages can produce corona effects which cause a continuous interfering spectrum with a large bandwidth.
8 In a gas discharge lamp a periodic discharge is generated producing an interference noise signal level above 1 MHz. A filtering element such as an induction coil can reduce this influence.
9 Finally, interfering sources with a very small frequency bandwidth should be mentioned, such as HF generators, magnetrons, welding equipment and all types of transmitters.

Interference of these kinds can be eliminated by two measures:

(a) the application of an isolating transformer;
(b) the use of shielding leads and cabinets.

You will not be surprised that a lot of mistakes can be made when applying these two methods, and that is what we will discuss in the following sections. At first glance it may seem very simple, but large mistakes can be made. Let us look at the mistakes and the solutions, in this sequence.

8.3 ASPECTS OF GUARDING

Applying a type of guard and/or shield is a good method to prevent interference. In this section we will discuss aspects of guarding qualitatively before going into more detail and making calculations. Guarding can be considered as the technique of applying one conductor only as the common, which may be grounded or earthed.

Many instruments have three input terminals, marked with a 'High' or '+'; a 'Low' or '−' sign; and an 'Earth' or '⊥' mark. The terminal marked with Earth is a safety earth to which all metal parts of an instrument are connected and which in its turn may be connected to earth or ground of the mains cable. In a proper installation this safety earth prevents any of the conducting parts of the system which can be touched from rising in potential above 42 V. A voltage of 42 V with respect to earth is the commonly accepted safety voltage for the human body.

Basically current is the limiting factor here and Figs 8.4 and 8.5 show this sensitivity as a function of frequency and current, and the consequences of the current on a human body.

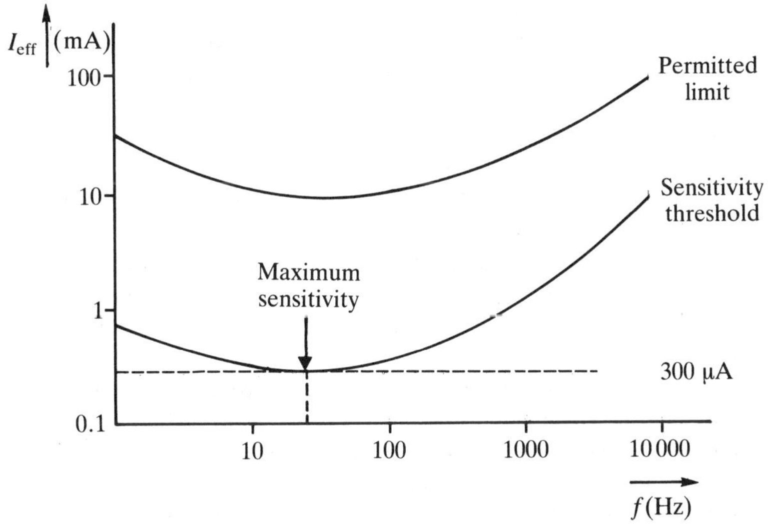

Figure 8.4 Current safety margins for the human body as a function of the frequency. Note the dip between 40 and 60 Hz

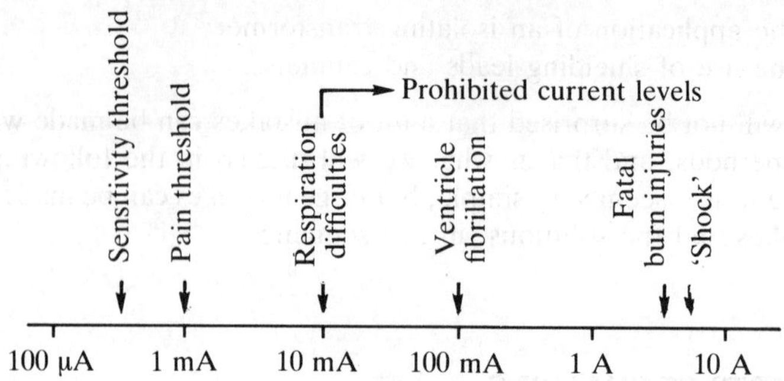

Figure 8.5 Survey of the consequences of a 50 Hz AC current flowing from hand to hand in a human body

In many instruments, a metal link or lead, the so-called 'Low' can be connected to the safety earth terminal, linking the common of the system to ground. If this is the only connection to earth the measurement can be performed straightforwardly, but great care must be taken to ensure that the system under test has no other connections to earth because this can introduce errors in sensitive measurements. In this case the 'Low' and earth have to be disconnected, resulting in a 'floating' common preventing ground loops.

8.3.1 Ground and earth

Another aspect to be considered in earth connections is the very common pollution of earth terminals in the mains plug. In many real circumstances the earth terminal is not clean at all, but polluted with large interference and can even have several volts with respect to ground level.

Ground can be defined as any reference conductor that can be used for a common return. Any ground should be an *equipotential* for the whole system involved. Earthing is just a particular case of grounding. On many instruments, there is a metal link that provides the possibility to make a connection between low and ground to ensure safety earthing.

That a ground is not always an equipotential and the care that must be taken before using a common can be seen from the following examples.

In a ship several hundreds of volts can exist between decks, superstructures and rigging. Across an aircraft up to 100 V difference can exist between structural points such as nose, wing and tail parts. In motor vehicles several volts can exist between different points.

When lightning occurs, earth stray currents can generate several kilovolts between different ground connections. A good earth connection can be made by driving copper- or cadmium-cladded iron rods into the ground, sometimes to a depth of up to 20 meters, depending on the transition impedance of the rods to earth. In most cases, this transition impedance should not exceed 0.2 Ω. Different values may be required and are application dependent. The transition impedance is to a large extent a function of the specific conductivity and humidity of the soil. It is not unlikely that more than one rod is necessary to obtain the required transition impedance. Used as a prescribed safety earth, a voltage of 42 V between the metallic housing and earth cannot be exceeded.

8.3.2 A grounded system

With this in mind, consider the case when two points in a measuring system are connected to ground as depicted in Fig. 8.6. It will be clear that a type of ground loop is created if the transducer and the measuring instrument are both connected by relatively long cables. You may think of thermocouples or strain gauges for the transducer and a chart recorder for the measuring instrument. In this circuitry, R_1 and R_2 represent the respective cable resistances. In general there are always stray or erratic currents in earth and these currents combined with the soil resistance form an imaginary voltage source V_{cm} causing a current to flow in R_1 and R_2.

It is obvious that the current in R_2 will be larger than in R_1 due to the relatively high input impedance Z_i of the measuring instrument; hence, due to V_{cm}, unequal series voltages are generated in R_2 and R_1 and will be added or subtracted to the differential voltage to be measured. A common-mode voltage, V_{cm}, is converted into a differential-mode voltage due to asymmetries in the input stage of the measuring instrument. The introduced error is called a series mode voltage error.

The same situation arises when a collection of instruments is mounted in a metal rack and each instrument is electrically connected to it. A current can flow through the rack via the signal ground terminals, no matter whether the instruments are connected to ground or not. This current can flow because in large racks several hundreds of millivolts may exist between different drawers. Another aspect is the feedback mechanism of interfering signals to the input likely to be coupled back into the circuit by capacitive or electromagnetic induction, and again a series mode error is introduced in the measuring system. The feedback problem will be discussed in more detail later on.

Suppose now the cable resistances R_1 and R_2 are neglected and a thermocouple transducer with given internal resistances R_{tr1}, R_{tr2} are connected to the

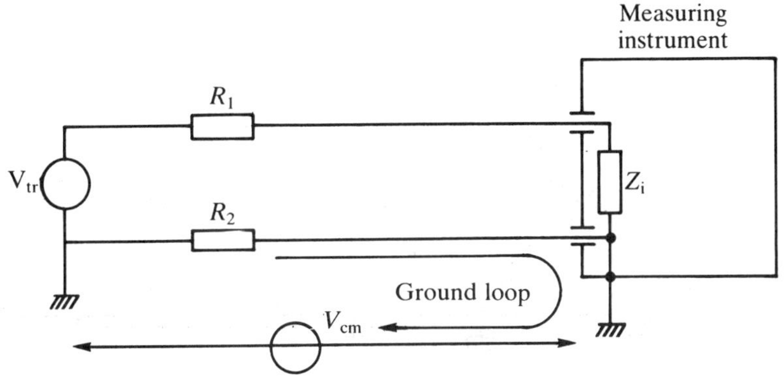

Figure 8.6 The creation of a ground loop by earthing at two different points

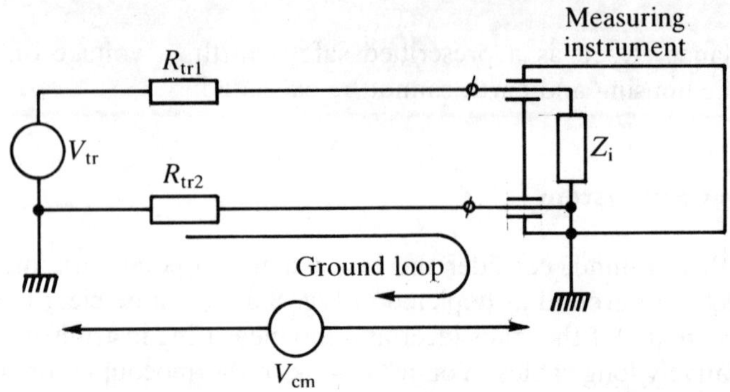

Figure 8.7 Conversion of a common-mode voltage into a series-mode voltage caused by the thermocouple impedance only

input of an amplifier with an input impedance Z_i, then the configuration in Fig. 8.7 will result. A thermocouple is a welded connection of two different metals with different specific resistances and if long thermocouple wires are involved, large differences in resistance can occur. Hence in this configuration the electrical equivalent of the thermocouple is considered as a voltage source with two equally divided but unequal resistors.

When this system is earthed at two points as shown, due to V_{cm} and Z_i two different ground loops are created, resulting in two currents different in magnitude and developing two different series mode voltage across the R_{tr}s. This results in a differential-mode voltage across Z_i, demonstrating again the conversion of a common mode voltage V_{cm} into a series mode voltage seen by the amplifier's input as a differential mode-signal.

In many measuring applications a type of bridge circuitry is applied which can be fed by a DC or AC voltage source V_{cc}. Very often, one of the bridge nodes is earthed and it will be obvious that the voltage levels of points A and B, with respect to earth, are ½V_{cc} and both are determined by the accuracies of the bridge branches. Hence, a bridge configuration will always deliver a so-called transducer common-mode voltage to the input of the measuring system. This is illustrated in Fig. 8.8.

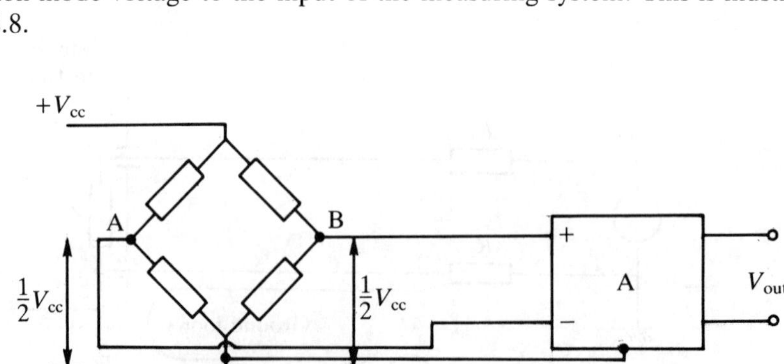

Figure 8.8 Bridge measuring configuration acting as a common-mode voltage source

In most cases we are not interested in this common-mode voltage, and the differential mode voltage generated by the bridge imbalance is the required signal. Hence a bridge measurement requires also a differential input stage to eliminate the common-mode voltage.

Great care must be taken because, if a fixed imbalance of the bridge is present, this results in a fixed differential signal, which should be avoided. A bridge-balancing procedure is easy to accomplish and a variety of good compensation techniques are available. But when the bridge signal varies with temperature or time, when a pressure should be measured for instance, considerable errors can be introduced and other precautions must be taken.

8.4 ASPECTS OF SHIELDING

As we said in the introduction, shielding can be defined as the placing of sensitive electronic parts and components in a metal casing to prevent electric and magnetic fields entering that casing.

8.4.1 A shielding cabinet

Suppose we have placed a non-inverting amplifier in a shielding cabinet as illustrated in Fig. 8.9. Inside the cabinet three parasitic capacitors C_1, C_2 and C_3 can be distinguished, of which C_1 and C_2 form a parasitic feedback path from output to input. This can result in undesired oscillations or jamming of the amplifier. The feedback path can be eliminated by connecting the shield to the common and thus short-circuiting C_3 as is shown in Fig. 8.10.

From this example it will be clear that a shield is only effective when the shield is connected to the common or zero-signal reference potential of the shielded circuitry, eliminating the undesired feedback path.

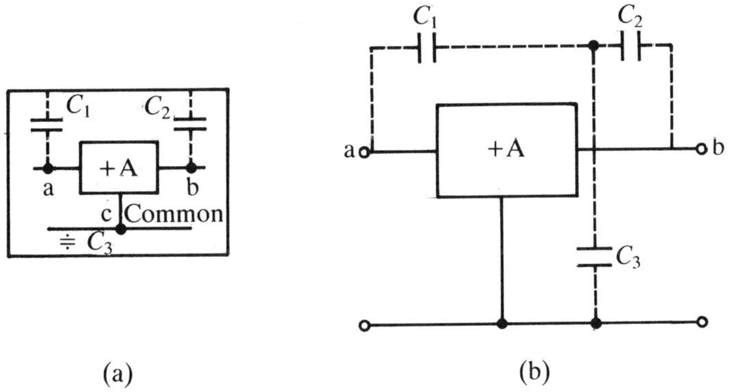

Figure 8.9 (a) Parasitic feedback of a shielded amplifier; (b) its electrical equivalent

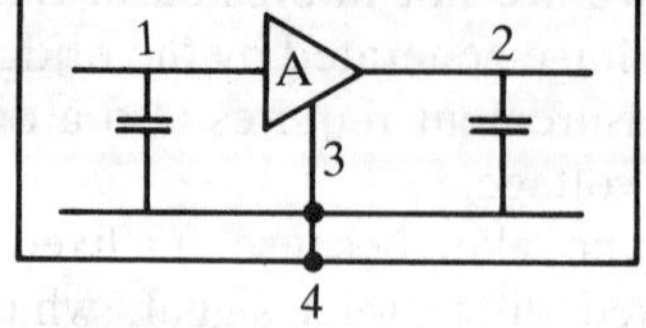

Figure 8.10 Eliminating the parasitic feedback path by short-circuiting C_3

8.4.2 A shielding cabinet as part of an electronic circuit

Fig. 8.11 illustrates the use of a shielding cabinet as a part of an electronic system. The shielding cabinet is coupled to the mains via a parasitic capacitor which will result in two currents i_1 and i_2 flowing through the cabinet to the earth connection. If within the cabinet different common connections are made, suppose at A and B, a voltage difference V_{AB} between A and B can be developed by V_{cm}. This results in a series mode voltage with V_{ss} forming an input signal for transistor Q_2 via R_l and clearly distortion will occur.

The solution to this problem is applying one common connection only to eliminate V_{AB}, as is illustrated in Fig. 8.12. Hence one point only of the circuitry's common should be connected to the shield.

It is interesting to make the following remark connected to this. Suppose you are walking outside in an open field and there is going to be a thunderstorm and you cannot find a hiding-place: there is one thing you should do and one thing you should never do. When lightning strikes the ground, large stray currents can be induced. If this is likely to happen, make one point of your body common as effectively as possible, hence sit on the ground with your legs drawn up and close together, preventing large voltage differences to be developed between your legs and resulting in stray currents passing your legs.

The thing you should never do is to hide under a large tree. When lightning

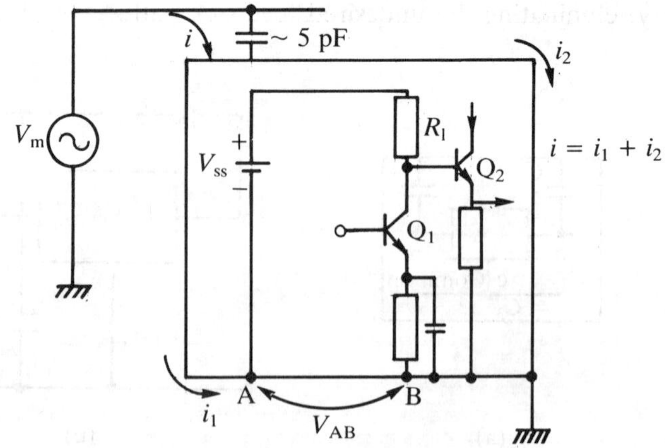

Figure 8.11 A shield as part of an electronic circuit

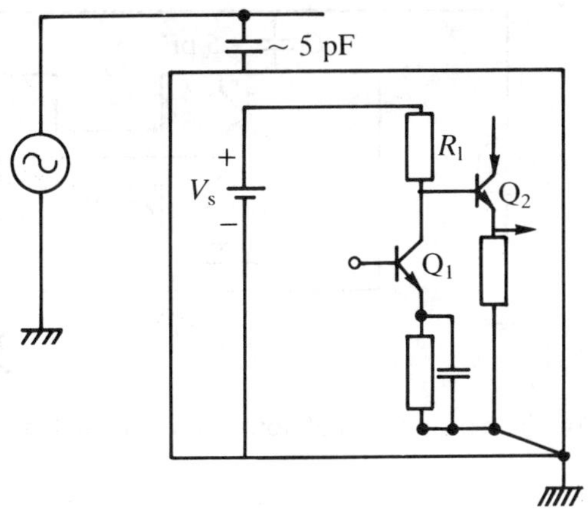

Figure 8.12 Solution to the hazard of Fig. 8.11

hits a tree, the tree will likely be split into two parts. This will be due to the huge amount of water, which especially in summertime is always present in a tree, hence a tremendous steam explosion can occur. Note that the developed power involved in a stroke of lightning can exceed 1 000 000 V by 20 000 A, which represents a power generation of 20 000 MW in a split second!

8.4.3 The use of more than one shielding cabinet

Suppose two instruments are required both of which need shielding, more than one shielding cabinet is involved and it is recommended that only one earth connection is used for the two cabinets. But sometimes a problem can arise, as illustrated in Fig. 8.13. If the right-hand cabinet is earthed and both cabinets are connected to the mains via parasitic capacitors, a current i_1 can flow through the parasitic capacitor C_1, the cabinet 1 and the shield of the coaxial cable to the ground connection of cabinet 2. When a rather long coaxial cable connects both cabinets a

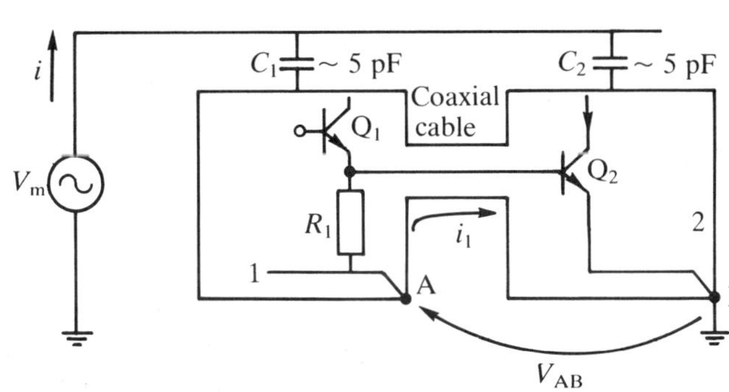

Figure 8.13 The cause of interference when two shielding cabinets are connected in the wrong fashion

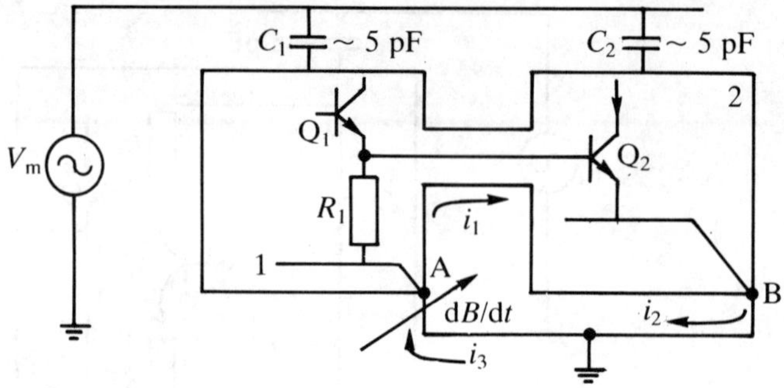

Figure 8.14 The grounding of both cabinets creates a magnetic-sensitive loop

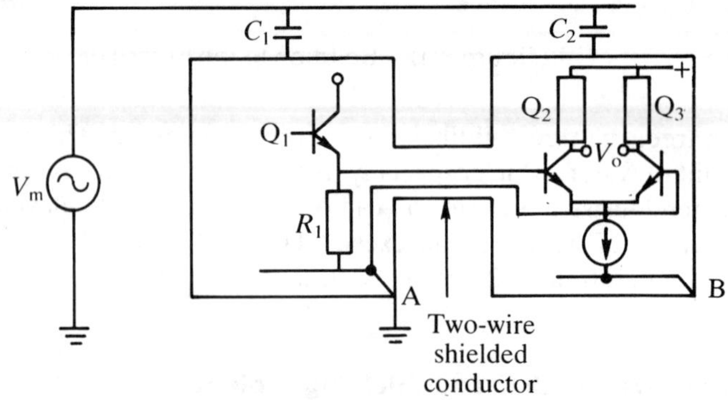

Figure 8.15 A multiple cabinet configuration avoiding the magnetic-sensitive ground loop and the influence of parasitic capacitors C_1 and C_2

considerable voltage difference can be developed. This voltage difference V_{AB} can be considered as a series mode voltage source with the voltage across R_1 and thus will be a part of the input signal for Q_2.

It will be clear that interference and consequently errors will result. The same situation arises when the left-hand cabinet is earthed, when a current is generated from right to left.

Now suppose you ground *both* cabinets, a voltage difference between the cabinets can still be developed due to current flows i_1, i_2 and i_3, different in magnitude in the loop ABA. But besides this mechanism another source of interference is introduced, as is depicted in Fig. 8.14. A type of loop is made in which an extra current can be generated if a changing magnetic field dB/dt is present.

This problem can be solved by implementing a two-wire shielded coaxial cable and applying a differential input stage at the input of the shielded circuitry in the second cabinet. This solution is depicted in Fig. 8.15, from which you can see that no groundloop exists and parasitic currents cannot influence any input or create a series-mode voltage.

In the following section another example of creating a ground loop by a wrong connection of the common is discussed.

8.4.4 The connection of a common in a multiple cabinet configuration

The system to be investigated is shown in Fig. 8.16 and consists of a three-cabinet configuration interconnected via two-wire shielded cables. At a first glance you may think the common connection seems to be right, but a more careful consideration shows otherwise.

This is clarified in Fig. 8.17 where the equivalent electrical circuit is drawn. It is presumed that rather long cables are installed, of which R_{lc} represents the equivalent resistance. A voltage difference will be generated over R_{lc}, which equals $V_{ad} + V_{dc}$. Then for the input signal V_{in} connected to the amplifier's input we have

$$V_{in} + V_{cd} + V_{da} - V_{sig} = 0 \quad (8.9)$$

then for V_{in} we find:

$$V_{in} = V_{sig} + V_{ad} - V_{cd} \quad (8.10)$$

Note the exchange in the indices of V_{ad}.

From this incremental model it can be seen what has to be done to avoid this source of interference.

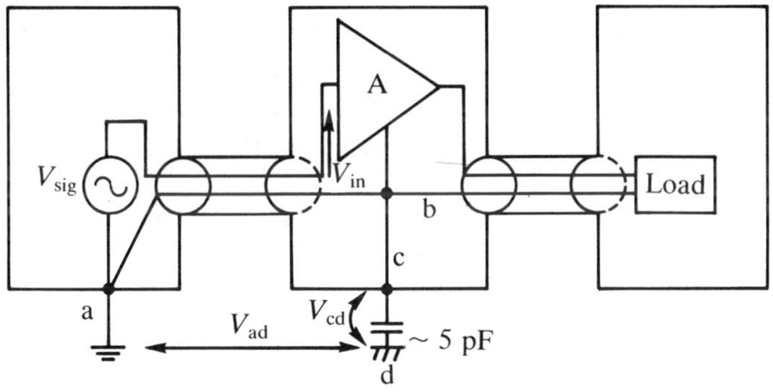

Figure 8.16 The connection of the common in a multiple cabinet configuration

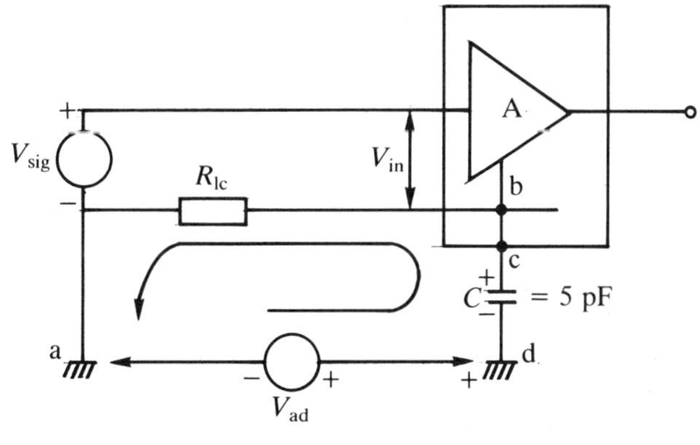

Figure 8.17 Equivalent incremental model for Fig. 8.16

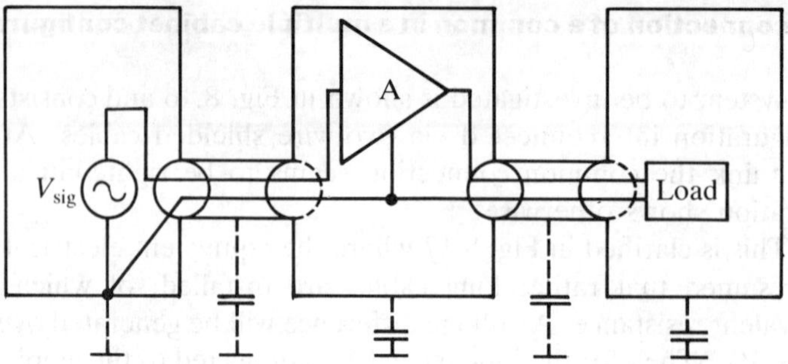

Figure 8.18 The correct common connection for the system

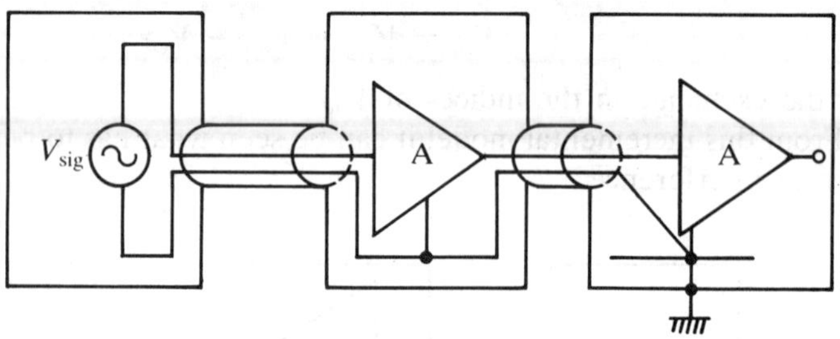

Figure 8.19 An alternative connection for the common when an oscilloscope is used as load

If we disconnect bc the influence of R_{lc} is eliminated and a suitable common connection for the circuit will remain. This is illustrated in Fig. 8.18.

From this example we can draw a general statement:

The common should be connected only to the shield at the ground connection of the zero-signal reference potential.

Hence this earth connection is made at the input of the system, but if this is not possible, the common connection may be made at the output as is often performed when the load is an oscilloscope. This is shown in Fig. 8.19, where the common of the CRT involved is connected to its own cabinet.

Another example is a digital voltmeter (DVM) conceived as a load for the signal source to be measured. This type of instrument can have a large capacitance between its common and the shield, e.g., about 10 nF. The situation is drawn in Fig. 8.20, where a grounded signal source is connected to the input terminals via a two-wire shielded cable. When the shield of the DVM is also grounded a current I_1 will flow into the common.

If the ohmic resistance R_{ac} is of considerable magnitude, a large series mode voltage V_{ac} can be generated across R_{ac}. This series mode voltage can be eliminated by bypassing R_{ac} applying a conductor between the points 1 and 2. The shunted current will flow now via the loop a–1–2–c–a with a smaller ohmic resistance, hence a reduced interference will result.

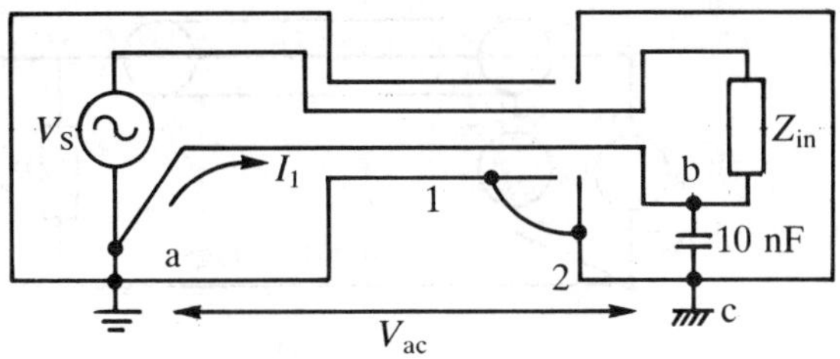

Figure 8.20 Signal source loaded with a large capacitance between common and shield

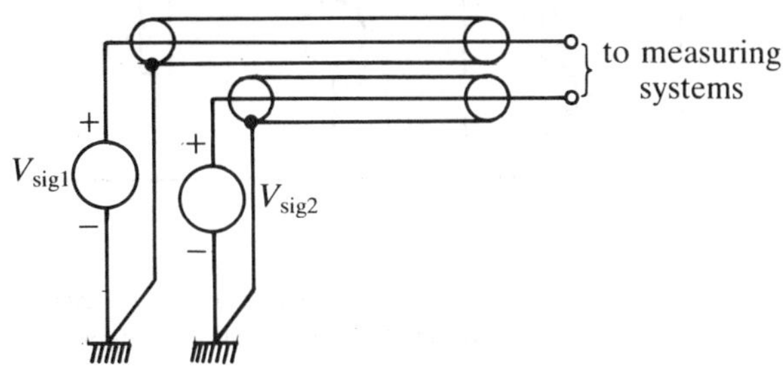

Figure 8.21 A shielding configuration when more than one independent signal source is present

When in a measuring system more than one independent signal source is present, each signal source should have its own shield, without any interconnection to other shields in the system except for a common reference ground potential. This principle is illustrated in Fig. 8.21.

It will be clear that the shield and the shielded conductor are always capacitively coupled by a (small) parasitic capacitor. If the shield has a voltage difference with respect to the signal source conductor, a current can flow through the parasitic capacitor between shield and conductor. This introduces an error signal and means that the shield is not allowed to be at a potential with respect to the signal source. To prevent this occurrence, shield currents should be avoided at all times. This problem is illustrated in Fig. 8.22.

We will return to this point when active guarding is treated. We can summarize our discussion so far with the following basic statements:

1 An electrostatic shield is only effective when it is grounded to the common of the circuitry.
2 The common should be connected to the shield at the signal–ground connection only.
3 When in a system more than one shield is applied, the shields have to be series connected and ultimately to the signal ground and/or common.
4 In general, shield currents must be avoided at all times.
5 A common which is used as a ground must be an 'equipotential'.

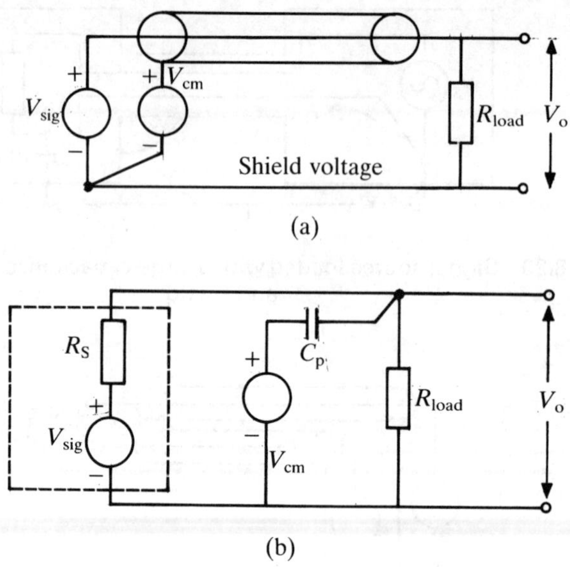

Figure 8.22 The consequences of a shield potential with respect to the signal source, due to shield currents

8.5 INTERFERENCE CAUSED BY A MAGNETIC FIELD

Interference caused by a magnetic field is much more difficult to eliminate than interferences coming from electrostatic fields. The main reason is that magnetic fields can penetrate conducting materials, and the moment there are changing magnetic fields with respect to time, induced voltages caused by the mechanism of a $d\phi/dt$ or a dB/dt are generated in conductors, resulting in currents flowing in the conductors.

For this type of interference two causes can be distinguished. Firstly, if wires in the neighborhood of a system are conducting high alternating currents, the wires will act as a transmitting antenna and are the source of a magnetic field. Secondly, if within a system wires are conducting high currents, the source is already present inside the system.

In the first case precautions should be taken to prevent the magnetic field entering the system. In the second case compensation techniques can be applied. But look first at the penetration depth of a magnetic field in three different kinds of very common materials. The penetration depth z of a magnetic field in a conductor can be described by the exponential equation

$$B_z = B_0 \exp(-\alpha z) \tag{8.11}$$

where

- B_0 is the field strength at the conductor's surface
- α is the absorption coefficient which in its turn is a function of the frequency and the applied material
- z is the penetration depth in the material

In Table 8.1 the influence of material and frequency is demonstrated for three different materials at different frequencies.

Table 8.1 Penetration depth as a function of material and frequency

Frequency (Hz)	*Penetration depth (z) (mm)*		
	Copper	*Aluminum*	*Steel*
50	8.97	11.5	0.91
60	8.5	10.9	0.86
100	6.6	8.5	0.66
1 000	2.1	2.7	0.20
10 000	0.66	0.84	0.08
1 000 000	0.08	0.08	0.008

From this table it can be understood that it is not so easy to shield against magnetic fields. There is one other material to be mentioned: that is mumetal, which is very effective at low frequencies. But at 100 kHz mumetal is less effective than steel, copper or aluminum. If this material is saturated by very strong fields it provides no shielding at all. Fig. 8.23 shows the effectiveness of magnetic shielding of several materials versus frequency.

The best way of minimizing magnetic interference is to minimize loop areas. When the direction of the magnetic field is known, the interception of the magnetic field can be avoided by a correct mounting of the loop area involved.

A good example of a compensation technique is illustrated in Fig. 8.24, where a correct and an incorrect compensation technique are depicted. If two equal and

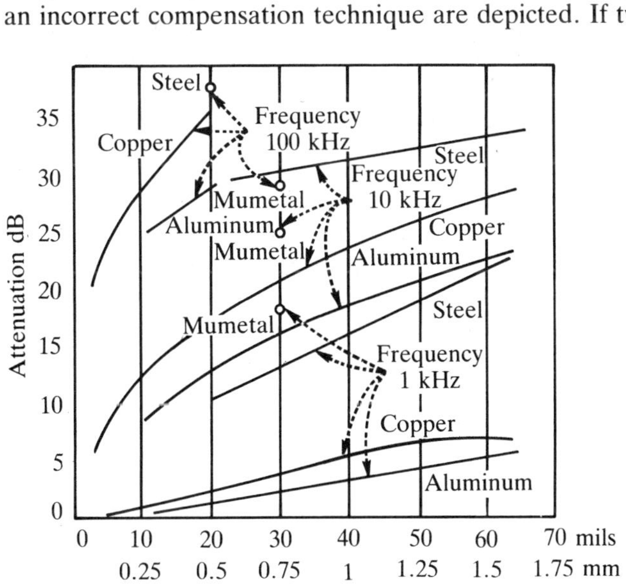

Figure 8.23 Magnetic shielding attenuation versus frequency of several materials (reprinted by permission, John Wiley & Sons Inc.)

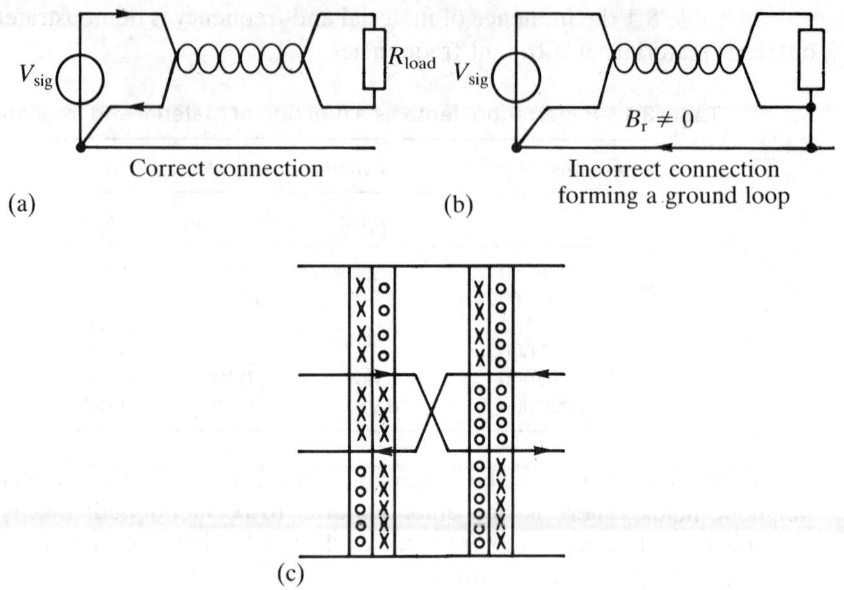

Figure 8.24 Applying a compensation technique with twisted wires

opposite currents flow through two twisted conductors, two equal and opposite fields are generated, producing a zero net field. This is illustrated in Fig. 8.24c.

Note again that wires can act either as a receiver or as a transmitter of interference. The compensation cancels the present magnetic fields under the condition that the conductors are completely symmetric and no new ground loop is introduced.

Another example of a good symmetrical conductor is a coaxial cable, which for reason of symmetry is completely insensitive to magnetic interferences.

Here are some basic rules for minimizing magnetic interference:

1 Use twisted pair cable for conductors carrying large currents. In this case, the wires are the source of the magnetic field.
2 Shield the circuit with an appropriate material depending on the applied frequency and field strength.
3 Avoid any wire running parallel to the magnetic field and cross magnetic fields at right angles.
4 Locate magnetic-sensitive parts as far as possible from magnetic field sources and at right angles.
5 Avoid ground loops and reduce any area which might be sensitive to magnetic fields.
6 Minimize cable length as far as possible.

In the following section we shall discuss some coupling techniques and related problems you may encounter when different measuring systems are interconnected.

8.6 TYPES OF COUPLING MECHANISMS

In many circumstances, systems are interconnected by a kind of coupling technique. We shall discuss several coupling techniques and possible errors which can occur when applying these techniques. In general, coupling between different measuring systems can be performed by different techniques, each of which has its own advantages and disadvantages. Coupling of measuring systems can be accomplished in the following different ways:

(a) direct coupling;
(b) magnetic coupling;
(c) r.f. coupling;
(d) microwave coupling;
(e) light coupling.

Suppose we have a system as illustrated in Fig. 8.25 consisting of two stages A_1 and A_2 and interconnected in one of the five ways mentioned above, denoted by the dashed arrow. Suppose further that the signal source V_s and the load R_l are at different ground potentials, then the parasitic capacitor C_{12} allows a current to flow through R_{s1}.

It will be clear that a common-mode voltage represented by the ground potential difference is converted into a series mode voltage in the same fashion as we have seen before. This results in a differential mode voltage for the first stage. We will discuss the consequences of several coupling mechanisms. Of the five methods of coupling mentioned we will only discuss direct coupling and magnetic coupling, because they are the most frequently used coupling methods.

8.6.1 Direct coupling

Direct coupling is a galvanic connection between different parts of the system to which for instance a differential amplifier is applied as illustrated in Fig. 8.26. In

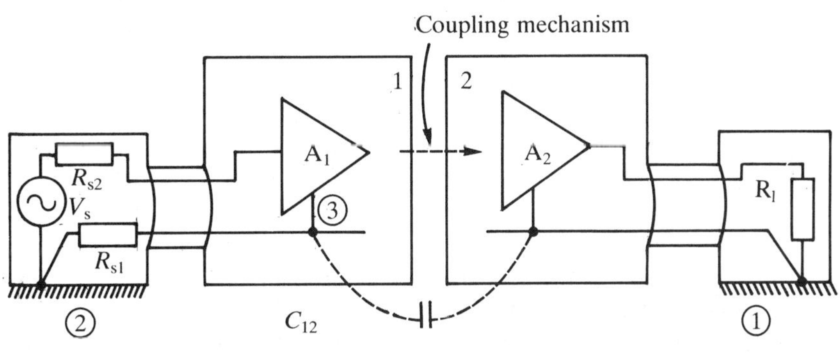

Figure 8.25 Types of coupling mechanism. The dashed arrow represents the coupling mechanism to be discussed

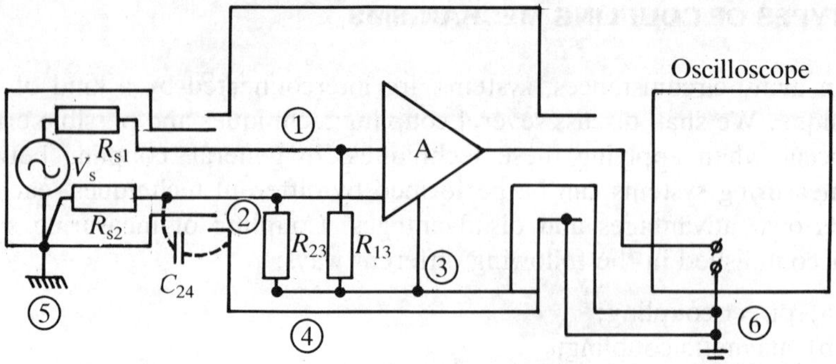

Figure 8.26 Principle of direct coupling applying a differential amplifier

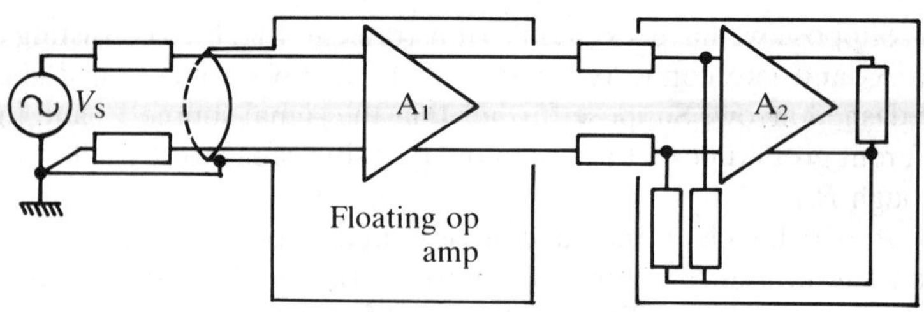

Figure 8.27 Direct coupling of high common-mode voltages via a balanced attenuator

this circuit a parasitic capacitor, C_{24}, between one lead and the shield 4, is present, over which a voltage difference is generated when a ground potential difference V_{56} is present.

This causes a current to flow through R_{s2}, when two ground connections are made. The leakage resistor R_{23} allows current to flow from point 5 to 5 via 2, the common 3 and point 6. In this figure R_{13} and R_{23} are the so-called leakage resistances with respect to common of both amplifier inputs. These leakage resistances can be interpreted as the isolation resistances of both inputs with respect to the common, and are to be detected in every input stage.

As will be shown, in a DVM these resistances should exceed a value of 1000 MΩ. The capacitor C_{24} can be understood as a cable or shield capacitor and, for instance, can have a capacity of 60–100 pF per meter. When high common-mode voltages are involved a balanced attenuator must be installed, because the maximum rating of the input voltage of an amplifier is limited. This is illustrated in Fig. 8.27.

In this way the common-mode voltage can be attenuated by a required factor. The second stage will reject the remaining common-mode voltage and reamplify the desired differential-mode signal. Most amplifiers cannot handle common-mode voltages exceeding 30 V, so when a common-mode voltage of 1000 V must be handled, in the attenuator an attenuation factor of 40 must be implemented. For this purpose, a floating amplifier for the first stage should be installed.

For direct coupling we can mention the following advantages:

(a) The coupling mechanism is accurate and stable.
(b) There are no limitations in bandwidth.
(c) A complete symmetrical input can be realized.

Of course there are some disadvantages:

(a) The common-mode voltages have to be processed by the circuitry.
(b) The level of the common-mode voltage is confined by the level of the power supply voltage.
(c) The common-mode rejection ratio is strongly frequency dependent.

8.6.2 Magnetic coupling

Magnetic coupling is the most widely used although the coupling mechanism has several disadvantages such as costs and bandwidth limitations. Applying a coupling transformer can serve two purposes, i.e.:

1 In numerous cases the transportation of energy is involved from the mains towards the system to supply.
2 Transportation of information. In this case a type of modulation and demodulation technique is involved by which the information is added to a carrier wave with a relatively high frequency.

The principle of operation of a transformer coupling for transporting information is drawn in Fig. 8.28 in which a voltage difference V_{14} due to a ground loop and a leakage capacitor C_{23} can be recognized. This can cause a current to flow round the input when an imperfect or wrong shielding technique is applied. As will be shown, this can require an input impedance of more than 1000 MΩ for accuracy reasons.

For magnetic coupling we can mention the following advantages:

(a) The maximum common-mode level is determined by the transformer's insulation quality.
(b) No external voltage can influence the shielded input, so the signal source may be floating.

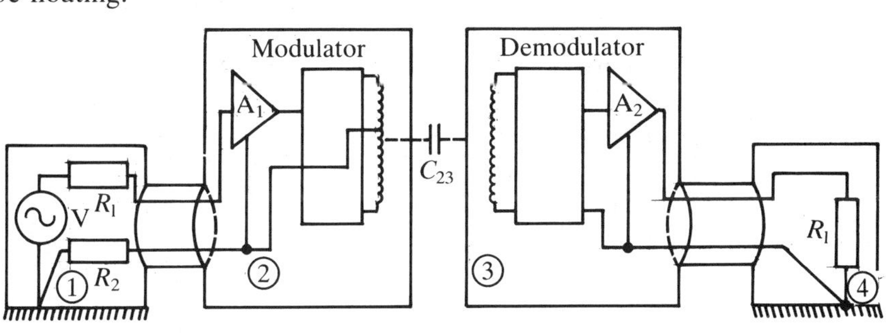

Figure 8.28 Principle of operation of magnetic flux coupling

(c) Common-mode signals are not amplified and are no longer a function of frequency and/or amplitude. The CMRR will remain independent of gain after the first stage.
(d) If at the input a type of modulation is used, within the shielding no electronic component is required. This means that at the input no power is necessary which might cause parasitic currents to flow into the input and shielding.
(e) A complete galvanic insulation between two stages in a system can be realized.

The disadvantages are:

(a) There is an increase in weight and costs of the system.
(b) There is a limitation in bandwidth determined by the quality of the transformer.

In the following section we shall discuss the interference of power supply transformers in more detail.

8.7 INTERFERENCE CAUSED BY POWER SUPPLY TRANSFORMERS

Most instruments use a mains supply transformer to deliver the required electric power. This transformer also provides a good electrical insulation from the mains, because only a magnetic coupling via the core material is involved, by which the electrical energy is transferred from the primary to the secondary coil via a magnetic field. Of course this is an important safety feature but this inherently involves a disadvantage too, as will be seen.

In normal circumstances the instruments are connected to the mains via a 3-core cable, of which one lead carries the power voltage and the other is the neutral or null lead. This null lead is connected to earth at the power distribution station where always at the secondary coil of the step-down transformer the starpoint is earthed. But simultaneously the safety guard wire is also earthed at the same point to which all metal housings are connected via the third lead of the mains cable.

Suppose no isolation transformer is installed at the instruments side, then the situation is as illustrated in Fig. 8.29. As can be seen from this figure, an extra low ohmic path to earth is created for all currents forced to flow by an isolation breakdown or an accidental short-circuit. Under normal conditions there is also a

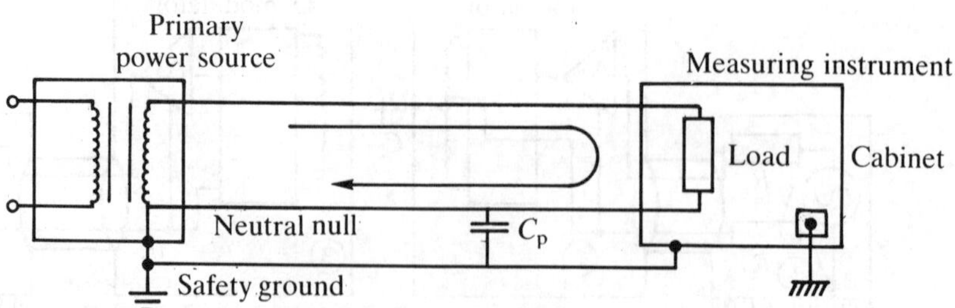

Figure 8.29 Circuit without a power-isolation transformer

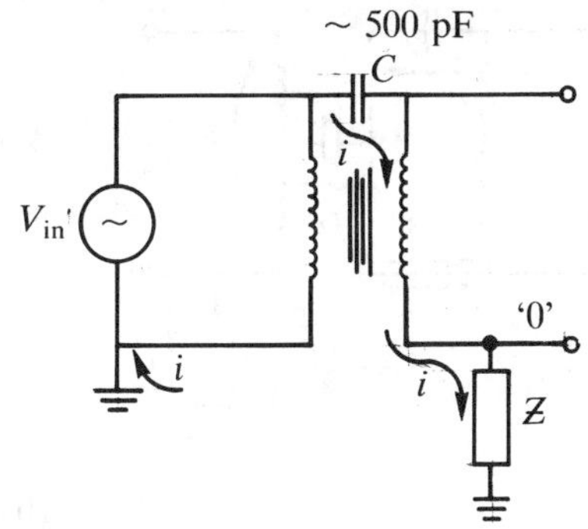

Figure 8.30 Stray capacitance in a power supply transformer

small current via a capacitive coupling of the conductors. This means a common-mode voltage will be formed between this power safety ground and the instrument ground and a lot of 'hum' at the input terminals will be the result in all circumstances. So, a most important reason for installing an isolation transformer is for safety.

When a power isolation transformer is used in a measuring system, another interference problem is introduced. In most countries the primary of this transformer is earthed. This is illustrated in Fig. 8.30 in which the stray capacitance C between primary and secondary coil is shown.

This capacitance exists between the primary and the secondary windings and is represented by one lumped capacity C ranging from several hundred picofarads up to 5 nF. The common of the circuitry at the secondary is normally grounded by a conductor which represents a certain impedance Z. A current, i, can flow from the primary to the secondary conductor whose magnitude is determined by $1/\omega C$ which represents the largest impedance value in the loop to be considered.

The result will be a voltage $V_p = iZ$ above ground level for the common of the secondary side. This need not be too serious, but when two or more instruments are used, each with its own power supply transformer, the commons can have different voltage levels which can cause large currents to flow when they are connected together.

This stray capacitance can be diminished considerably applying two shields for each transformer coil at both sides. The fact that, in a transformer, only one shield cannot eliminate stray capacitances is illustrated in Fig. 8.31. It can be seen that capacitive currents will flow through the ground connection at the secondary side in both cases.

The solution is shown in Fig. 8.32, with two shields, one around the primary side and one around the secondary side which are open windings connected to one side only of their respective windings. If closed windings are used, large eddy currents will flow, totally spoiling the desired effect.

Applying both shields results in a large capacitance between the shield and the cable core, giving a low impedance path for the common-mode generator V_{cm}. In this way, the stray capacitance between the primary and the secondary can be reduced to a value of 0.1 pF.

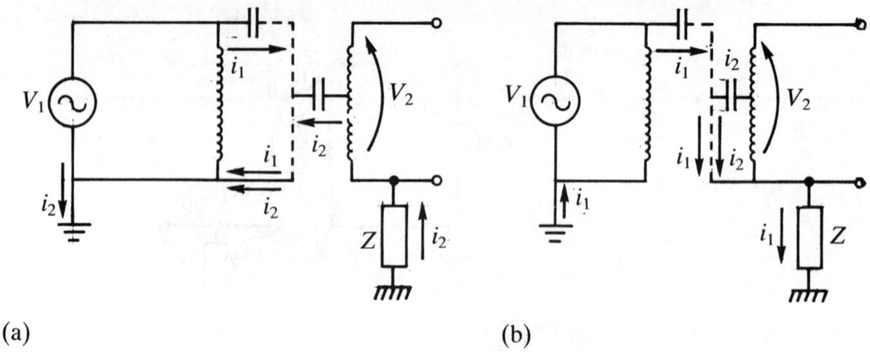

Figure 8.31 The application of one shield does not eliminate parasitic currents in the secondary common

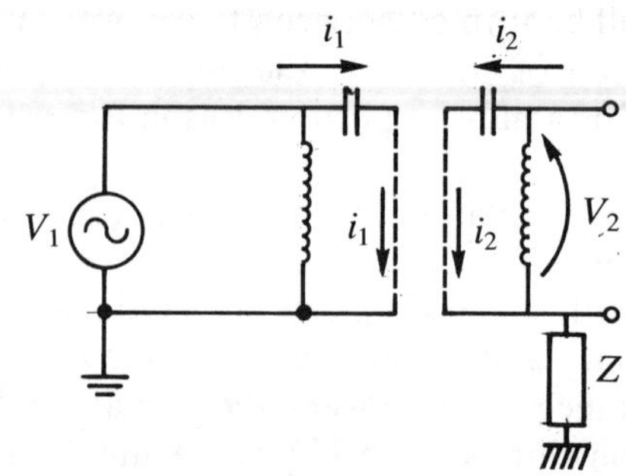

Figure 8.32 Method for avoiding large stray capacitances in a power supply transformer

Of course this technique has some disadvantages such as the high costs of manufacturing these types of transformers. The shields are hand-made and careful manufacturing is required to prevent short-circuiting.

Finally, we will give some other examples in which such a type of transformer is applied.

* * *

Example 8.4

Fig. 8.33a illustrates the application of the shielding principle. The shields around the primary and the secondary windings are connected to one side of the winding, so that stray capacitances between the windings and the shields are eliminated at one end. Moreover, a large capacitance is achieved between the shield and the cable core, resulting in a low-impedance path for the common-mode generator V_{cm}. The secondary is often connected to a center tap of the secondary winding, reducing the voltage difference. This also reduces the stray capacitance between the primary and secondary windings and a value of 0.1 pF can be obtained, while without shielding the stray capacitance between the two windings may exceed 2000 pF.

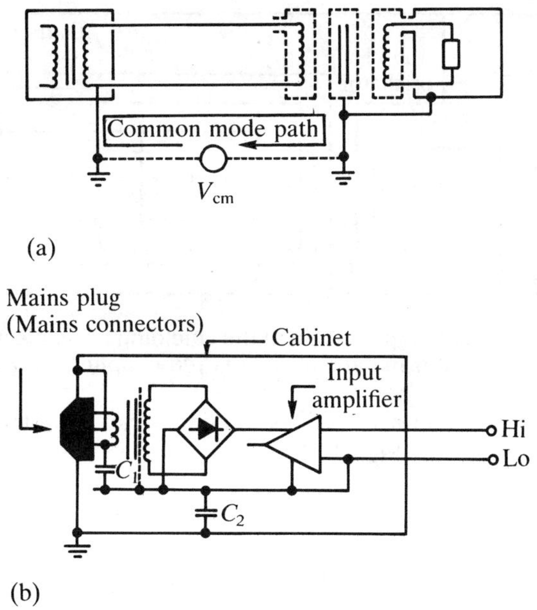

Figure 8.33 (a) The shielding principle applying a power-isolation transformer; (b) its use in a digital multimeter input configuration (courtesy Philips)

This principle is applied in a digital multimeter as is illustrated in Fig. 8.33b, for which the specifications are:

Capacitance from Mains to Lo is $C_1 = 50$ pF,
Capacitance from Lo to chassis is $C_2 = 1.5$ nF

It will be clear that 50 pF is quite a good separation between the mains and the measuring circuitry.

* * *

When more stringent input specifications are required, a solution like that in Fig. 8.34 can be applied, which is discussed in the following example.

* * *

Example 8.5

In order to achieve a further reduction of the influence of the mains, an extra transformer T_2 can be added. One of the secondary windings of T_1 is used to feed a 100 kHz oscillator which forms a part of the measuring system. The output of amplifier A is fed to the remaining part of the instrument's amplifier via a light-coupled diode device. Hence a perfect galvanic separation is obtained. A floating guard which may be connected to an extra terminal on the instrument is placed around the input amplifier A and its supply unit to eliminate the influence of external parasitic voltages. With this type of configuration an AC common-mode

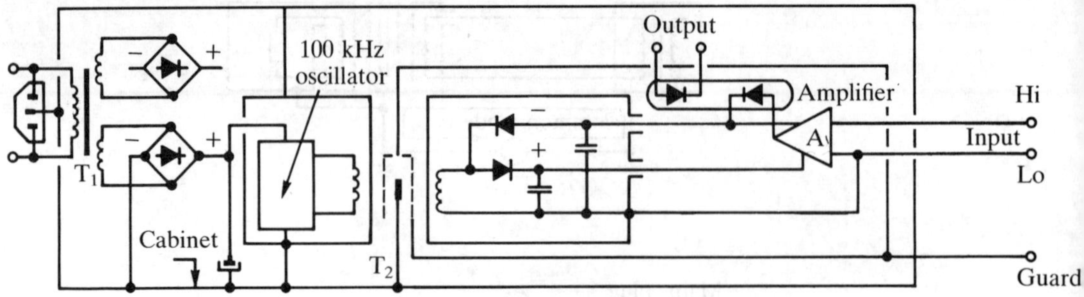

Figure 8.34 Application of the shielding principle applying a second transformer and a type of opto-coupling (courtesy Philips)

rejection of 126 dB for the mains frequency can be obtained. The principle of this set-up is depicted in Fig. 8.34.

* * *

In the next example, a multiple cabinet configuration is discussed.

* * *

Example 8.6

Suppose a multiple instrument configuration is designed as illustrated in Fig. 8.35. The instruments A and B are connected to the signal source V_s as shown. Suppose further that $C_{45} \approx 1$ nF and a ground potential of 10 V is present, then the current in the loop 1, 2, 4, 5, 8, 1 can be calculated as

$$I = 10/(1/\omega C) = 10/(3 \times 10^6) = 3.2\ \mu A$$

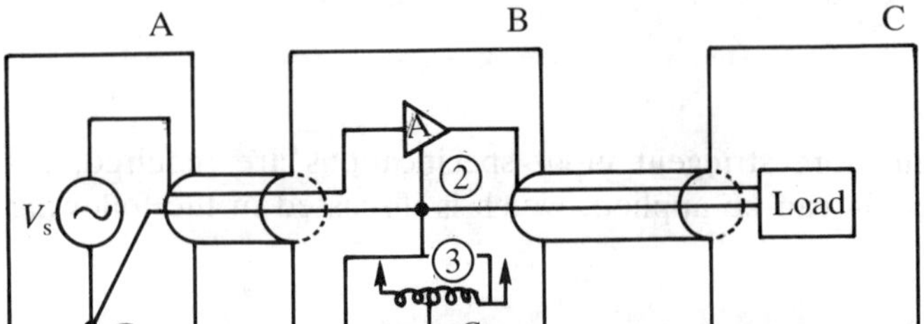

Figure 8.35 Earthing the primary of the shielding transformer in a multiple instrument configuration

If the common conductor has a 1 Ω impedance, a voltage difference between point 1 and 2 is developed of about 3 μV. This parasitic voltage can seriously destroy the performance of a sensitive measurement set-up.

The solution to prevent this parasitic current is by grounding the primary shield at point 1 as is indicated by the dashed line, and disconnecting the shield at point 8, as shown in Fig. 8.35.

* * *

In the following section the application of differential amplifiers in measuring set-ups and guarded environments will be discussed in more detail.

8.8 THE DIFFERENTIAL AMPLIFIER AND GUARDING

To discuss the possibilities of differential amplifiers in guarding applications we will repeat some basic concepts which are necessary for a good understanding of the common-mode rejection ratio (CMRR). For a thorough treatment of the differential amplifier reference is made to the relevant textbooks.

In measuring applications, it very often occurs that differential amplifiers are implemented because this type of amplifier is able to discriminate between a common-mode signal and a differential mode signal. In general these types of signals are always present. The quality of this kind of amplifier determines in most cases the total quality of the measuring system, because in numerous measuring applications this type of amplifier is used and can be found directly at the input stage of a measuring instrument.

If we consider the different types of active elements then a single-input amplifier such as a transistor, with a single-ended output is a device which cannot be considered as an ideal active element. An ideal active element can be defined as a controllable power source with no limitations in polarity, in voltage and current at both the input and the output. A good example of such an active element is the operational amplifier which is called a nullor in its most ideal form. In a nullor, the input current, i_i, and the input voltage, v_i, by definition are claimed to be zero.

There is some confusion between the different amplifier configurations. For example, sometimes an operational amplifier is called an instrumentation amplifier, but this is just a special configuration in which three operational amplifiers are involved. Such a three-amplifier configuration has improved features with respect to a single operational amplifier.

An operational amplifier can be considered as just a building block which can be used either as a differential amplifier or as a single active element, depending on how a few external passive components are connected to the respective connections.

For our purposes it will be sufficient to present only some basic concepts which describe the quality of the differential amplifier. Consider the general schematic layout for a grounded circuit as illustrated in Fig. 8.36.

In this figure the following quantities can be defined. V_s represents the signal

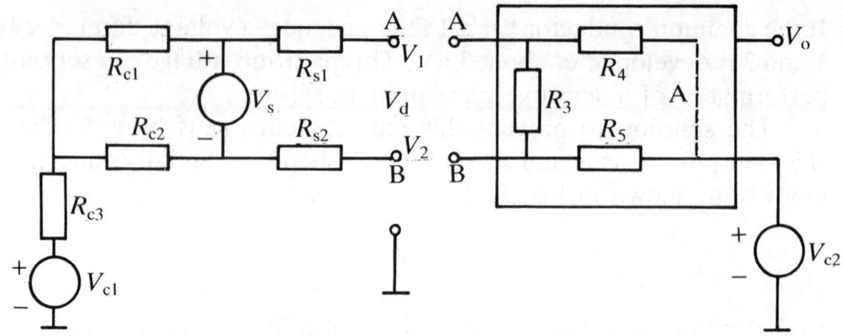

Figure 8.36 General layout of a grounded source circuit

source with source impedances R_{s1} and R_{s2}, and R_3 is the amplifier's input resistance. The leakage resistors of the amplifier from input to output ground are represented by R_4 and R_5. The common-mode voltages of input and output with respect to ground are V_{c1} and V_{c2} respectively. Finally, R_{c1}, R_{c2} and R_{c3} are the respective transition impedances to be distinguished and are inherently present with the common-mode signal.

In the same way as is in equations (8.7) and (8.8), the differential-mode and the common-mode voltage can be defined. With different indexes we define for the differential-mode voltage

$$V_d = V_1 - V_2 \tag{8.12}$$

and for the common-mode voltage at the amplifier's input side,

$$V_c = \tfrac{1}{2}(V_1 + V_2) \tag{8.13}$$

Suppose the input resistance is infinite; then no input current will flow and the following equation holds:

$$V_s = V_d = V_1 - V_2 = 0 \tag{8.14}$$

The average value of the input voltage at the amplifiers's input side equals

$$\tfrac{1}{2}(V_1 + V_2) = \tfrac{1}{2}(V_{c1} + V_{c2} + V_s) \tag{8.15}$$

When $V_{c1} = V_{c2}$ equation (8.15) yields

$$\tfrac{1}{2}(V_1 + V_2) = \tfrac{1}{2}V_s + V_{c1} = \tfrac{1}{2}V_s + V_{c2} \tag{8.16}$$

This equation contains the unwanted value V_{c1} or V_{c2}. This need not be a problem, but when for instance $V_{c1} \neq V_{c2}$ then $V_{c1} - V_{c2} \neq 0$ and a current will flow in R_{s1} or R_{s2}.

The following example demonstrates the consequences.

* * *

Example 8.7

Suppose a digital voltmeter is applied to execute a measurement on a voltage source delivering a signal output of $V_s = 10$ mV. The required accuracy is 0.1%, which means an allowed $V = 10\ \mu\text{V}$ as common-mode voltage. Suppose further for the moment that there is an imbalance involved between R_{s1} and R_{s2} of 1 kΩ, then the input current is

$$I_{in} = \frac{V}{R_{s2}} = \frac{10 \times 10^{-6}}{1000} = 10\ \text{nA}$$

If $V_{c1} - V_{c2} = 10$ V then the leakage impedance must exceed

$$R_{leakage} = \frac{10}{10 \times 10^{-9}} = 1000\ \text{M}\Omega$$

which is a normal figure of merit for input impedances with respect to ground. When for instance in a 3.5 digit digital voltmeter the *lowest* digit is 1 μV with a maximum of 1999 mV, this value is too small. Hence for the desired accuracy an error of 10 μV means that the last two digits are meaningless. This is again an example of how at the input of an amplifier a common-mode voltage is converted into a differential-mode voltage affecting the accuracy to a large extent. The solution is guarding, as discussed in the next section.

* * *

8.8.1 The common-mode rejection ratio

As claimed already in Chapter 3, a differential amplifier can be used to supress common-mode voltages and to amplify differential mode voltages. To be able to treat this chapter more independently and to emphasize relevant aspects concerning guarding, we will repeat some characteristics already discussed. The characteristic features of a differential amplifier can be expressed into two figures of merit, of which one is called the discrimination factor F and the other is called the rejection factor H. The rejection factor is the most important one, because it expresses to what extent a common-mode signal can be supressed by the applied amplifier. Once a common-mode voltage is converted into a differential-mode voltage, any further supression will fail. The discrimination factor F is just the ratio between the differential-mode gain and the common-mode gain of the same amplifier and can always be improved by a series connection of more stages. For a good amplifier both figures of merit should be high, for instance 120 dB.

To clarify the above statements reference is made to Fig. 8.37 where a single-

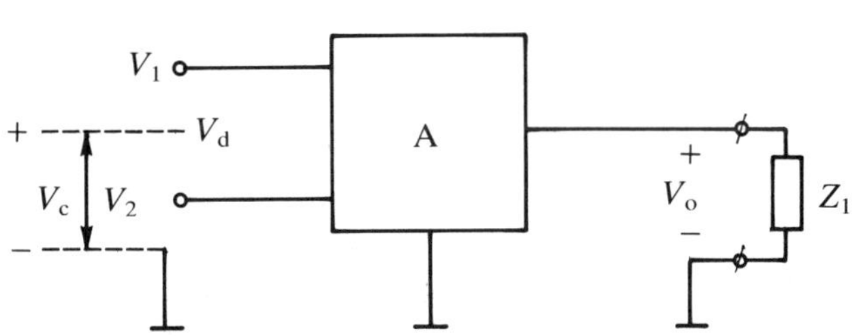

Figure 8.37 A single-ended output differential amplifier

ended output amplifier is depicted. In general the behavior of a differential amplifier can be described with the following set of equations:

$$V_{od} = A_{dd} V_{id} + A_{cd} V_{ic} \tag{8.17}$$

$$V_{oc} = A_{dc} V_{id} + A_{cc} V_{ic} \tag{8.18}$$

where, when any sign is neglected,

A_{dd} is the differential–to–differential signal gain factor
A_{cd} is the common–to–differential signal gain factor
A_{dc} is the differential–to–common signal gain factor
A_{cc} is the common–to–common signal gain factor

The gain factors are defined as

$$\begin{aligned} A_{dd} &= V_{od}/V_{id} \\ A_{cd} &= V_{od}/V_{ic} \\ A_{dc} &= V_{oc}/V_{id} \\ A_{cc} &= V_{oc}/V_{ic} \end{aligned} \tag{8.19}$$

with the other respective quantities kept constant, and where V_{od} and V_{id} respectively are the differential output and input voltages. In the same way, V_{oc} and V_{ic} are the common-mode output and input voltage respectively. Note that the double index in A is usually reserved for double input/double output stages, but in this fashion it generalizes the problem.

As can be seen from these equations, four different gain factors can be distinguished in which the indices ascribe the considered gain factor, for instance A_{cd} describes the extent to which a common-mode signal is converted into a differential-mode signal, etc. With the help of these equations, it is a simple affair to define the rejection factor, H, and the discrimination factor, F. The following definition holds for the rejection factor:

$$H \stackrel{\text{def}}{=} \frac{A_{dd}}{A_{cd}} = \frac{V_{od}/V_{id}}{V_{od}/V_{ic}} = \left.\frac{V_{ic}}{V_{id}}\right|_{V_{od}=\text{const}} \tag{8.20}$$

Usually this quantity is expressed in dB and is called the common-mode rejection ratio factor, CMRR, and so the manufacturer would claim for instance $20 \log|H| = x$ dB for a given amplifier, where 60 dB is a poor and 120 dB is a good figure of merit. But an important note must be made here. For a good judgment of the qualifications of an amplifier the measurement conditions should be specified by the manufacturer, such as the applied load and frequency. It will be clear that for an ideal amplifier H is infinite.

The discrimination factor F is defined as

$$F \stackrel{\text{def}}{=} \frac{A_{dd}}{A_{cc}} = \frac{V_{od}/V_{id}}{V_{oc}/V_{ic}} \tag{8.21}$$

Note that the denominator is the ratio V_{oc}/V_{ic}. You can see from these equations that both H and F can be calculated from measured voltages. As said, H is more important and describes how perfect the amplifier is. Very often the causes are symmetry faults as a result of design errors and/or applied technology faults during manufacturing.

To improve the CMRR two important measurements can be applied:

1 The input has to be isolated with respect to ground.
2 The amplifier's input must be balanced perfectly.

Isolation gives best results; balancing gives a further improvement. In the next sections we will see how this can be made effective.

8.8.2 The CMRR in shielding circumstances

So far different possibilities have been discussed of shielding and guarding to reduce the influence of ground loop currents, without looking *inside* the instrument. But inside the instrument shielding techniques too can give a large improvement of the CMRR.

Consider for instance an instrument with a metal shield applied inside the instrument as is depicted in Fig. 8.38. You may think of a DVM in which a metal shield is applied. Very often this shield is referred to as a guard if it is floating on the low-potential side, which here is supposed to be the common.

Although in the DVM, and thus the input amplifier itself, neither the 'Hi' or the 'Lo' terminal of the amplifier is connected to ground, the measuring leads may be grounded at the transducers side with its symmetrical source resistances.

Although there is no galvanic connection between the amplifier and earth, an AC current I_{cm} can flow through the stray capacitance C_1 owing to the series mode voltage V_{cm} over R_{t1} and R_2 when a ground loop is made. Bear in mind that for simplicity the input impedance of the amplifier is supposed to be infinite so no input current will flow. In the following example a calculation is executed and, for this purpose, the equivalent electrical circuit for the input is depicted in Fig. 8.39.

* * *

Example 8.8

Suppose the following practical values are given: the power supply frequency is 50 Hz, $C_1 = 10$ nF, $R_{t1} = 1$ kΩ and $V_{cm} = 2$ V, then the input current can be calculated to be

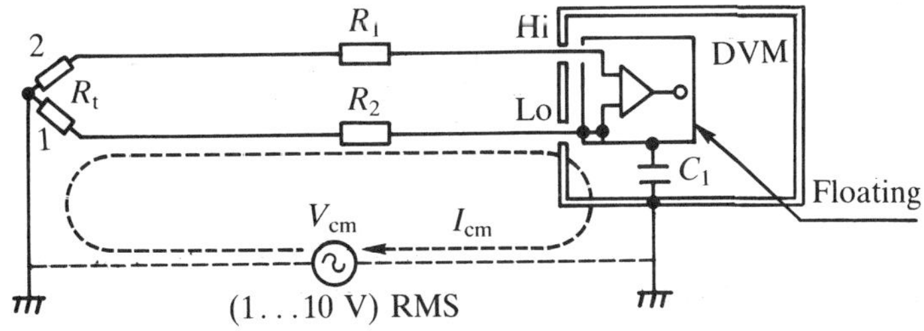

Figure 8.38 Digital voltmeter with floating guard

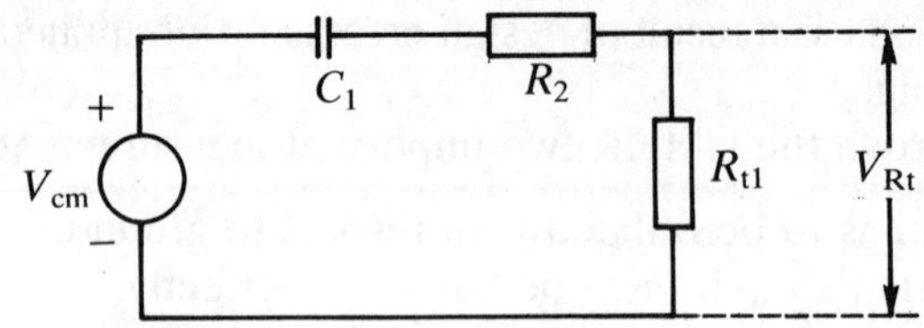

Figure 8.39 Equivalent input circuit of the DVM in Fig. 8.38

$$I_{cm} = \frac{V_{cm}}{X_{c1} + R_2 + R_{t1}} \tag{8.22}$$

Substituting the values for X_{c1} it is found that

$$X_{c1} = 1/(2\pi 50 \times 10 \times 10^{-9}) = 318\ \text{k}\Omega$$

and this value is in most cases much larger than R_2 and R_{t1}. Hence I_{cm} is found to be

$$I_{cm} = 2/318\,000 = 6.28\ \mu\text{A}$$

and thus

$$V_{Rt} = I_{cm} R_t = 6.28\ \text{mV}$$

Note that the series-mode voltage over R_t is determined by the ratio of R_t and X_{c1} and is directly added to the result of the measurand when AC measurements are performed. If, for instance, this series-mode voltage is compared with the output voltage of a chromel–alumel thermocouple of 40 µV/K, an unacceptable error is introduced.

This error can be reduced by filtering and/or integration techniques, because the magnitude of the error is mainly determined by the magnitude of the parasitic capacitance and the magnitude of the frequency.

In this case the CMRR can be defined as the ratio between the fault potential caused by the common-mode voltage and the magnitude of the common-mode voltage itself, both determined at the input side, hence

$$\text{CMRR} = \left.\frac{\text{present value of } V_{cm}}{\text{fault value caused by } V_{cm}}\right|_{\text{input side}} \tag{8.23}$$

If this relationship is applied for the above given values, by substitution we obtain

$$\text{CMRR} = \frac{V_{cm}}{V_{fault}} = \frac{2000}{6.28} = 318$$

or in dB,

$$\text{CMRR} = 20 \log 318 = 50\ \text{dB}$$

which of course is a rather poor specification.

* * *

For the CMRR we can conclude that its magnitude is strongly dependent on:

1 the applied frequency;
2 the value of the parasitic capacitance C_1;
3 the impedance of the transducer;

and, what is most important, for an instrument these conditions should be mentioned by the manufacturer when a CMRR is specified. The conditions to be mentioned are standardized by the IEC subcommittee 13B, April 1972, section 23.5.

In the next section we will investigate how the CMRR can be improved.

8.9 IMPROVEMENT OF THE COMMON-MODE REJECTION RATIO

As you have seen in the last section, one way to improve the CMRR is to increase the value of X_{c1}, or conversely, to reduce the parasitic capacitance C_1. To realize a reduced influence of the parasitic capacitance two principles of operation can be applied:

1 The application of a second or floating guard.
2 The application of double-shielded cables or active guarding as it is sometimes called.

Both principles will be discussed in detail.

8.9.1 The application of a floating guard

The principle of applying a second or floating guard is illustrated in Fig. 8.40, which shows a shielded cable of which the shield is connected to the guard at the instrument side and to the ground at the transducer side. Further, in this configuration a cabinet, a guard and a floating housing can be recognized, of which

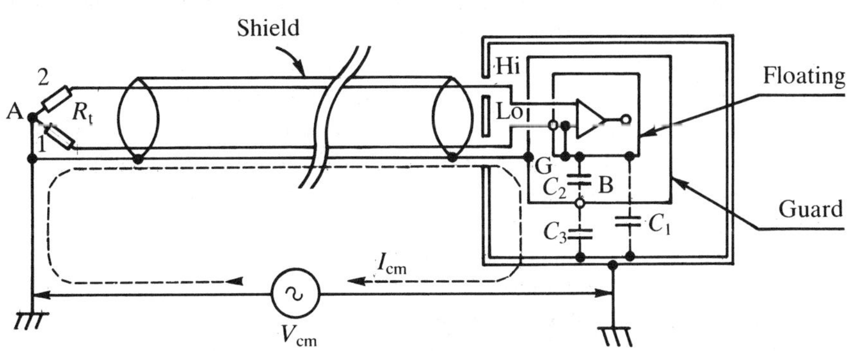

Figure 8.40 Application of a second shield to improve the CMRR

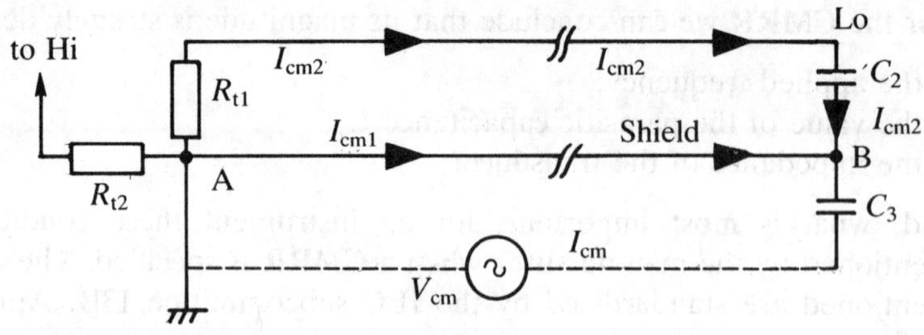

Figure 8.41 Equivalent circuit for Fig. 8.40 showing the two current loops

the floating housing is the one placed innermost. Besides these points, note how the three terminals 'Hi', 'Lo' and 'G' are connected to their respective housings.

If a common-mode voltage V_{cm} is present a current I_{cm} will be produced, but this current now splits up into two parts at point A, at the tip of the transducer. The transducer's impedance is represented by R_{t1} and R_{t2} and the amplifier's input impedance is supposed to be infinite, so no current will flow in R_{t2}.

The equivalent electrical input circuit is depicted in Fig. 8.41, in which two currents loops have to be distinguished.

According to Kirchhoff's current law for node B in Fig. 8.41, the following equation holds:

$$I_{cm} = I_{cm1} + I_{cm2} \tag{8.24}$$

The respective current loops are:

1 For I_{cm1} via the cable shield, C_3 and ground, back to V_{cm}, not affecting the result at all.
2 For I_{cm2} via R_{t1}, the 'Lo' terminal, C_2, C_3 and ground back to V_{cm}.

If it is assumed now that R_{t1} has a very small value with respect to X_{c2} then the voltage difference over C_2 will be determined by $I_{cm2}R_{t1}$ independently of the magnitude of X_{c2}. Even if C_2 is of the order of some nanofarads the *effective* capacitance of C_2 can be of the order of some picofarads because there is almost no potential accoss the plates of C_2. Hence, the current through C_2 will be very small. Now you can see how the trick works: if a voltage across a capacitor is very small, the capacitor is not effective at all regardless of its magnitude.

The assumption must be made that the ratio between R_{t1} and X_{c2} is very large, or more precisely $R_t \ll X_{c2}$, and a high impedance must be present between the 'Lo' and the guard. Very often a low source impedance is an advantage, as is the case here. The consequence of this is that via the shield, C_2 is short-circuited and in this loop, no current due to V_{cm} will flow. Hence all the current generated by V_{cm} will flow in the lower loop only.

The effective capacity is strongly reduced by a factor of, say, 1000 which then equals an improvement of another 60 dB. In practice the ultimately effective capacitance between Lo and earth, here denoted by C_1, can have an order of magnitude of 0.5–5 pF, while the capacitance between guard and earth is in the order of some nanofarads. Of course it is true that all three mentioned capacitances are bypassed by the isolation resistances.

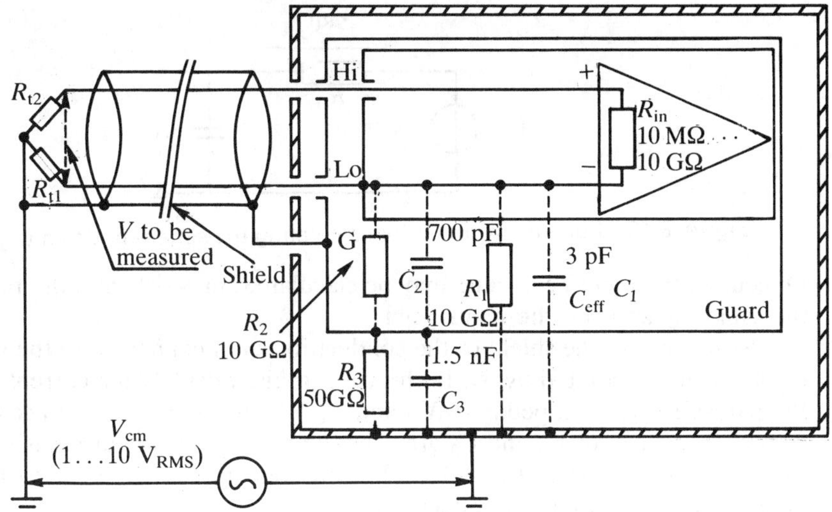

Figure 8.42 Complete diagram of a floating guard configuration. Note that the effective resistance is $R_{eff} = R_1 + R_2$

A more complete diagram with relevant practical values for the respective quantities is given in Fig. 8.42.

In the following example it is shown how this can work out. Again the series-mode voltage due the common-mode voltage V_{cm} must be calculated.

* * *

Example 8.9

Suppose again the frequency of the mains is 50 Hz and the other values are as depicted in Fig. 8.42, then the capacitive impedance of X_{C2} equals

$$X_{C2} = \frac{1}{2\pi 50 \times 700 \times 10^{-12}} \approx 4.5\ \text{M}\Omega$$

If this value is compared with the respective isolation resistances R_2 and R_3, then it will be clear that X_{C2} is much smaller than these. Hence we may consider the capacitive impedance only. Suppose further that the input impedance of the amplifier is again infinite and no current flows through C_2, due to an effective short-circuiting of C_2 by R_{t1}, which implies that the voltage difference developed across C_2 is almost zero. The equivalent electrical circuit of Fig. 8.43 can then be applied to calculate the series-mode voltage V_{Rt}.

$$V_{Rt} = \frac{R_{t1}}{R_{t1} + X_{c1}} V_{cm} \tag{8.25}$$

If for instance $R_{t1} = 1\ \text{k}\Omega$, $C_1 = 3$ pF and the common-mode voltage $V_{cm} = 5$ V, then applying equation (8.25) we find

$$\text{CMRR} = 20 \log \left| \frac{V_{cm}}{V_{Rt}} \right| = 20 \log \left| \frac{R_{t1} + X_{c1}}{R_{t1}} \right| = 120\ \text{dB}$$

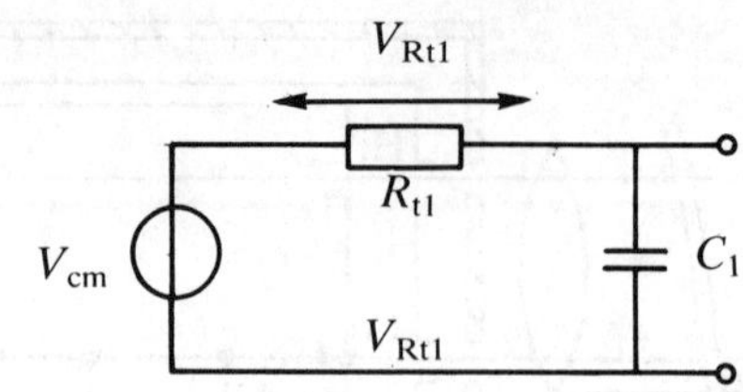

Figure 8.43 Equivalent circuit for calculating the series-mode voltage V_{Rt}

Of course, the reciprocal value may be calculated, in which case the result will be the same, apart from the minus sign.

Now suppose the shield of the conductors is not connected to the guard; then C_2 is not short-circuited by R_t. In this case, in the calculation a current flowing via the parasitic series, impedance X_{c2} and X_{c3} should be taken into account. If the respective values for C_2 and C_3 are substituted, the equivalent parasitic capacitor for C_2 and C_3 becomes $C_v = 478$ pF, and applying equation (8.25) the CMRR experiences a considerable reduction:

$$\text{CMRR} = 20\log\frac{1/2\pi 50 \times 478 \times 10^{-12}}{1000} = 76.5\ \text{dB}$$

For DC measurements, we have to account for the isolation resistances R_2 and R_3 only, because these resistances now determine the total current. Under the condition of DC measurement the CMRR can be calculated as

$$\text{CMRR}_{\text{DC}} = 20\log\frac{R_2 + R_3}{R_{t1}}\ \text{dB} \tag{8.26}$$

Substituting the known values in equation (8.26) yields

$$\text{CMRR}_{\text{DC}} = 20\log[(10 + 50) \times 10^9/1000] = 156.5\ \text{dB}$$

* * *

In the next example it will be shown once more why the CMRR is an important figure of merit.

* * *

Example 8.10

Suppose the configuration of Fig. 8.42 is applied in a digital voltmeter set-up and an accuracy of 0.001% is claimed at a resolution of 1 μV. For the lowest measuring range of, say, 1 V the claimed accuracy means a value of 10 μV. Suppose further a series-mode voltage of 50 μV is calculated, then you will understand that the last two digits make no sense at all. If a high accuracy combined with a high resolution is needed, a floating guard system can provide the right solution. Of course the costs of such an expensive system must be compared with the requirements when low-level measurements have to be executed.

* * *

8.9.2 Active guarding or the application of double-shielded cables

The configuration used to realized active guarding is illustrated in Fig. 8.44. In this circuit a buffer amplifier or so-called voltage follower is applied. The probe cable is a double-shielded one of which the inner shield is connected to the output of the buffer amplifier and the outer shield is connected to the non-inverting input terminal of the amplifier. Besides these connections, one of the signal source terminals is connected to the outer shield. The result is that the voltage of the inner floating shield equals exactly the voltage of the signal source. The result is a totally ineffective parasitic capacitor C_1 between source terminal and inner floating shield.

The remaining capacitive load is formed by C_2, but any current flowing through C_2 is supplied by the output of the voltage follower and will not result in a current through R_t, which might cause a substantial series-mode voltage error. The source voltage is present between the non-inverting terminal and the inner shield, and hence over the plates of the parasitic capacitor C_2.

Basically the same principle is applied as discussed in the previous example, because a parasitic capacitor is made ineffective by a kind of short-circuiting involving the plates of the capacitor present. Here this short-circuiting is realized by an active element such as a buffer amplifier. Measuring the signal source voltage is performed by a second amplifier not depicted in Fig. 8.44.

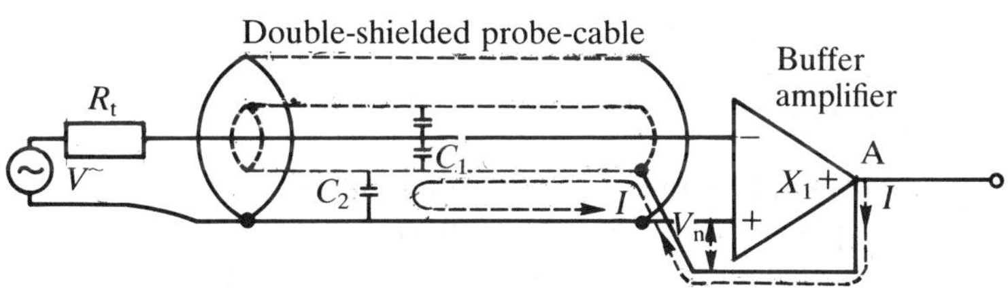

Figure 8.44 Principle of active guarding by applying a double-shielded cable

8.10 WORKED EXAMPLES

* * *

Example 8.11

The configuration in Fig. 8.45 can be considered as a floating attenuator. This means that the leakage resistances of both input terminals can be considered to be infinite. Then the output voltage V_o due to the source signal V_s can be calculated as

$$V_o = V_s R_p/(Z_{s1} + Z_{s2} + R_{s3} + R_{s4} + R_p) \tag{8.27}$$

The transfer function is given by

$$A_{dd} = \frac{V_o}{V_s} = R_p/(Z_{s1} + Z_{s2} + R_{s3} + R_{s4} + R_p) \tag{8.28}$$

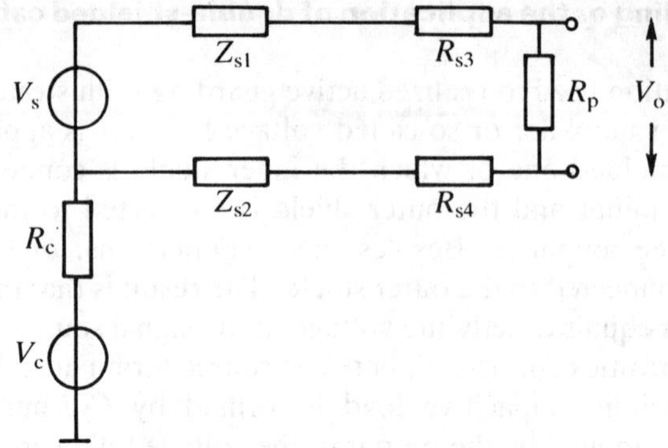

Figure 8.45 Input circuit with floating attenuator

In this configuration it will be clear no series-mode voltage or error voltage can be developed by the common-mode voltage V_c, hence $V_{fault} = A_{cd} V_c$ is zero and by definition for the CMRR we find

$$\text{CMRR} = \frac{A_{dd}}{A_{cd}} = \frac{V_{cm}}{V_{fault}} = \infty$$

Sometimes the reciprocal value is calculated; then the result is $1/H = 0$.

* * *

Example 8.12

Assume a digital voltmeter with a floating guard as shown in Fig. 8.42 has the following specifications. The display contains 4½ digit; this is equivalent to a maximum range of 19 999. The minimum range is 20 mV and the resolution is 1 μV. The input impedances are respectively:

High-Low	1 GΩ // 500 pF in the 20 mV range
Low–Guard	10 GΩ // 700 pF
Guard–Ground	50 GΩ // 1.5 nF

When the imbalance in the input circuit equals the standardized value of 1 kΩ, the CMRR equals 150 dB at DC and 120 dB at 50 Hz/AC.

A measurement is performed on a voltage source with an intrinsic impedance R_s, of 100 Ω in one lead, the other lead is supposed to be zero ohm. The source voltage is 100 μV which should be measured with an accuracy of 1%.

Demanding the specified accuracy, we will investigate how frequency and a wrong connection of the guard influence the CMRR, and therefore we need to calculate the following cases:

(a) The maximum allowed common-mode voltage at an interfering frequency of 20 Hz when the guard is properly used.
(b) The maximum allowed common-mode voltage with an interfering frequency of 20 Hz when an error is made connecting the guard to the low.

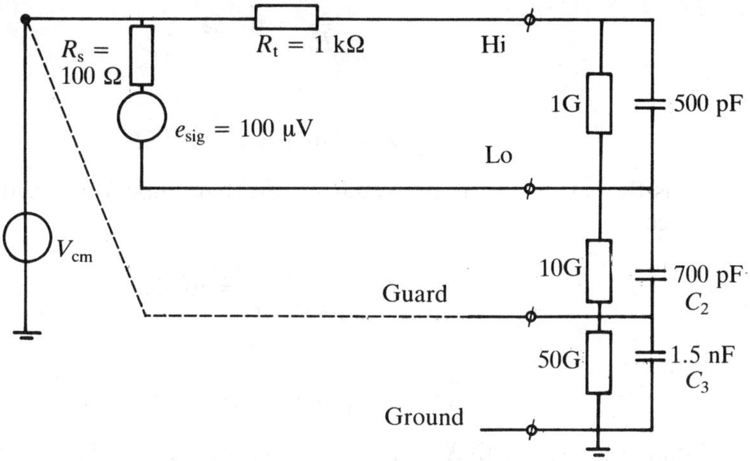

Figure 8.46 Equivalent measuring input circuit of the DVM when the guard is properly used. Note that the guard is connected to V_{cm} (dashed lines).

This can be achieved by the following means:

(a) Consider the equivalent circuit for the input measurement configuration in Fig. 8.46. The guard is properly used, which means that it is:
(i) connected to the screen surrounding the input leads; and
(ii) via the screen connected to the common-mode voltage V_{cm} at the transducer side, and hence to earth, in order to make C_2 ineffective by short-circuiting C_2 with R_s.

In this circuit the imbalance of 1 kΩ is drawn in the Hi terminal lead. Again no current is presumed to flow due to the high input impedance of the amplifier. This means that any error voltage will be developed across the signal source resistance R_s of 100 Ω. The series mode voltage V_{cm} is converted into a differential mode (error) voltage.

It is shown previously that the CMRR can be expressed as a measure of imbalance present in the input circuit. Note that the 1 kΩ imbalance is used as a standardized value at which the CMRR should be specified by the manufacturer. Once the CMRR is specified, the resulting effective impedance Z of the measuring system can be derived from the given impedance ratio. This procedure can be expressed as

$$\text{CMRR}|_{50\,\text{Hz}} = 20\log\frac{R_t + Z}{R_t} \tag{8.29}$$

Here also the condition holds that $R_t \ll |Z|$, and thus equation (8.29) simplifies as

$$\text{CMRR}|_{50\,\text{Hz}} = 20\log\frac{Z}{R_t} = 120\ \text{dB}$$

Substituting the specified data yields for Z

$$\frac{Z}{R_t} = 10^6\ \Omega$$

or

$$Z = (10^6)(R_t) = 10^9\ \Omega$$

Thus this value equals the resulting effective impedance of the input circuit, hence

$$Z = \frac{1}{2\pi f C_{eff}} = 10^9\ \Omega$$

or,

$$C_{eff} = 1/(2\pi 50 \times 10^9) = 3.18\ pF$$

Comparing X_{Cin} with X_{Ceff} it can be seen that $X_{Cin} \ll X_{Ceff}$ and therefore X_{Ceff} will dominate the total effective impedance described by Z. Neglecting the current through R_{in} of the input amplifier, and keeping in mind that C_2 is made ineffective in Fig. 8.47a, Fig. 8.47b represents the remaining equivalent circuit.

Now the effective capacitance C_{eff} is known, the CMRR can be calculated with the source impedance of 100 Ω (!) at the specified frequency of 20 Hz, instead of 50 Hz. To calculate the CMRR the same equation (8.29) as before can be applied, thus with $R_t = R_s = 100\ \Omega$, we find

$$\begin{aligned} \text{CMRR}|_{20\,\text{Hz}} &= 20\log\frac{Z}{R_t} = 20\log\frac{1/(2\pi 20 \times (3.2) \times 10^{-12})}{100} \\ &= 20\log 25 \times 10^6 = 148\ \text{dB} \end{aligned}$$

A larger CMRR is obtained such as can be expected from an applied lower frequency of 20 Hz.

From the specifications we know that the resolution is 1 μV, the source voltage is 100 μV and the required accuracy is 1%, hence the allowable voltage error is 1 μV. Applying a proper guarding the magnitude of the allowed voltage error will again be determined by the ratio of the now present imbalance $R_s = 100\ \Omega$ and the ultimately effective impedance Z, or

$$V_{error} = V_{cm}\frac{R_s}{Z} \tag{8.30}$$

Rearranging the equation will deliver the first desired answer

$$V_{cm} = V_{error}\frac{Z}{R_s} = (10^{-6})(25 \times 10^6) = 25\ \text{Volts}$$

This result equals the maximum allowed common-mode input voltage at an interfering frequency of 20 Hz maintaining the specified accuracy.

(b) When the guard is connected to the Low, the equivalent circuit of Fig. 8.48 will illustrate the actual situation.

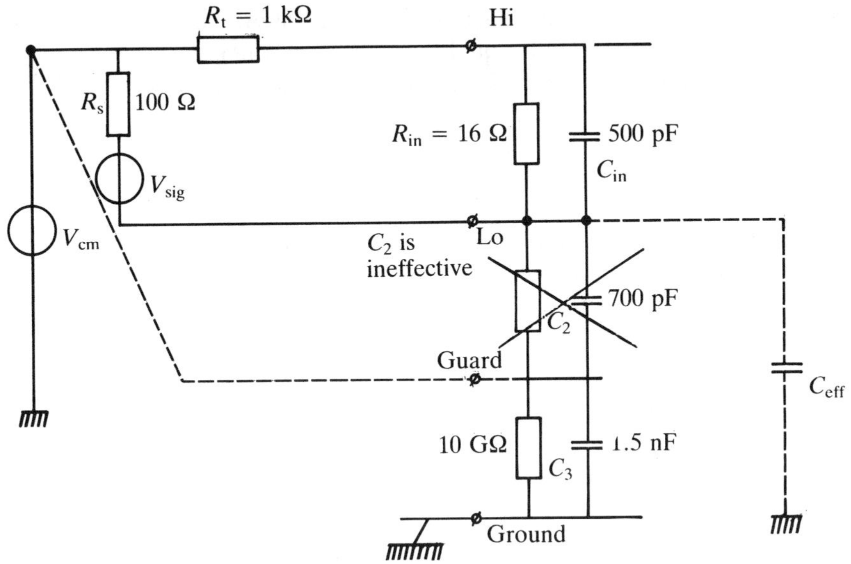

(a)

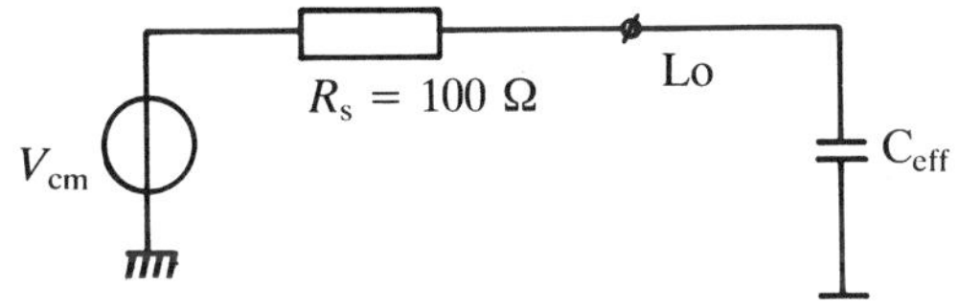

(b)

Figure 8.47 (a) Equivalent circuit when C_2 is made ineffective; (b) simplified equivalent circuit when the guard is properly used

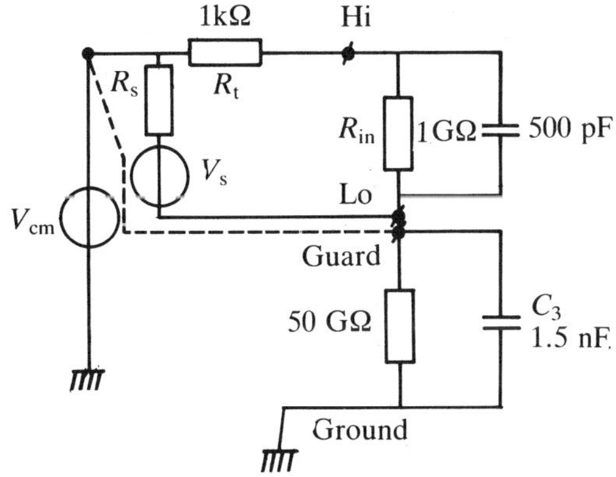

Figure 8.48 The equivalent circuit when the guard is connected to the Low

In this case the remaining effective capacitance C3 is the capacitance between low and ground and equals 1.5 nF. Again neglecting any input current, the maximum allowable V_{cm} at an interfering frequency of 20 Hz and with the same specified accuracy, we find

$$V_{cm} = \frac{Z}{R_s} \times V_{error} = \frac{1/(2\pi 20 \times 1.5 \times 10^{-9})}{100}(10^{-6}) = 53 \text{ mV}$$

It will be clear that a wrong connection of the guard influences accuracy, CMRR, or the allowed V_{cm} in large extent.

*   *   *

8.11 PROBLEMS

8.1 Explain why in the configuration depicted in Fig. 8.49, the common-mode voltage V_{cm} is converted into a series-mode voltage.

8.2 Determine the common-mode voltage at both input terminals of the operational amplifier drawn in Fig. 8.50.

8.3 Of the measuring circuit illustrated in Fig. 8.51 the following specifications are known. The supply voltage is $V_{cc} = V$, the CMRR of the op.amp. for DC is 100 dB and the Wheatstone bridge is in a balanced state. Determine the output voltage V_{out}.

8.4 Determine by which measurement technique in the configuration of Fig. 8.51 the influence of V_{cm} can be eliminated.

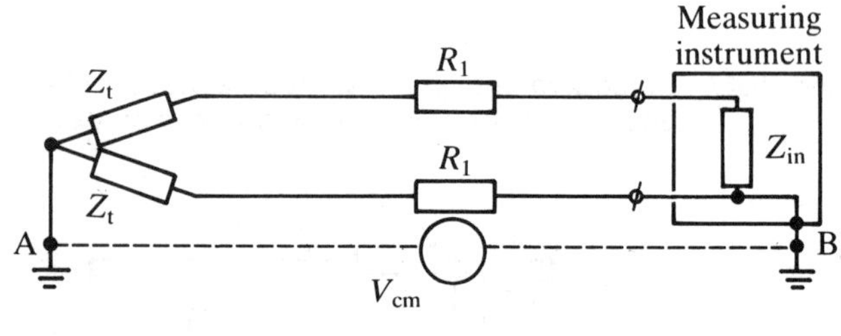

Figure 8.49 A common-mode voltage is converted into a series-mode voltage

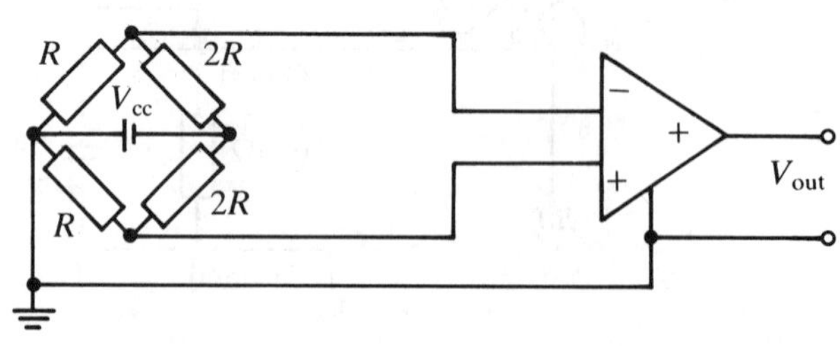

Figure 8.50 Bridge measuring circuit

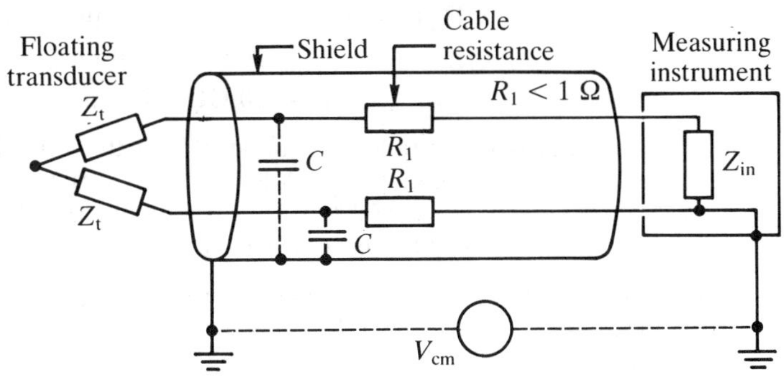

Figure 8.51 Configuration with floating transducer

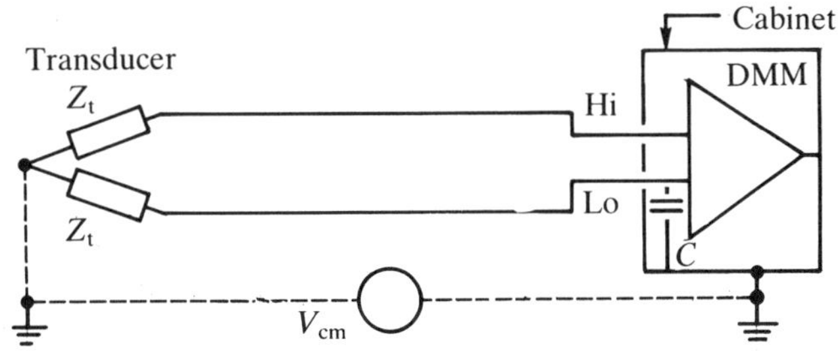

Figure 8.52 Configuration with earthed transducer

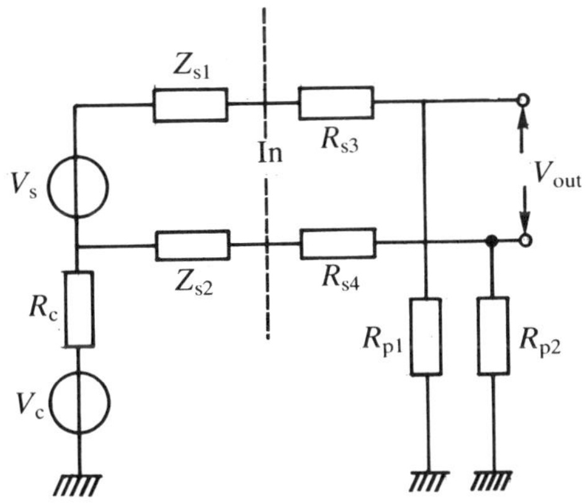

Figure 8.53 An earthed attenuator

8.5 Determine for the configuration depicted in Fig. 8.52 the CMRR for AC measurements in a country with a mains supply frequency of 60 Hz. The stray capacitance between the low terminal and the cabinet equals 2.6 nF, the transducer impedance is 1 kΩ.

8.6 Determine the CMRR of the earthed attenuator depicted in Fig. 8.53.

8.7 A digital DC voltmeter has the following specification impedances:

High–Low	1 GΩ // 500 pF
Low–Guard	10 GΩ // 700 pF
Guard–Ground	50 GΩ // 1.5 nF

The common-mode rejection ratio is 130 dB at 50 Hz with 1 kΩ imbalance in the input leads. The series-mode rejection ratio (SMRR) is 60 dB at 50 Hz. A DC measurement is executed on a source signal of 200 μV having an imbalanced source resistance of 10 kΩ. The following quantities have to be calculated:

(a) The error voltage due to a DC common-mode voltage of 100 V when
   (i) the guard is connected to the ground;
   (ii) the guard is properly used.
(b) The common-mode rejection ratio in both cases.
(c) The maximum allowable series-mode voltage with a frequency of 50 Hz when an accuracy of 10% is required.
(d) The maximum allowable common-mode voltage with a frequency of 50 Hz when the guard is properly used and the required accuracy again is 10%.

Nine

NOISE CALCULATIONS

9.1 INTRODUCTION

This chapter is devoted to noise calculations. Noise is one of the most discussed topics in electronics. When very low signal levels are involved, noise plays a vital role in signal handling. Noise can have a great variety of causes but they all have one feature in common with respect to the signal being measured. Noise can cause the signal to be drowned so it cannot be detected any longer. This is mainly determined by the signal-to-noise ratio, but, apart from this, the applied measuring technique has an important influence on the threshold value that can be detected. Of course the ultimate limit will be a physical limit and cannot be surpassed.

There is an increasing need to control signal-to-noise ratios because in communications and telecommunications applications the distance to be bridged is growing. Any increase in distance means a loss in signal-to-noise ratio. Despite an increasing shift to digital signal handling when very large distances are involved, the noise level still comes into scope. But an even more serious noise problem arises when biological signal sources have to be measured. Very often these sources combine a very high source impedance with very low signal level.

A good example of such a source are our heart-muscle cells. The electrical phenomenon is due to a flow of K^+, Ca^{++}, or Na^+ ions in the so-called canals in the cell membranes. These membranes show an inherently high source resistance with an order of magnitude of GΩ. The currents flowing through these membranes have an order of magnitude of picoamperes and so the signal power is in the same order as the noise power. Another example you met in Chapter 8 is the measurement of an electroencephalogram in which very low signal levels (μV) have to be measured.

This chapter offers you a practical approach with only one objective – to make noise calculations possible with all the relevant concepts required for calculation, without discussing noise mechanisms. The following chapter is devoted to a more fundamental physical approach and discusses the nature of several noise mechanisms. It can be treated separately or skipped in a first reading.

9.2 NOISE VOLTAGE, NOISE CURRENT AND NOISE FIGURE

The physical aspects of noise are discussed in Chapter 10, so here we can skip all sorts of terms like spot noise, thermal noise, broadband noise, shot noise, white noise, fluctuation noise, pink noise, $1/F$ noise, flicker noise, popcorn noise, bipolar spike noise, low noise, no noise and loud noise. This can make it all very confusing but only two things are important here: there is noise and there is signal.

The whole idea is to have the noise level very small compared to the signal level; or stated conversely a high signal-to-noise ratio (S/N) is desired always. Now it happens that the S/N is related to the noise figure, noise factor, noise power, noise voltage, and noise current. To simplify matters further it also happens that any noisy channel or amplifier can be completely specified for noise in terms of two noise *generators* $\overline{e_n}$ and $\overline{i_n}$ as is illustrated in Fig. 9.1.

All we really have to understand are the quantities NF, $\overline{e_n}$, and $\overline{i_n}$.

The *noise voltage*, $\overline{e_n}$, or more precisely the equivalent short-circuit input RMS noise voltage, is defined as that noise voltage which would appear to originate at the input of the noiseless amplifier if the input terminals were shorted. It is expressed in nanovolts per root Hertz (nV/$\sqrt{\text{Hz}}$), at a specified frequency, or in microvolts (μV) in a given frequency band.

This noise voltage can be determined by shorting the input terminals, measuring the output RMS noise voltage, dividing by the amplifier gain, and referencing to the input. This explains the term equivalent short-circuit input RMS noise voltage.

In the measurements, an output bandpass filter of known characteristics is used and the measured value is divided by the square root of the bandwidth, $\sqrt{B}$, if the data has to be expressed per unit bandwidth or per root Hertz. It appears that the level of $\overline{e_n}$ is not a constant over the frequency band but very often increases at lower frequencies as is illustrated in Fig. 9.2. Very often for every specified amplifier or transistor this type of graph is provided by the manufacturer and is easy to use in calculations, as will be seen later on.

The *noise current*, $\overline{i_n}$, or, more precisely, the equivalent open-circuit RMS noise current, is that noise which appears to be present at the input of the noiseless amplifier due only to noise currents. It is expressed in picoamps per root Hertz (pA/$\sqrt{\text{Hz}}$) at a specified frequency, or in nanoamps in a given frequency band.

This noise current can be measured by shunting a capacitor or resistor across the input terminals in such a way that the noise current will give rise to an additional noise voltage which then equals $\overline{i_n}\ R_{in}$, or $\overline{i_n}\ X_{Cin}$, depending on what is

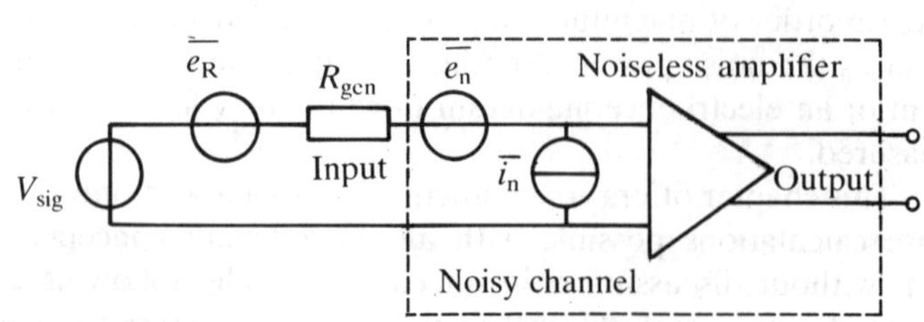

Figure 9.1 The noise characterization of an amplifier

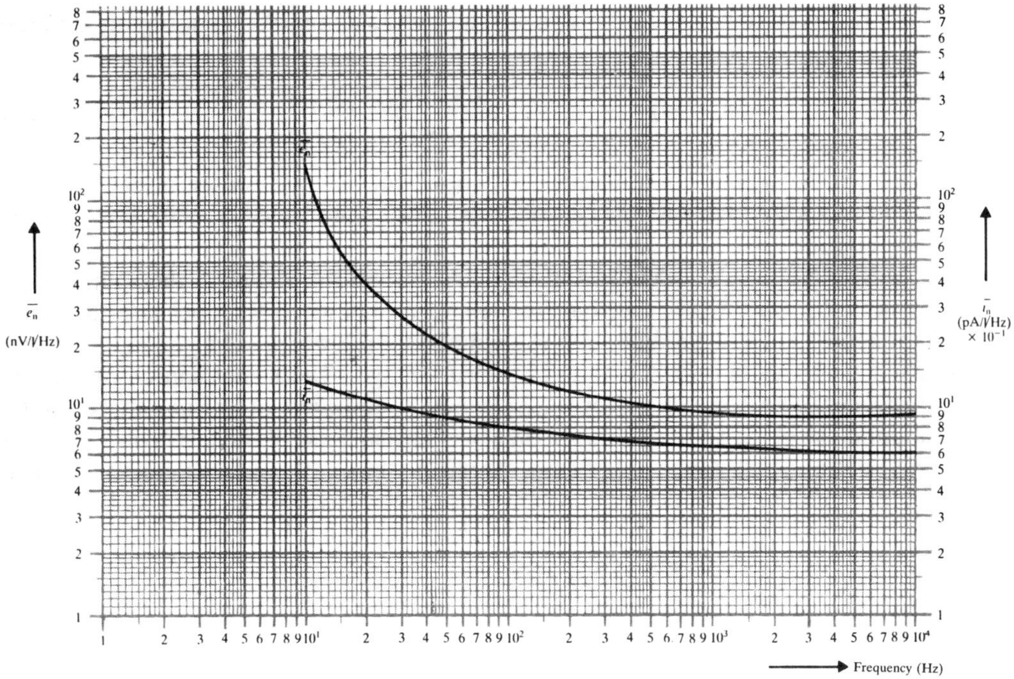

Figure 9.2 Noise voltage and noise current as functions of frequency for a given op amp (courtesy National Semiconductors Ltd)

connected to the input terminals. The output voltage is measured, divided by the amplifier gain, referenced to the input, and the contribution known to be due to the noise voltage $\overline{e_n}$ and the resistor noise is subtracted from the total measured noise.

If a capacitor is used at the input there is only a noise voltage $\overline{e_n}$ and a noise voltage $\overline{i_n}X_{Cin}$, because a capacitor does not generate noise so an $\overline{e_c}$ cannot exist. Again the noise current $\overline{i_n}$ is measured with a bandpass filter of known characteristics and divided by the square root of the bandwidth, converting to pA/$\sqrt{B}$ if appropriate. Noise current increases at lower frequencies for operational amplifiers and bipolar transistors, but increases at higher frequencies for field-effect transistors. These aspects will be discussed later on in more detail. The current noise behavior for the same amplifier is depicted in Fig. 9.2.

Finally we have to define the noise figure. The *noise figure*, NF, is defined as the logarithm of the ratio of the input signal-to-noise ratio and the output signal-to-noise ratio. This is expressed in the equation

$$\mathrm{NF} = 10 \log \frac{(S/N)_{\mathrm{in}}}{(S/N)_{\mathrm{out}}} \tag{9.1}$$

where S and N are power levels or $(\mathrm{Volt})^2$ levels. It is measured by first determining the S/N at the input with the amplifier disconnected and second determining the S/N at the output with the signal source connected at the amplifier, and finally dividing both measured quantities.

Derived from this description we can put equation (9.1) in a slightly different way:

$$\mathrm{NF} = 10 \log \frac{\text{total noise power}}{\text{noise power of the source}} \tag{9.2}$$

This will be shown in the following section, but, as you can see already, the NF expresses a comparison between the total noise power and the noise power of the source itself.

The NF can be calculated in real terms when the values of the source impedances R_{gen} and X_{gen}, as well as frequency, are known. This is because the amplifier's noise current produces a noise voltage source which equals $\overline{i_n}\, Z_{\text{gen}}$ and R_{gen} itself produces a so-called thermal noise voltage. The signal source in Fig. 9.1 produces some noise; however, V_{sig} is generally considered to be noise free and the input noise of the source is only present as thermal noise of the resistive component R_{gen} of the signal source impedance. As will be shown, this thermal noise is white in nature because it contains a constant *noise power density* per unit bandwidth.

In the next section we will discuss the relationship between the noise voltage source, the noise current source and the noise figure NF.

9.3 THE RELATIONSHIP BETWEEN THE NOISE VOLTAGE, NOISE CURRENT AND NOISE FIGURE

We will examine the relationship between $\overline{e_n}$ and $\overline{i_n}$ at the amplifier's input. When the signal source is connected to the input of the amplifier, the $\overline{e_n}$ appears in series with the source voltage V_{sig} and e_R, where e_R represents the thermal noise voltage developed across R_{gen}. The $\overline{i_n}$ flows through R_{gen}, thus producing another noise voltage with magnitude $\overline{i_n}\, R_{\text{gen}}$. It will be clear that this noise voltage is dependent on the value of R_{gen}.

All of these noise voltages add at the input in RMS fashion; this means that the square root of the sum of the squares of the noise voltages needs to be considered. If we neglect possible correlations between $\overline{i_n}$ and $\overline{e_n}$, the total *input noise density* can be calculated from

$$\overline{e_N^2} = \overline{e_n^2} + \overline{e_R^2} + \overline{i_n^2} R_{\text{gen}}^2 \tag{9.3}$$

So, in general, three different sources of noise can be distinguished: the first is delivered by the amplifier only, the second comes from the signal source, and the third is determined by the product of the noise current of the amplifier and the generator source resistance.

Further examination will reveal the relationship between the different quantities. For this purpose we refer to equation (9.1). Note that this equation is expressed in power densities, and can be rewritten in the following way:

$$\mathrm{NF} = 10 \log \frac{S_{\text{in}} \times N_{\text{out}}}{S_{\text{out}} \times N_{\text{in}}} \tag{9.4}$$

and for N_{out} we can write, when, G_p equals the power gain of the amplifier,

$$N_{\text{out}} = G_p\overline{e_N^2} \tag{9.5}$$

Substituting equation (9.5) in (9.4) yields

$$\text{NF} = 10 \log \frac{S_{\text{in}} G_p \overline{e_N^2}}{S_{\text{in}} G_p \overline{e_R^2}} \tag{9.6}$$

and this yields for the noise figure

$$\text{NF} = 10 \log \frac{\overline{e_N^2}}{\overline{e_R^2}} \tag{9.7}$$

Compare the result of equation (9.7) with equation (9.2). Equation (9.7) can be written as

$$\text{NF} = 10 \log \frac{\overline{e_n^2} + \overline{e_R^2} + \overline{i_n^2} R_{\text{gen}}^2}{\overline{e_R^2}} \tag{9.8}$$

or

$$\text{NF} = 10 \log \left[1 + \frac{\overline{e_n^2} + \overline{i_n^2} R_{\text{gen}}^2}{\overline{e_R^2}} \right] \tag{9.9}$$

From this equation it can be seen that when R_{gen} is small, noise voltages will dominate, and when a large source resistance R_{gen} is involved noise current will become important. A clear advantage accrues here to FET input amplifiers especially at high values of R_{gen} because, in an ideal FET, there is no $\overline{i_n}$ at all. Note that for a meaningful NF value it must be accompanied by a value for R_{gen} as well as frequency and/or bandwidth.

In the following section a first approach will be given to calculating total noise voltage.

9.4 CALCULATING TOTAL NOISE VOLTAGE

As stated already, always keep in mind that we have to reckon with power levels or V^2 levels before the total noise voltage level can be determined, taking the square root. The consequence of this is that every noise term must be squared prior to addition.

Consider again equation (9.3), then each term for a given amplifier and known signal source must be fixed. Very often the manufacturer provides graphs concerning the noise behavior of the specified amplifier. Sometimes the total input noise voltage $\overline{e_N}$ versus frequency is given for different values of source resistances $R_s = R_{\text{gen}}$, of which Fig. 9.3 is an example. But very often the manufacturer provides two separate graphs, one depicting the noise voltage of the amplifier and the other depicting the noise current versus frequency, and occasionally they are combined in one graph. An example of such a graph is Fig. 9.2. So you have to consider carefully the applied coordinate axis and units. Here are some examples.

* * *

Example 9.1

Determine the total equivalent input noise per unit bandwidth for an amplifier operating at 1 kHz from a source resistance of 10 kΩ. The ambient temperature is 300 K.

When the manufacturer has provided us with a graph depicting the total noise voltage versus frequency with R_S as parameter, we can draw the required value directly from that graph. Assume here that Fig. 9.3 is applicable, then at 1 kHz with parameter $R_s = 10$ kΩ the total noise voltage has to be read from the graph. This delivers $\overline{e_N} \approx 17$ nV/$\sqrt{\text{Hz}}$. If the source resistance, R_s, is not one of the depicted values an estimation can be made by interpolation. But when this total noise graph is not available you have to perform a little more effort, as outlined in the next example.

* * *

Example 9.2

Determine the total equivalent input noise voltage per unit bandwidth for an amplifier operating at 1 kHz from a source resistance of 10 kΩ when the noise data of Fig. 9.2 are to be applied. The ambient temperature is again 300 K.

The first thing to do is to determine the value of the equivalent noise voltage for R_s; thus $\overline{e_R}$ must be fixed. This can be performed in two different ways.

As mentioned already, the nature of noise generated by R_s is white or thermal noise and can be derived from the following equation, which describes the available thermal noise power at a given temperature and bandwidth of the resistance involved:

$$P = 4kTB \qquad \text{(Watts)} \tag{9.10}$$

where:

k is Boltzmann's constant (1.38×10^{-23} J/K)
T is the temperature in degrees Kelvin (K)
B is the considered bandwidth in Hz

Two source representations can be derived from this equation, one as an equivalent RMS noise voltage source and one as an equivalent RMS noise current source, i.e.

$$P = \frac{\overline{e_{Rs}^2}}{R_s} = \overline{i^2} R_s = 4kTB \qquad \text{(W)} \tag{9.11}$$

then the equivalent voltage power is

$$\overline{e_{Rs}^2} = 4kTBR_s \qquad (\text{V}^2) \tag{9.12}$$

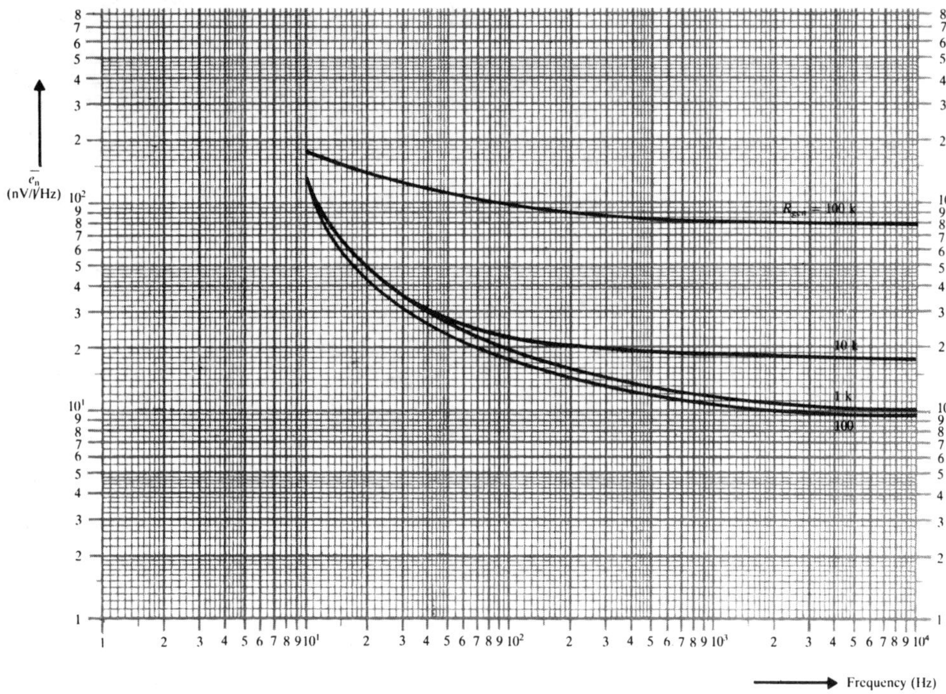

Figure 9.3 Total noise voltage versus frequency with source resistance R_s as parameter (courtesy National Semiconductors Ltd)

and the equivalent current power is

$$\overline{i_{Rs}^2} = \frac{4kTB}{R_S} \qquad (A^2) \tag{9.13}$$

Then the equivalent RMS noise voltage source is

$$\overline{e_{Rs}} = \sqrt{4kTBR_s} \qquad (V) \tag{9.14}$$

and the equivalent RMS noise current source is

$$\overline{i_{Rs}} = \sqrt{\frac{4kTB}{R_S}} \qquad (A) \tag{9.15}$$

This calculation can be executed for a bandwidth of 1 Hz and an ambient temperature of 300 K for the equivalent RMS noise voltage source as well as the equivalent RMS noise current source. The result is depicted in Fig. 9.4 and is very convenient for noise calculations when the reference temperature is taken as 300 K, which is very usual.

So every time an equivalent RMS noise voltage must be determined you can apply equation (9.10), or you can use Fig. 9.4 if the referred standard conditions are met. If not, then you have to use equation (9.10). To answer the question the following steps have to be taken:

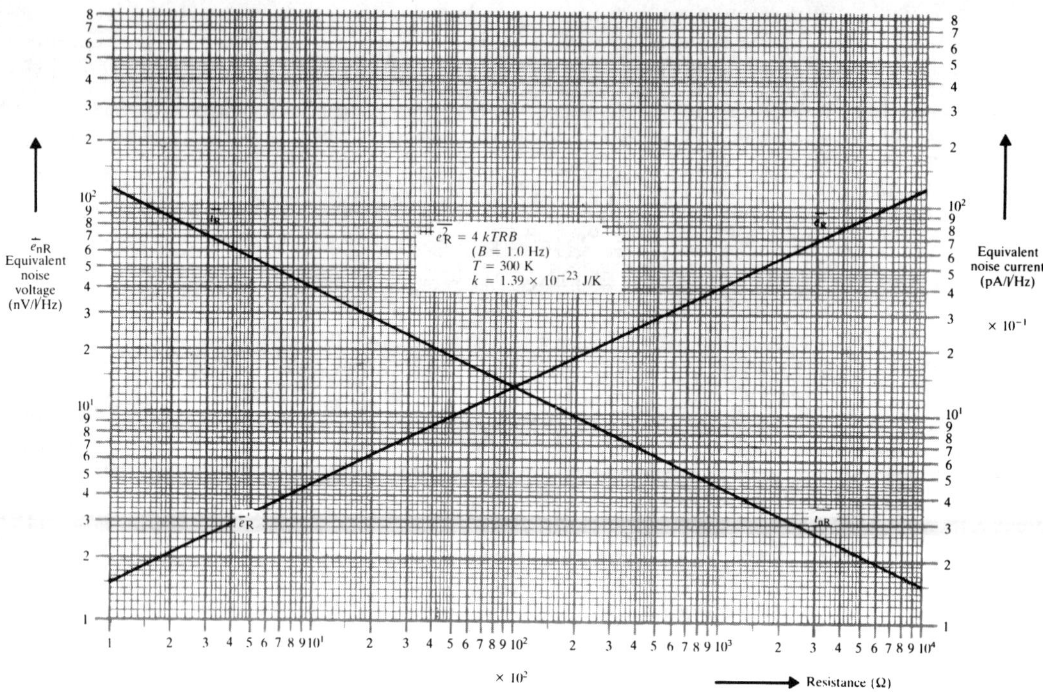

Figure 9.4 Thermal equivalent RMS noise voltage and RMS current versus resistance at 300K and 1 Hz bandwidth (courtesy National Semiconductors Ltd)

1 Read $\overline{e_R}$ from Fig. 9.4 at 10 kΩ; the value is 13 nV/√Hz for 1 Hz bandwidth.
2 Read $\overline{e_n}$ from Fig. 9.2 at 1 kHz; the value is 9.5 nV/√Hz.
3 Read $\overline{i_n}$ from Fig. 9.2 at 1 kHz; the value is 0.65 pA/√Hz, then multiply by 10 kΩ to obtain 6.5 nV/√Hz.
4 Square each term individually and substitute into equation (9.3), add all terms and take the square root of the result:

$$\overline{e_N} = \sqrt{\overline{e_n^2} + \overline{e_R^2} + \overline{i_n^2} R_s^2}$$
$$= \sqrt{(9.5)^2 + (13)^2 + (6.5)^2} = \sqrt{301.5}\ \text{nV}/\sqrt{\text{Hz}}$$

which slightly differs from the result of Example 1 due to reading inaccuracies in the graph.

* * *

The result is the total RMS noise voltage at the input in 1 Hz bandwidth at a frequency of 1 kHz. If the total noise voltage is desired in a given bandwidth one must integrate the noise over the specified bandwidth. This is most easily performed in a noise measurement set-up, but may be approximated as follows:

1 If the frequency range of interest is in the flat band between 1 kHz and 10 kHz in Fig. 9.2 it is simply a matter of multiplying $\overline{e_N}$ by the square root of the bandwidth. In this case with the specified bandwidth range, the total noise voltage is

$$\overline{e_N} = 17.4\sqrt{9000} = 1.65\ \mu V$$

2 If the frequency band of interest is not in the flat band of Fig. 9.2, the band must be broken into sections, calculating the noise in each section, then squaring and multiplying by section bandwidth, summing all sections and finally taking the square root of the sum. This complete procedure is expressed in the following equation where i denotes the total number of sub-blocks.

$$\overline{e_N} = \sqrt{\overline{e_R^2} B + \sum (\overline{e_n^2} + \overline{i^2} R_{gen}^2)_i B_i} \tag{9.16}$$

For most purposes a sub-block of one or two octaves may be sufficient. This procedure is demonstrated in the next example.

* * *

Example 9.3

Determine the RMS noise level in the frequency band from 50 Hz to 10 kHz for the amplifier of Fig. 9.2, operating with a voltage source with an impedance of R_{gen} = 2 kΩ. The ambient temperature is 300 K.

The easiest way to answer this question is to construct a table.

1 Read $\overline{e_R}$ from Fig. 9.4 at 2 kΩ. This gives 6 nV/$\sqrt{\text{Hz}}$; square this and multiply by the entire bandwidth. Write the result (358 200 nV^2) in the last column of Table 9.1.
2 Read the median value of $\overline{e_n}$ in a relatively small frequency band, say 50–100 Hz, from Fig. 9.2. This gives 17.5 nV/$\sqrt{\text{Hz}}$; square this and write the result (306 nV^2) in the table. For the median read the value for frequency 75 Hz.
3 Read the median value of $\overline{i_n}$ in the 50–100 Hz band from Fig. 9.2 (0.85 pA/$\sqrt{\text{Hz}}$), multiply by R_{gen} = 2 kΩ square the result ($1.7^2 = 3$ nV^2/Hz) and enter it in the table.
4 Sum the squared results from steps 2 and 3, multiply by the bandwidth and enter the result (309 × 50 = 15 450 nV^2) in the table.
5 Repeat steps 2 to 4 for frequency bands of 100–300 Hz, 300–1000 Hz, and 1–10 kHz. Note that the median values are taken for frequencies 200, 650 and 5500 Hz respectively.

Table 9.1 Noise calculations for Example 9.3

Frequency band (Hz)	*Bandwidth B* (Hz)	$\overline{e_R^2}$ (nV)^2/Hz	$\overline{e_n^2}$ (nV)^2/Hz	$\overline{i_n^2} R_{gen}^2$ (nV)^2/Hz	*Sum × B* (nV)^2
50–10 000	9950	$6.0^2 = 36$			36 × 9950 = 358 200
50–100	50		$17.5^2 = 306$	1.7^2	309 × 50 = 15 450
100–300	200		$12.5^2 = 156$	1.5^2	158 × 200 = 31 600
300–1000	700		$10.0^2 = 100$	1.3^2	102 × 700 = 71 400
1000–10 000	9000		$9.0^2 = 81$	1.2^2	82 × 9000 = 738 000
				$\overline{e_N^2}$	1 214 650

6 Sum the last column of the table and take the square root to obtain the total RMS noise $\overline{e_N}$ in the 50 Hz–10 kHz band range. The total noise voltage is found to be 1.1 μV.

* * *

As you see in this example the contribution of the noise current is very modest because of the relatively small value of R_{gen}. When the total noise voltage is calculated it is very easy to determine the NF and the signal-to-noise ratio.

9.5 CALCULATING THE NOISE FIGURE AND THE SIGNAL-TO-NOISE RATIO

The noise figure and the signal-to-noise ratio can easily be calculated once the total RMS noise voltage in the required frequency band is determined.

* * *

Example 9.4

Applying equation (9.7) for the noise figure and substituting the values of Example 9.3, we find

$$\begin{aligned} NF &= 10\log\frac{\overline{e_N^2}}{e_R^2} \\ &= 10\log\frac{1\,214\,650}{358\,200} \\ &= 10\log 3.9 = 5.3\text{ dB} \end{aligned}$$

which thus expresses a power ratio between the total present noise power and the noise power developed by the signal source with $R_{gen} = 2\text{ k}\Omega$ in the frequency band of 50 Hz–10 kHz.

* * *

Example 9.5

Determine the S/N for an RMS $V_{sig} = 4$ mV connected to the input of the amplifier of the previous example for the specified bandwidth and $R_{gen} = 2\text{ k}\Omega$.

The signal-to-noise ratio S/N is defined as the logarithm of the ratio of the signal source power and the total noise power referenced across the same resistance:

$$S/N = 10\log\frac{V_{sig}^2}{e_N^2} \qquad (9.17)$$

or

$$\text{S/N} = 20 \log \frac{V_{\text{sig}}}{\bar{e}_{\text{N}}} \tag{9.18}$$

Applying equation (9.18) and substituting the calculated value for $\overline{e_{\text{N}}} = 1.1\ \mu\text{V}$ and $V_{\text{sig}} = 4$ mV yields

$$\begin{aligned} \text{S/N} &= 20 \log \frac{4 \times 10^{-3}}{1.1 \times 10^{-6}} = 20 \log 3.64 \times 10^3 \\ &= 20[\log 10^3 + \log 3.64] \\ &= 20 \times [3 + 0.56] = 71.2 \text{ dB} \end{aligned}$$

* * *

It is also possible to plot NF versus frequency at various values of R_{gen} for any given plot of $\overline{e_{\text{n}}}$ and $\overline{i_{\text{n}}}$. However, there is no all-purpose conversion plot relating NF, $\overline{e_{\text{n}}}$, $\overline{i_{\text{n}}}$, R_{gen} and frequency f. If either $\overline{e_{\text{n}}}$ or $\overline{i_{\text{n}}}$ is neglected, a reference chart can be constructed. An example of such a chart is given in Fig. 9.5 when only $\overline{e_{\text{n}}}$ is considered. It is applicable for most op amps if R_{gen} is less than about 200 ohms and of course for FETs at any R_{gen}, because no significant $\overline{i_{\text{n}}}$ exists for FETs. However, the actual NF for op amps with $R_{\text{gen}} > 200$ ohms is higher than is indicated on the chart.

The graph of Fig. 9.5 can be used to find NF if $\overline{e_{\text{n}}}$ and R_{gen} are known, or to

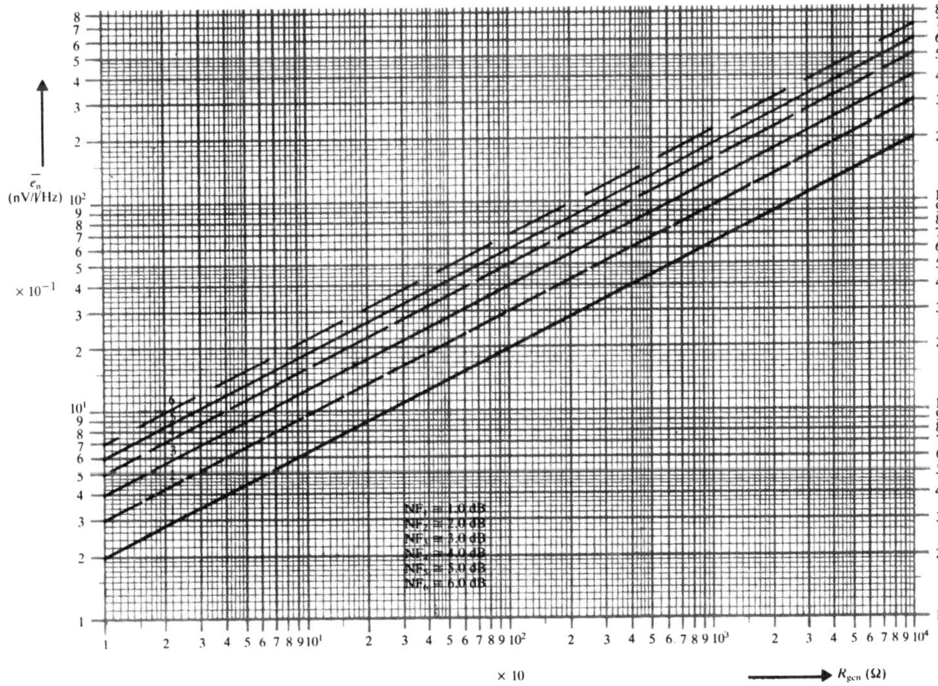

Figure 9.5 Spot noise NF versus R_{gen} when considering only $\overline{e_{\text{n}}}$ and $\overline{e_{\text{R}}}$, and $\overline{i_{\text{n}}} R_{\text{gen}}$ may be neglected (courtesy National Semiconductors Ltd)

find $\overline{e_n}$ if NF and R_{gen} are known. Another possibility is to find the maximum allowed R_{gen} for a given max NF when $\overline{e_n}$ is known. But keep in mind that these values are only valid on the conditions that $\overline{i_n}$ is negligible, at the specific frequency of interest for the NF and $\overline{e_n}$, and a bandwidth of 1 Hz. If a deviating bandwidth is involved the plot is valid so long as $\overline{e_n}$ is multiplied by $\sqrt{B}$.

* * *

Example 9.6

Determine the RMS noise voltage $\overline{e_n}$ for a bandwidth of 9000 Hz with R_{gen} = 1000 Ω, the NF is 1 dB and the chart of Fig. 9.5 is applicable.

1 Read at $R_{gen} = 1000\ \Omega$ on the graph of 1 dB; $\overline{e_n} = 1.9\ \text{nV}/\sqrt{\text{Hz}}$
2 Multiply $\overline{e_n} = 1.9\ \text{nV}/\sqrt{\text{Hz}}$ by $\sqrt{9000}$
3 Then the RMS noise voltage in the specified frequency band is $1.9 \times \sqrt{9000} = 185\ \text{nV} = 0.185\ \mu\text{V}$.

* * *

In the following section we will discuss the fact that a low NF does not always mean a low noise.

9.6 THE NOISE FIGURE MYTH

The noise figure is easy to calculate because the signal level of the source need not be known because V_{sig}, denoted by S_{in} in equation (9.6), drops out. Because NF is so easy to handle you might think that NF is important, but never lose sight of the fact that the S/N_{out} is what remains most important in the final analysis, be it in audio, video, or digital data systems. You can in fact choose a high R_{gen} to reduce NF to almost zero if $\overline{i_n}$ is very small.

In this case $\overline{e_R}$ will be the major source of noise, overshadowing $\overline{e_n}$ completely. The result will be a very low NF but a very low S/N as well because of the very high noise generated by R_{gen}. So don't be misled into thinking that a low NF means low noise.

A maximum S/N is only achieved when $R_{gen} = 0$. As you know already, $\overline{e_R} = \sqrt{(4kTB)}$ is strongly temperature dependent, and one way to reduce $\overline{e_R}$ considerably is to cool the signal source with liquid helium, which has a temperature of 4 K. This may be costly but in satellite land-communication receivers this technique is applied very often to improve the signal-to-noise ratio. Suppose the distance to Uranus has to be bridged, which equals a distance of about 3×10^9 km, then a signal has to travel with the speed of light for $t = 3 \times 10^9/3 \times 10^5 = 10^4$ seconds which equals 2.78 hours before it will reach Earth. Generally, such a distance and time delay will cause a large reduction in the signal-to-noise ratio.

9.6.1 The optimum source resistance

Another term worth considering is the optimum source resistance R_{opt}, which is defined as the value of R_{gen} which produces the lowest NF in a given system at a given frequency. This can be described by the equation

$$R_{opt} = \frac{\overline{e_n}}{\overline{i_n}} \tag{9.19}$$

This can be proven by differentiating equation (9.8) with respect to R_{gen} and equating it to zero. Note again that this does not mean the lowest noise. The derivation is as follows. First conversion of equation (9.8) into a natural logarithm (ln) gives

$$\text{NF} = \frac{10}{\ln 10} \ln \left[\frac{\overline{e_R^2} + \overline{e_n^2} + \overline{i_n^2} R_{gen}^2}{\overline{e_R^2}} \right] \tag{9.20}$$

Put $R_{gen} = R$, note that equation (9.20) is like

$$\partial\text{NF} = 4.34 \ln \left(\frac{T}{N} \right) \tag{9.21}$$

and with $\overline{e_R^2} = 4kTBR$ and differentiating equation (9.21) with respect to R gives

$$\begin{aligned} \partial\text{NF} &= 4.34 \frac{1}{T/N} \text{d}(T/N) \\ &= 4.34 \frac{1}{T/N} \frac{N\text{dT} - T\text{d}N}{N^2} \end{aligned}$$

This yields

$$\frac{\text{dNF}}{\text{d}R} = 4.34 \frac{1}{TN} [\overline{e_R^2}[4kTB + 2R\overline{i_n^2}] - [\overline{e_R^2} + \overline{e_n^2} + \overline{i_n^2} R^2]4kTB] \tag{9.22}$$

For a maximum or minimum value this equals zero, yielding

$$\begin{aligned} 4kTBR[4kTB + 2R\overline{i_n^2}] &= [\overline{e_R^2} + \overline{e_n^2} + \overline{i_n^2} R^2]4kTB \\ 2R^2\overline{i_n^2} &= \overline{e_n^2} + \overline{i_n^2} R^2 \\ R^2\overline{i_n^2} &= \overline{e_n^2} \end{aligned}$$

and thus

$$R_{opt}^2 = \overline{e_n^2}/\overline{i_n^2}$$

or

$$R_{opt} = \frac{\overline{e_n}}{\overline{i_n}} \tag{9.23}$$

If this value for optimum source resistance is substituted in equation (9.9) an expression for optimum NF is found (see Problem 9.7). In general, the noise figure is a function of source resistance and operating frequency and plotting this for

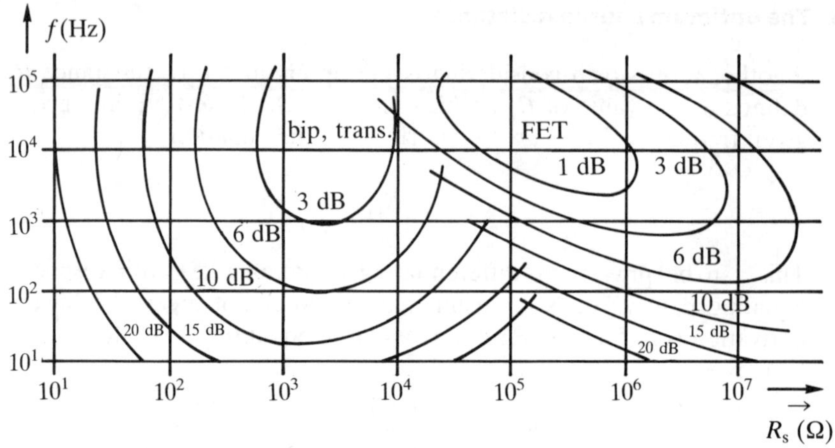

Figure 9.6 Constant noise figure contours for a bipolar transistor and junction field-effect transistor (courtesy National Semiconductors Ltd)

constant noise figure the so-called noise figure contours are obtained. This is illustrated in Fig. 9.6 where for comparison the bipolar and the juction field-effect transistor are depicted in the same graph. It is interesting to note that a junction field-effect transistor allows greater source resistance for the same NF. It is striking and not well understood that the MOSFET shows a much worse noise behavior.

* * *

Example 9.7

Determine R_{gen} for the lowest NF at a frequency of 600 Hz when the op amp of Fig. 9.2 is applicable.

1 Read from the chart at 600 Hz $\overline{e_n} = 10\ \text{nV}/\sqrt{\text{Hz}}$
2 Read from the chart at 600 Hz $\overline{i_n} = 0.66\ \text{pA}/\sqrt{\text{Hz}}$
3 Substitute these values in equation (9.23); the result is $R_{opt} = 15.15\ \text{k}\Omega$.

Now compare this result with Fig. 9.3, then you see the following values for $\overline{e_N}$:

1 $\overline{e_N} = 20\ \text{nV}/\sqrt{\text{Hz}}$ for R_{gen} of about 15 kΩ and
2 $\overline{e_N} = 10\ \text{nV}/\sqrt{\text{Hz}}$ for R_{gen} in the range of 0–100 Ω.

* * *

From this example a most important conclusion can be drawn: applying R_{opt} does not mean lowest noise *per se*, it only means the lowest NF! There is only one condition under which applying R_{opt} means lowest noise. Lowest noise, applying $R_{opt} = R_{gen}$, is guaranteed when $\overline{e_R^2} = mR_{gen}$ and this can only be realized when a type of transformer coupling is applied. This will be investigated in more detail in the following section.

9.6.2 The optimum signal-to-noise ratio

Here we will investigate the conditions for R_{gen} on which a maximum signal-to-noise ratio can be realized.

The signal-to-noise ratio is already defined in equation (9.17) and can be written as

$$\mathrm{S/N} = 10\log\frac{V_{sig}^2}{B(\overline{e_R^2} + \overline{e_n^2} + \overline{i_n^2}R^2)} \tag{9.24}$$

Again conversion to the natural logarithm and differentiating with respect to R gives the following expression for the differential equation:

$$\mathrm{d(S/N)} = \frac{4.34}{TN}(N\mathrm{d}T - T\mathrm{d}N) \tag{9.25}$$

where

$T = V_{sig}^2$
$N = B(\overline{e_R^2} + \overline{e_n^2} + \overline{i_n^2}R^2)$
$R = R_{gen}$ and B is the considered bandwidth
$\overline{e_R^2}$ is the noise power of source for 1 Hz bandwidth

The result for the differential quotient is

$$\frac{\mathrm{d(S/N)}}{\mathrm{d}R} = \frac{4.34}{V_{sig}N}[\overline{e_R^2} + \overline{e_n^2} + \overline{i_n^2}R^2]2\,\mathrm{d}(V_{sig})/\mathrm{d}R - V_{sig}[4kT + 2\overline{i_n^2}R] \tag{9.26}$$

From this equation, an extreme value for R can be derived when a certain relationship between V_{sig} and R is given. Notice that only the right-hand term in equation (9.26) is of interest. This means that as long as this term is larger than zero the S/N increases with R and

$$\frac{\mathrm{d(S/N)}}{\mathrm{d}R} > 0 \tag{9.27}$$

Three conditions for V_{sig} will be considered:

1 $V_{sig}^2 = m^2R$
2 $V_{sig} = mR$
3 $V_{sig} < m\sqrt{R}$

1 $V_{sig}^2 = m^2R$

This condition can be realized when a transformer coupling is involved; then

$$\mathrm{d}(V_{sig})/\mathrm{d}R = \frac{m}{2\sqrt{R}} \tag{9.28}$$

Considering only the right-hand term in equation (9.26) and substituting the given relationship for V_{sig} yields

$$\frac{2m}{2\sqrt{R}}[\overline{e_R^2} + \overline{e_n^2} + \overline{i_n^2}R^2] > m\sqrt{R}[4kT + 2\overline{i_n^2}R] \tag{9.29}$$

Rearranging this equation gives

$$\overline{e_R^2} + \overline{e_n^2} + \overline{i_n^2}R^2 > 4kTR + 2\overline{i_n^2}R^2 \tag{9.30}$$

Note again that $4kTR = \overline{e_R^2}$, thus

$$\overline{e_n^2} > \overline{i_n^2}R^2$$

or

$$R < \overline{e_n}/\overline{i_n} = R_{opt} \tag{9.31}$$

Thus the conclusion appears to be that as long as $R < R_{opt}$ the S/N ratio will increase with increasing R_{gen} on the condition that $V_{sig} = m^2R$.

2 $V_{sig} = mR$
In this case the signal level is proportional to R_{gen} and

$$d(V_{sig})/dR = m$$

Again following the same philosophy and substituting now $V_{sig} = mR$ we find for the right-hand term of equation (9.26)

$$2m[\overline{e_R^2} + \overline{e_n^2} + \overline{i_n^2}R^2] > mR[4kT + 2\overline{i_n^2}R] \tag{9.32}$$

Rearranging this equation gives

$$\overline{e_R^2} + \overline{e_n^2} > 0 \tag{9.33}$$

Thus on the condition that $V_{sig} = mR$ the S/N increases with R_{gen} for any amplifier because equation (9.33) holds, and it makes sense to apply the highest possible practical value for R_{gen}. However this condition cannot always be fulfilled.

3 $V_{sig} < m\sqrt{R}$
Here also an optimum may be determined. If for instance $V_{sig} = mR^{0.4}$ then $dV_{sig}/dR = 0.4mR^{-0.6}$. Substituting in equation (9.26) the given value for V_{sig} gives

$$0.8m/R^{0.6}[\overline{e_R^2} + \overline{e_n^2} + \overline{i_n^2}R^2] > mR^{0.4}[4kT + 2\overline{i_n^2}R]$$

$$0.8[\overline{e_R^2} + \overline{e_n^2} + \overline{i_n^2}R^2] > 4kTR + 2\overline{i_n^2}R \tag{9.34}$$

$$0.8\overline{e_n^2} > 0.2\overline{e_R^2} + 1.2\overline{i_n^2}R^2 \tag{9.35}$$

The conclusion appears to be that the S/N will increase with increasing value of R_{gen} as long as equation (9.35) holds, thus until

$$\overline{e_n^2} = 0.25\overline{e_R^2} + 1.5\overline{i_n^2}R^2 \tag{9.36}$$

Summarizing this all, it does not make sense to change the R_{gen} of an existing signal source in an attempt to make $R_{gen} = R_{opt}$. In particular, do not add series resistance to a source for this purpose. This only makes sense when a transformer coupling is applied manipulating the turns ratio or to redesign the R_{gen} to operate with pre-amps where R_{opt} is known.

It does make sense to increase the designed resistance of signal sources to match or to exceed the value of R_{opt} so long as the signal voltage increases with R_{gen}. We will illustrate this with an example.

* * *

Example 9.8

An amplifier is required to boost a signal coming from a transducer with a resistive impedance $R_{gen} = 10\ \text{k}\Omega$. The amplifier requirements and specifications are respectively:

gain $A_v = 100$
frequency range 10 Hz–10 kHz
noise voltage $\overline{e_n} = 12\ \text{nV}/\sqrt{\text{Hz}}$ at 1 kHz
noise current $\overline{i_n} = 0.01\ \text{pA}/\sqrt{\text{Hz}}$ at 1 kHz

The noise characteristics of the amplifier may be considered linear. A non-inverting amplifier configuration will be used.

Determine R_{opt} and calculate the NF and the total noise at the input of the amplifier when the transducer is connected to the input without affecting the gain and the noise of the feed-back resistors may be neglected, in the following two cases:

(a) The optimum source resistance is installed.
(b) The transducer is directly connected to the amplifier.

The optimum source resistance at 1 kHz can be calculated by applying equation (9.19) and substituting the given values:

$$R_{opt} = \frac{\overline{e_n}}{\overline{i_n}} = \frac{12 \times 10^{-9}}{0.01 \times 10^{-12}} = 1200 \times 10^3 = 1.2\ \text{M}\Omega$$

In case (a) the circuit involved is depicted in Fig. 9.7. The total noise $\overline{e_N}$ at the

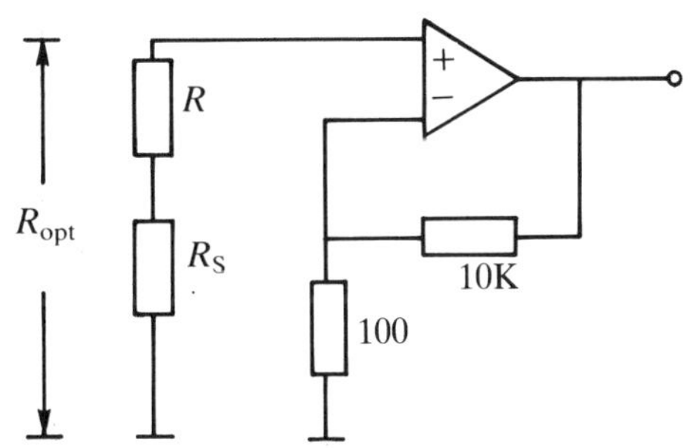

Figure 9.7 Configuration where the optimum source resistance is implemented

input can be found applying equation (9.3), replacing R by R_{opt}, and taking the square root of equation (9.37). This is executed in Table 9.2.

1 Read $\overline{e_R} = 130$ nV/$\sqrt{}$Hz from Fig. 9.4, square and put the result into Table 9.2.
2 Given the value $\overline{e_n} = 12$ nV/$\sqrt{}$Hz, square it and enter into the table.
3 Given the value $\overline{i_n} = 0.01$ pA/$\sqrt{}$Hz, multiply by $R_{opt} = 1200$ kΩ, square the result and enter into the table.
4 Sum the squared results from steps 1, 2 and 3 and multiply the sum by $f = B = 9990$ Hz.
5 Take the square root of the sum; this gives the total noise voltage $\overline{e_N} = 13.1$ µV for the specified bandwidth.
6 The NF can be calculated from equation (9.7):

$$\text{NF} = 10\log\frac{\overline{e_N^2}}{\overline{e_R^2}} = 10\log\frac{171\,708\,120}{168\,831\,120} = 10\log 1.02$$

$$= 0.08 \text{ dB}$$

which is very low indeed.

Table 9.2 Noise voltage calculation for R_{opt}

Frequency bandwidth B (Hz)	$\overline{e_R^2}$ (nV²/Hz)	$\overline{e_n^2}$ (nV²/Hz)	$\overline{i_n^2}R^2$ (nV²/Hz)	*Sum × B* (nV²)
9990	16 900	$(12)^2 = 144$	$(0.01 \times 1200)^2 = 144$	2 877 120
				168 831 000
				171 708 120

In case (b) the current circuit when the transducer is directly connected to the input of the non-inverting amplifier is depicted in Fig. 9.8.

1 Follow the same procedure as in case (a), but now substitute the transducer impedance R only. From Fig. 9.4 $\overline{e_R} = 13.5$ nV/$\sqrt{}$Hz.
The respective values are depicted in Table 9.3.

Table 9.3 Noise performance calculations for the source directly connected to the amplifier

Frequency bandwidth B (Hz)	$\overline{e_R^2}$ (nV²/Hz)	$\overline{e_n^2}$ (nV²/Hz)	$\overline{i_n^2}$ (nV²/Hz)	*Sum × B* (nV²)
9990	$(13.5)^2 = 182.25$	$(12)^2 = 144$	$(0.01 \times 10)^2 = 0.01$	1 438 659
				1 820 678
				3 259 337

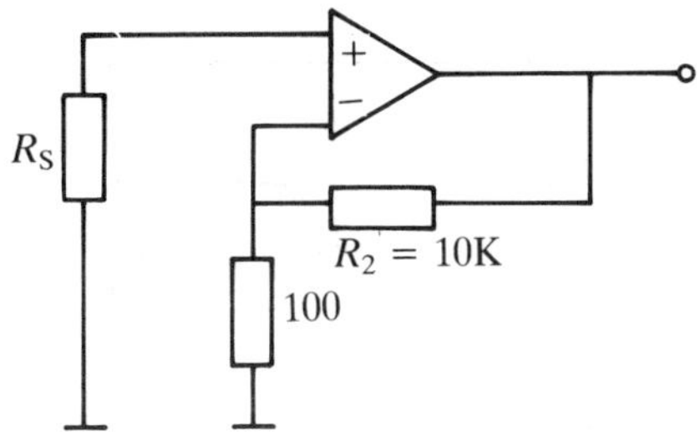

Figure 9.8 Configuration when the transducer is connected directly to the amplifier

2 Taking the square root of the sum find $\overline{e_N}$ = 1805 nV = 1.8 μV, which is considerably smaller than in the case of $R = R_{opt}$.
3 Calculating the NF gives

$$\text{NF} = 10\log\frac{3\,259\,337}{1\,820\,678} = 10\log 1.79$$
$$= 2.5 \text{ dB}$$

Note that it is not necessary to have multiplied with B to find the NF. for comparison the results are repeated below:

Case (a) $\overline{e_N}$ = 13.1 μV NF = 0.08 dB
Case (b) $\overline{e_N}$ = 1.8 μV NF = 2.5 dB

* * *

It will be clear that noise figure is only a measurement of the amplifier noise with respect to the source noise. From this we can draw a very important conclusion. *Never* add a series resistance with the source to improve the NF. The noise figure will improve but the noise performance is degraded considerably. So live with the existing R_{gen} or apply a transformer coupling to make sense.

In the following example it will be shown that for an op amp with ideal feedback the equivalent input noise is independent of gain.

* * *

Example 9.9

Determine the equivalent input noise per unit bandwidth for the feedback configuration shown in Fig. 9.9. The following specifications are known. The amplifier operates at 1 kHz from a source with a source resistance of 1 kΩ. The values of the feedback resistors are R_1 = 100 kΩ and R_2 = 1 kΩ. Fig. 9.2 represents the noise specifications of the amplifer.

Here the thermal noise generated by the feedback resistors has to be taken into account, which can be best performed by moving outside the feedback loop the

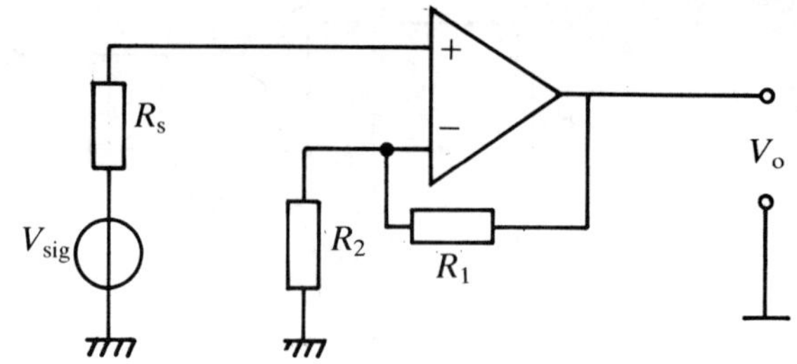

Figure 9.9 Inverting amplifier configuration with ideal feedback

noise influence of the parallel connected resistors R_1 and R_2. In addition to this, the noise generated by $\overline{i_n} \times (R_1//R_2)$ must be added even though the minus input is a virtual ground.

The easiest way to include these effects is to consider $R_1//R_2$ in series with R_S. We then get the equivalent configuration shown in Fig. 9.10 and we construct a table as follows

1 Read $\overline{e_R}$ at 1 kΩ from Fig. 9.4 (4.55), square it, and enter the result in Table 9.4.
2 Read $\overline{e_{R_2//R_1}}$ at 1 kΩ from Fig. 9.4 (4.55), square it and enter the result in the table.
3 Read $\overline{e_n}$ from Fig. 9.2 (9.5), square it and enter the result in the table.
4 Read $\overline{i_n}$ from Fig. 9.2 (0.65), multiply by $(R_s + R_1//R_2)$, square the result and enter in the table.
5 Sum all entries, take the square root to obtain the total RMS noise at the input as $\overline{e_N} = 11.5$ nV/$\sqrt{}$Hz at 1 kHz in a 1 Hz bandwidth.

* * *

As you see, the gain is not involved in the total noise calculation at the input of the amplifier.

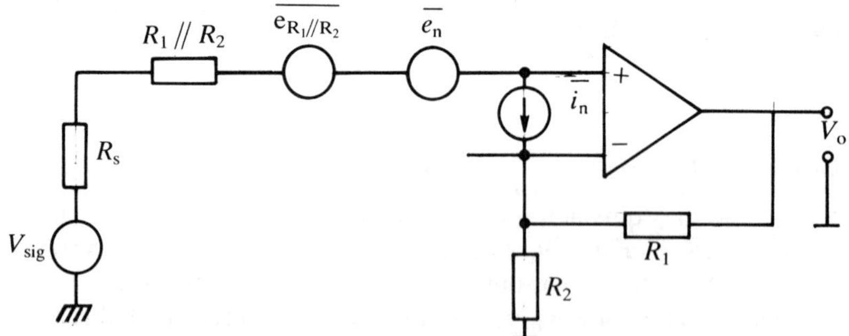

Figure 9.10 Equivalent noise circuit with noise generators moved outside the feedback loop for Fig. 9.9

Table 9.4 Noise calculations for Example 9.9

Source	$(\mathrm{nV}/\sqrt{\mathrm{Hz}})^2$	$(\mathrm{nV})^2/\mathrm{Hz}$
$\overline{e_R^2}$	$(4.55)^2$	20.7
$\overline{e_{R_1//R_2}^2}$	$(4.55)^2$	20.7
$\overline{e_n^2}$	$(9.5)^2$	90.2
$\overline{i_n^2}(R_S + R_1//R_2)^2$	$(0.65 \times 2.0\mathrm{k})^2$	1.69
Total $\overline{e_N^2}$		133.29

9.7 OTHER NOISE CALCULATING TECHNIQUES

We will conclude the chapter with three more examples to demonstrate some other techniques of noise calculations.

* * *

Example 9.10

Determine the effective (RMS) noise voltage and noise current for a resistor $R = 1\ \Omega$ at an ambient temperature of 1 K in a bandwidth of 0–10 kHz.

The graph of Fig. 9.4 will fail to solve this problem because it is constructed for an ambient temperature of 300 K. This means we have to apply the basic formula (9.10). Entering the given values yields

$$P = 4kTB = \frac{\overline{e_R^2}}{R}$$

$$\overline{e_R^2} = 4kTBR = 4 \times 1.38 \times 10^{-23} \times 1 \times 10^4$$
$$= 5.52 \times 10^{-19}\ (\mathrm{V})^2$$

and the equivalent RMS noise voltage is

$$\overline{e_N} = \sqrt{(5.52 \times 10^{-19})} = 0.74\ \mathrm{nV}$$

The RMS noise current is calculated as follows:

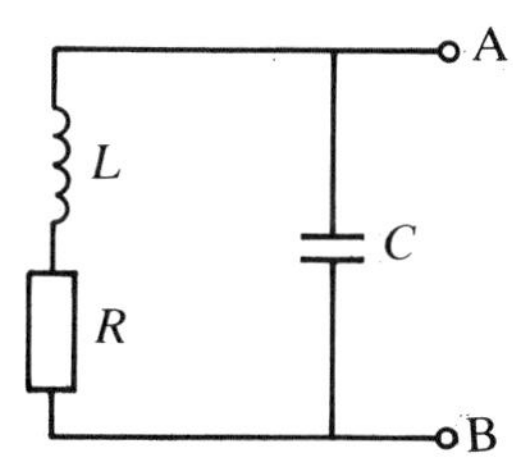

Figure 9.11 Circuit for Example 9.11

$$\overline{i_R^2} = \frac{4kTB}{R} = \frac{4 \times 1.38 \times 10^{-23} \times 10^4}{1}$$
$$= 5.52 \times 10^{-19}\ (\mathrm{A})^2$$

The equivalent RMS noise current is

$$\overline{i_R} = 0.74\ \mathrm{nA}$$

* * *

Example 9.11

Determine the equivalent noise voltage source e_{AB} between the points A and B of the circuit in Fig. 9.11.

In this configuration only one noise voltage source can be distinguished, i.e., R, because, as is known, an ideal coil or capacitor does not generate any noise contribution. Keep in mind that, here again, power densities must be considered. Thus we first calculate

$$\overline{e_{AB}^2} = \left|\frac{Z_1}{Z_1 + Z_2}\right|^2 \overline{e_R^2} \tag{9.37}$$

where

$$Z_1 = 1/j\omega C \text{ and } Z_2 = R + j\omega L$$

and again

$$\overline{e_R^2} = 4kTR$$

Entering these quantities yields

$$\overline{e_{AB}^2} = \left|\frac{1/j\omega C}{1/j\omega C + R + j\omega L}\right|^2 \overline{e_R^2} \tag{9.38}$$

Multiplying by $j\omega C$,

$$\overline{e_{AB}^2} = \frac{1}{|1 - \omega^2 LC + j\omega RC|^2} \overline{e_R^2} \tag{9.39}$$

or

$$\overline{e_{AB}^2} = \frac{1}{\{1 - \omega^2 LC\}^2 + \{\omega RC\}^2} \overline{e_R^2} \tag{9.40}$$

and finally taking the square root of equation (9.40) gives the required result. Note that the left-hand factor of $\overline{e_R^2}$ is dimensionless.

* * *

Example 9.12

Determine the equivalent noise voltage source $\overline{e_{N12}}$ for the circuit in Fig. 9.12.

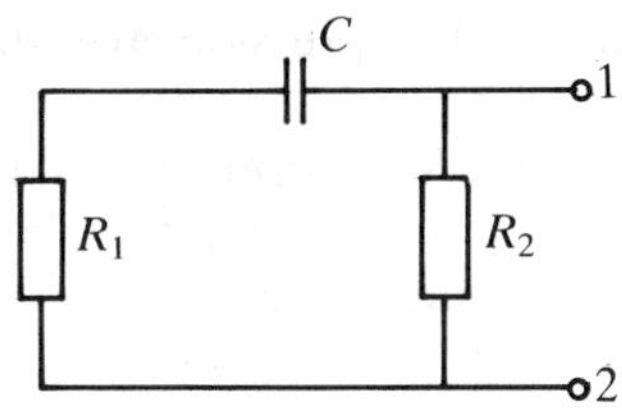

Figure 9.12 Circuit for Example 9.12

The easiest way to solve this problem is to apply the superposition theorem, taking into account both noise sources separately. This is allowed when it is assumed that there is no correlation of sources.

The two noise sources are respectively

$$\begin{aligned} \overline{e_{R1}^2} &= 4kTR_1 \\ \overline{e_{R2}^2} &= 4kTR_2 \end{aligned} \tag{9.41}$$

To see more clearly what is happening in the calculation the equivalent noise circuit can be split into two circuits just as the superposition theorem dictates. This is shown in Fig. 9.13.

In Fig. 9.13a the equivalent noise voltage source $\overline{e_{R1}^2}$ generates a noise power $\overline{e_{12a}^2}$ which equals

$$\overline{e_{12a}^2} = \frac{R_2^2}{|R_1 + R_2 + 1/j\omega C|^2}\,\overline{e_{R1}^2} \tag{9.42}$$

In the same fashion, for the circuit of Fig. 9.13b the equivalent noise power at points 1, 2 is

$$\overline{e_{12b}^2} = \frac{|1/j\omega C + R_1|^2}{|R_1 + R_2 + 1/j\omega C|^2}\,\overline{e_{R2}^2} \tag{9.43}$$

Summing both expressions gives the required result, thus

$$\overline{e_N^2} = \overline{e_{12a}^2} + \overline{e_{12b}^2} \tag{9.44}$$

$$= \overline{e_{R1}^2}\,\frac{(\omega R_2 C)^2}{1 + [(R_1 + R_2)\omega C]^2} + \overline{e_{R2}^2}\,\frac{1 + (\omega R_1 C)^2}{1 + [(R_1 + R_2)\omega C]^2}. \tag{9.45}$$

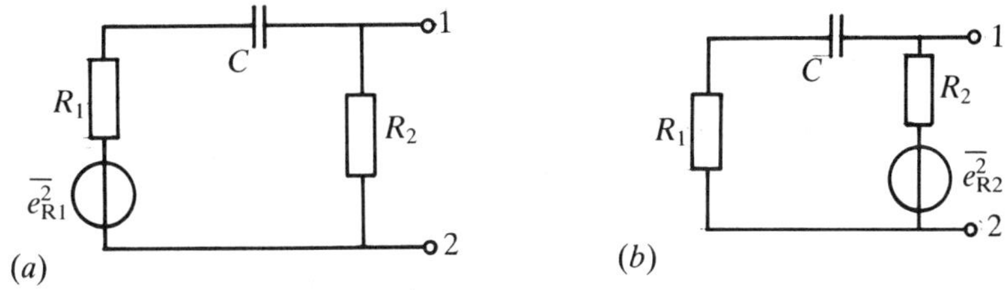

Figure 9.13 Equivalent noise circuit applying the superposition theorem to Fig. 9.12

Taking the square root of this expression gives the equivalent RMS noise voltage $\overline{e_N}$ between the points 1 and 2.

Note that in Fig. 9.13a, R_2 determines finally the fraction of noise power which will appear between the points 1, 2 caused by $\overline{e_{R1}^2}$, and in Fig. 9.13b the term $(1/j\omega C + R_1)$ determines the fraction of noise power which will finally appear between the points 1 and 2, generated by $\overline{e_{R2}^2}$.

* * *

This concludes our chapter on noise calculations. In the next chapter we will treat noise in a more physical fashion concerning active and passive components. But as you have seen it is not necessary to understand all the ins and outs of the physical details to perform noise calculations. Normally it suffices to use the noise specifications delivered by the manufacturer and to use these for the required noise calculations.

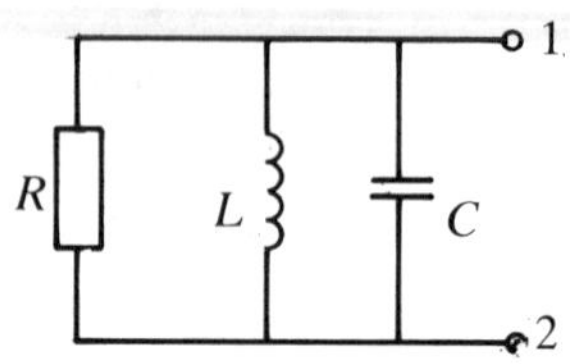

Figure 9.14

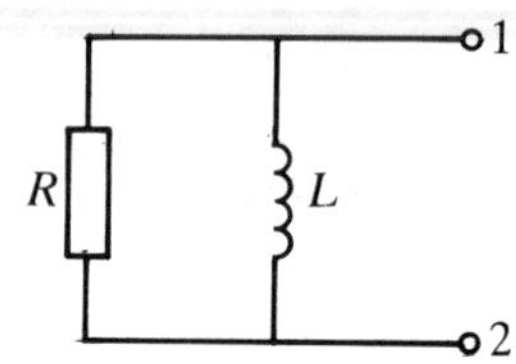

Figure 9.15

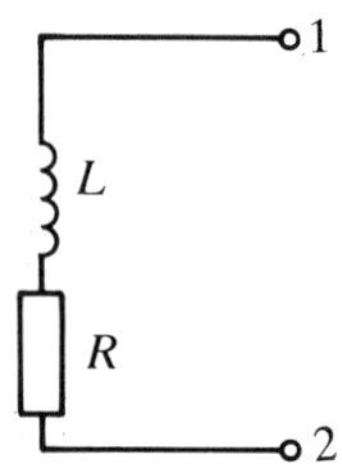

Figure 9.16

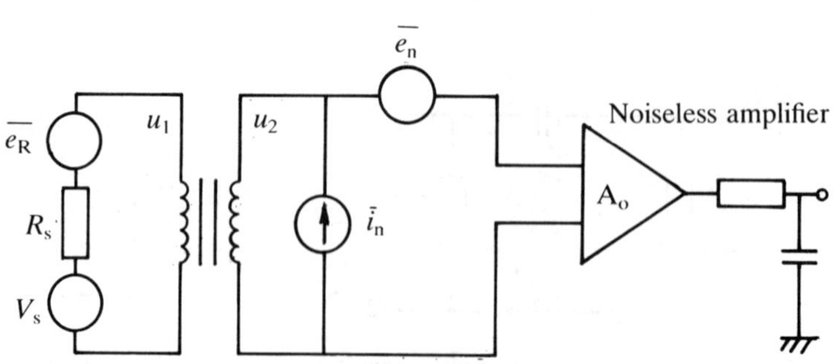

Figure 9.17

9.1 Calculate the noise voltage source and the noise current source for the following resistances in the frequency range as specified:
(a) $R = 1\ \text{M}\Omega$, 100–110 kHz, $T = 300$ K
(b) $R = 100\ \text{k}\Omega$, 10–11 MHz, $T = 300$ K

9.2 Determine the signal-to-noise ratio for an amplifier with $\overline{e_N} = 17.7\ \text{nV}/\sqrt{\text{Hz}}$ in the bandwidth from 1 kHz to 10 kHz where $e_{sig} = 100$ mV.

9.3 Derive an expression for the equivalent noise voltage source $\overline{e_{12}}$ for the circuit depicted in Fig. 9.14 at points 1, 2.

9.4 Derive an expression for the equivalent noise voltage source at the terminals 1, 2 for the circuit depicted in Fig. 9.15.

9.5 Derive an expression for the equivalent noise voltage source at the terminals 1, 2 for the circuit depicted in Fig. 9.16.

9.6 Calculate the shot noise current and the equivalent thermal noise voltage in a diode in a bandwidth of 2 MHz and a temperature of 300 K.

9.7 Derive an expression for the noise figure NF for optimum source resistance.

9.8 Figure 9.17 represents a low-noise pre-amplifier with $\overline{e_n} = 10^{-8}\ \text{V}/\sqrt{\text{Hz}}$ and $\overline{i_n} = 0.1\ \text{pA}/\sqrt{\text{Hz}}$. Connected to it is a voltage source with a source resistance $R_S = 20\ \Omega$ via a matching transformer. The transformer may be presumed to be ideal with a transformer ratio of $a = n_2/n_1$. Find (a) the optimum source resistance, (b) the unmatched noise figure without transformer, (c) the matched noise figure, with matching transformer.

Ten

PHYSICS OF NOISE

10.1 INTRODUCTION

In this chapter, we will continue our discussion of noise by looking at several frequently occurring noise mechanisms. An exhaustive treatment can be found in more theoretical textbooks devoted to noise only. Here we discuss only the most important mechanisms required for the description of noise that are met in commonly used electronic components.

Noise can be defined as a random movement of energy in solid-state material, conductors and components due to the finite temperatures of the molecules of the material. This thermal energy is transported at random by electrons as one of the possible energy carriers, but lattice vibrations can also transport energy at random. For this type of electron transport Ohm's law is no longer valid; the transport of electrons is not subjected to an outside electric field but is caused, for instance, by a thermal activation of the electrons completely at random. This means that the noise mechanism can be described from a statistical point of view and is a stochastic process. Other mechanisms show a certain relationship with frequency, but very often the absolute temperature is one of the main causes involved. Three mechanisms shall be discussed.

10.2 THERMAL NOISE OR JOHNSON NOISE IN CONDUCTORS

One of the most frequent causes of noise in a conductor is the random movement of electrons due to their finite temperature resulting in a certain noise power. This noise power can be described by relation (9.10), and was theoretically derived by Nyquist from a thermodynamic point of view:

$$P = 4kTB \qquad \text{Watts}$$

where again

P = noise power in watts
k = Boltzmann's constant, 1.38×10^{-23} J/K

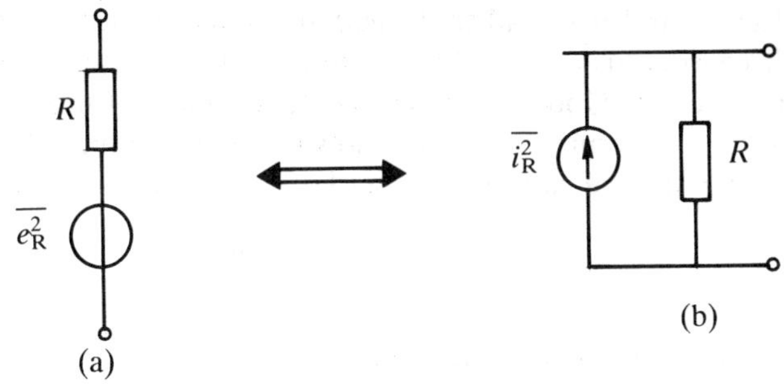

Figure 10.1 Equivalent thermal noise voltage and current source circuit of a resistor

T = absolute temperature in K
B = bandwidth in Hz

In this equation the frequency bandwidth, the absolute temperature and a physical constant are involved. If this expression is divided by the bandwidth B a power density per Hertz is obtained, which at T = 300 K equals

$$P/B = 4kT = 4.14 \times 10^{-21} \text{ W/Hz}$$

As you can see from this equation, it expresses a certain amount of available noise power to which a voltage or current source can be assigned to be connected in series or in parallel with a resistor by the known relationships

$$P = \overline{e_R^2}/R \tag{10.1}$$

$$P = \overline{i_R^2}R \tag{10.2}$$

If this noise power is expressed in an equivalent noise voltage or noise current, the following equations hold:

$$\overline{e_R^2} = PR = 4kTRB \qquad (V^2) \tag{10.3}$$

$$\overline{i_R^2} = P/R = 4kTB/R \qquad (\text{A}^2) \tag{10.4}$$

This is depicted in Figs 10.1 a and b. Note again that mean squares are involved based on statistical considerations. In standardized circumstances an ambient temperature of 300 K is assumed and on this condition the graphs of Fig. 9.4 are expressed in the root mean square (RMS) of the above expressions.

The values are given for a 1 Hz bandwidth which results in dimensions of nV/$\sqrt{}$Hz or pA/$\sqrt{}$Hz. Because there is a linear relationship between noise power and bandwidth, this type of noise is often called *white noise*.

10.3 SHOT NOISE

In many electronic components we have to deal with flows of charge carriers which cross potential barriers independently of each other. The character of this flow is rather discrete and it is said that these flows show a quantum-like character and show a discontinuous current flow behavior. The current as a function of time is a stochastic function. Good examples are the crossing of holes and electrons over a pn-junction or the emission of electrons in vacuum tubes. For a pn-junction this type of flow of charge carriers is described by the following expression:

$$\overline{i_n^2} = 2\,q\,I_D B \tag{10.5}$$

where

$\overline{i_n^2}$ = mean square noise current (A^2)
q = specific charge of an electron (1.6×10^{-19} C)
I_D = the DC current flowing through the junction
B = considered bandwidth in Hz

In the same fashion as we have seen for thermal noise, *shot noise* can be represented as a current *density* as follows:

$$\overline{i_n} = \sqrt{\frac{\overline{i_n^2}}{B}} \tag{10.6}$$

Here again a linear relationship exists between the noise power and the bandwidth, so this type can be assigned as white noise too. The DC current I_D can be found with the well-known diode equation,

$$I_D = I_0\{\exp(qV_D/kT) - 1\} \tag{10.7}$$

where

I_D is the DC current in the pn-junction
I_0 is the saturation current
q is the specific charge of an electron
V_D is the voltage over the pn-junction
k is Boltzmann's constant
T is the absolute temperature

Differentiating this expression yields the differential diode resistance. The result is often denoted by r_d to express its differential character. The following equation expresses the relationship between the differential resistance, r_d, and the current, I_D:

$$r_d = \frac{kT}{qI_D} \approx \frac{26}{I_D(\mathrm{mA})} \tag{10.8}$$

If T equals 300 K then with I_D substituted in mA the rightmost term is valid and is easy to remember.

Rearranging equation (10.8) and substituting I_D in equation (10.5) yields for the noise current

$$\overline{i_n^2} = 2q\frac{kTB}{qr_d} = \frac{4kT}{2r_d}B \tag{10.9}$$

To derive an expression for the equivalent noise voltage $\overline{e_n^2}$, equation (10.9) can first be multiplied by r_d, then we find the total noise power. Multiplying this expression by r_d gives

$$\overline{e_n^2} = 4kTB\tfrac{1}{2}r_d \tag{10.10}$$

The factor ½ in (10.10) is for convenience to get the well-known factor $4kTB$ in our expression.

10.4 FLICKER NOISE

Flicker noise or $1/f$ noise, sometimes called low-frequency noise, excess noise, contact noise, semiconductor noise, or pink noise, can be characterized as caused by a flow of charge carriers in a discontinuous medium. Good examples of these are tubes, carbon composite resistors, diodes, transistors, thermistors, thin film devices and light sources. Flicker noise is roughly inversely proportional to frequency and starts to be of importance at 1 kHz in both junction FETs and bipolar transistors. As can be seen from the following equation the equivalent noise voltage increases as frequency decreases. Flicker noise dominates thermal noise at frequencies below 100 Hz and is related to a flow of direct current.

$$\overline{e_n^2} = KI_D^a(1/f)^\alpha R^2B \tag{10.11}$$

where

- K is a constant and device dependent
- I_D is the direct current
- a is a constant ranging from 0.5 to 2
- α is a constant (see Table 10.1)
- B is the bandwidth
- R is the equivalent resistance through which the current flows

Some authors prefer to describe the noise flicker current, in which case equation (10.11) should be divided by the equivalent source resistance R^2 and $\overline{i_n^2}$ will result.

The constant α can vary between 0.8 and 1.3 depending on the process to be considered. Some processes are mentioned in Table 10.1.

To get an impression of the level of this type of noise, Fig. 10.2 shows the relative level of different types of noise as a function of frequency.

In semiconductors, $1/f$ noise is mainly determined by surface properties, and

Table 10.1 Survey of values for α as a function of source mechanism

$\alpha = 1$	for *pink* noise characterized by an equal power per octave
$\alpha = 2$	*red* noise, which can be characterized by the fluctuation in the earth's rotational speed
$\alpha = 2.7$	for galactic radiation

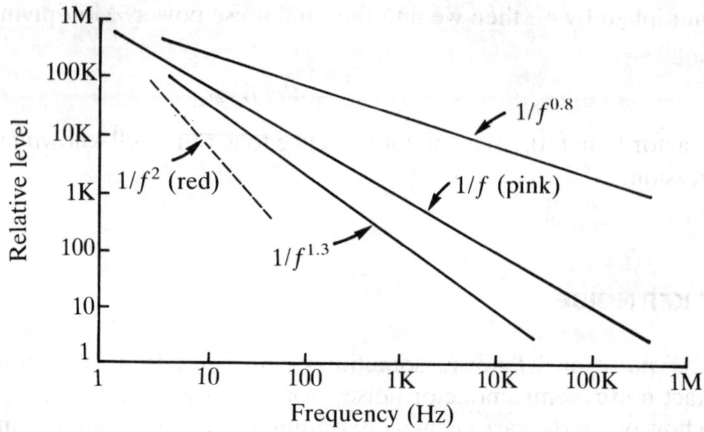

Figure 10.2 Relative noise level as a function of frequency for different noise processes

the generation and recombination of charge carriers in surface states. So far we have discussed different types of physical mechanism of noise source.

In the next section the noise behavior of three semiconductor components, the bipolar, the field-effect transistor and the diode, are discussed, to which all integrated circuits can be related.

10.5 NOISE IN SEMICONDUCTOR COMPONENTS

10.5.1 Noise behavior of the bipolar transistor

As you have already seen, any active element can be characterized with respect to noise by an equivalent noise voltage source and an equivalent noise current source referred to the input. If a bipolar transistor is considered, the following noise sources can be distinguished:

1 The thermal noise source of the internal base resistor r_{bb},

$$\overline{e_b^2} = 4kTr_{bb}B \tag{10.12}$$

2 A shot noise source in the collector current,

$$\overline{i_c^2} = 2qI_cB \tag{10.13}$$

3 A shot noise source in the base – emitter junction,

$$\overline{i_b^2} = 2qI_BB \tag{10.14}$$

4 A flicker noise component $\overline{i_{bf}}$ in the base emitter region. This term becomes important when frequencies are below ≈100 Hz and will be neglected in our calculations of the equivalent noise sources.

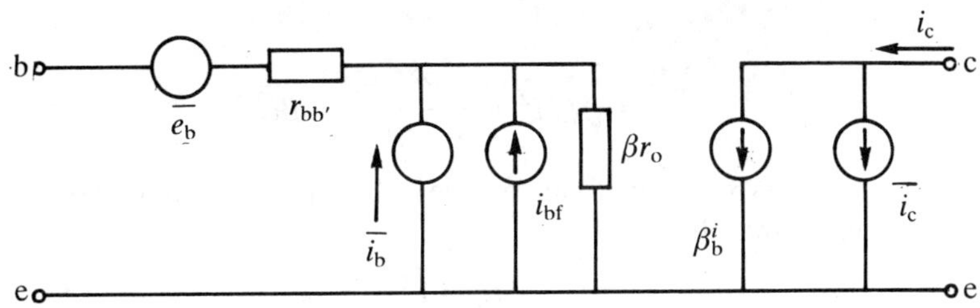

Figure 10.3 Bipolar transistor model with noise sources

These noise sources are depicted in Fig. 10.3, and to find the equivalent noise sources this circuit should be equated to an equivalent noise circuit.

If the correlation between i_c and i_b and the Early effect are neglected, and we further suppose that r_{bb} is much smaller than βr_0, the circuit for the bipolar transistor can be drawn as in Fig. 10.3. In this respect, no correlation means that the noise sources are independent. As said already, this circuit should be equated to an equivalent circuit which is depicted in Fig. 10.4 in which $\overline{e_n}$ and $\overline{i_n}$ are drawn. If both circuits are put equivalent with their respective input and output terminals, the following equations will hold under the circumstances as stated. First, short-circuiting the input terminals of both circuits gives

$$\overline{e_b^2} + r_{bb'}^2\overline{i_b^2} + r_0^2\overline{i_c^2} = \overline{e_n^2} \tag{10.15}$$

Note that $\beta i_b = i_c$ and, further, it is assumed that $r_{bb'} \ll \beta r_0$.

If, in the equivalent circuits, both inputs are opened, then, equating both output currents, this yields the expression

$$\beta^2\overline{i_b^2} + \overline{i_c^2} = \beta^2\overline{i_n^2} \tag{10.16}$$

Because $I_C = \beta I_B$ from equations (10.13) and (10.14) it can be derived that

$$\overline{i_b^2} = \frac{\overline{i_c^2}}{\beta} \tag{10.17}$$

Substituting equation (10.17) in (10.16) yields

$$\beta i_c^2 + i_c^2 = \beta^2\overline{i^2}_n \tag{10.18}$$

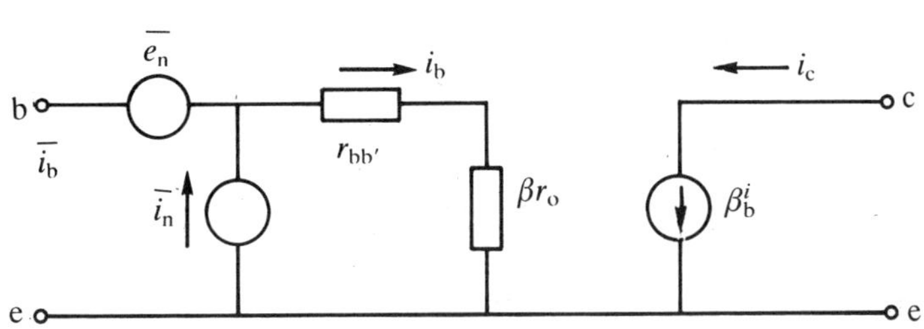

Figure 10.4 The equivalent noise circuit for the bipolar transistor

or if $\beta \gg 1$,

$$\overline{i_c^2} = \frac{\beta^2}{1+\beta}\,\overline{i_n^2} = \beta\overline{i_n^2} \tag{10.19}$$

and from this expression it can be seen that the equivalent noise current source equals approximately

$$\overline{i_n^2} = \overline{i_c^2}/\beta = \overline{i_b^2} \tag{10.20}$$

In accordance with (10.13) and (10.14), in this equation the current $\overline{i_c^2}$ and $\overline{i_b^2}$ can be replaced by the respective DC currents I_B or I_C; this yields the following expression for the equivalent noise current source referred to the input:

$$\overline{i_n^2} = 2qI_B B = 2q\,\frac{I_C}{\beta}\,B \tag{10.21}$$

Finally in this expression, the DC value I_C can be replaced by kT/qr_0 in accordance with equation (10.8), thus

$$\overline{i_n^2} = \frac{4kT}{2\beta r_0}\,B \tag{10.22}$$

If required, $\overline{i_n}$ can be found by taking the square root of (10.22).

In the same way, we can find an expression for the equivalent noise voltage source $\overline{e_n}$. This can be performed by substituting equations (10.12), (10.13) and (10.14) in (10.15) which results in

$$\overline{e_n^2} = 4kTR_{bb'}B + r_{bb'}^2 2qI_B B + r_0^2 2qI_C B \tag{10.23}$$

and substituting the respective DC current values in accordance with equation (10.8),

$$\overline{e_n^2} = (4kTr_{bb'} + \frac{4kT}{2\beta r_0} + \frac{4kT}{2r_0}\,r_0^2)\,B \tag{10.24}$$

and if $2\beta r_0 \gg r_{bb'}$, which in most cases may be presumed, then

$$\overline{e_n^2} = 4kT(r_{bb'} + \tfrac{1}{2}r_0)\,B \tag{10.25}$$

and again, if required the square root of this expression delivers the equivalent noise voltage source $\overline{e_n}$.

The result of this derivation is shown in Fig. 10.5, where a noise-free transistor is drawn with the respective noise sources. For every transistor the manufacturer provides extensive data concerning noise behavior.

10.5.2 Noise behavior of the field-effect transistor

The same procedure can be followed to describe the noise behavior of the FET as in the case of the bipolar transistor. Two circuits are to be drawn, one of which depicts the physical noise sources, and the second represents the circuit with the

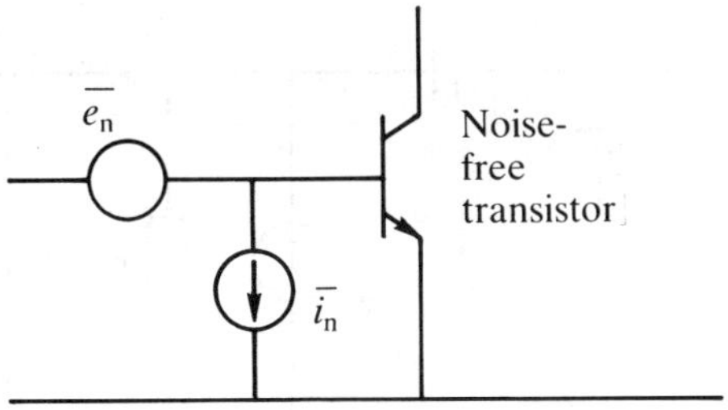

Figure 10.5 The noise-free transistor model with equivalent noise sources

equivalent noise sources. First, the physical noise sources are to be mentioned; mainly these sources are:

1 The thermal noise of the resistive part of the input impedance, which can be expressed in

$$\overline{e_{R1}^2} = 4kTR_1B \qquad (10.26)$$

where R_1 represents the input impedance

2 The thermal channel noise source, which can be expressed in the equation

$$\overline{i_{ch}^2} = 4kT\tau g_m B \qquad (10.27)$$

where τ is a constant related to the operating point
g_m is the transconductance defined as

$$g_m = \frac{\partial I_D}{\partial V_{gs}} = \frac{i_d}{v_{gm}} \qquad (10.28)$$

for a given operating point and $\tau = 1$ below the pinch-off voltage and $\tau = 2/3$ above the pinch-off voltage.

3 The gate noise, which can be considered as shot noise source due to its random behavior of the current flow in the pn-junction, expressed in

$$\overline{i_{gn}^2} = 2qI_G B \qquad (10.29)$$

where I_G is the DC gate current.
In MOSFETs, the gate current is in the order of nA and is usually negligible.

4 For low frequencies, the flicker noise current $\overline{i_{df}}$, generated in the drain section, comes into scope. This addition will also be neglected in the calculation of the equivalent noise sources.

In Fig. 10.6 the three noise sources are depicted: a noise voltage source in series with the input impedance, a noise current source in parallel with the input terminals and two noise current sources in parallel with the output terminals representing the channel noise and the $1/f$ noise.

The equivalent circuit with the equivalent noise sources is depicted in Fig. 10.7. The magnitude of these noise sources can be determined by equating the equivalent input and output terminals of both circuits with respect to voltage and impedance.

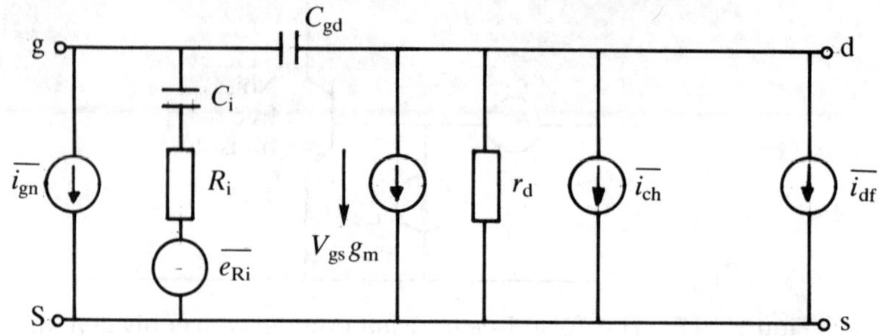

Figure 10.6 The field-effect transistor with noise sources

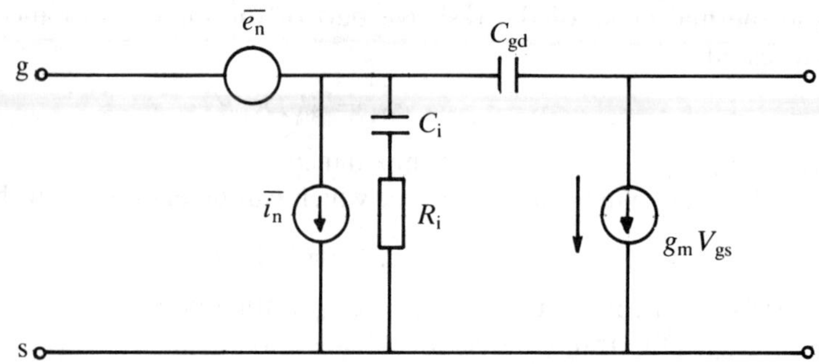

Figure 10.7 The equivalent noise circuit with equivalent noise sources $\overline{i_n}$ and $\overline{e_n}$ of an FET

Suppose in Fig. 10.6 the drain source is not loaded and V_{gs} is put equal to $\sqrt{\overline{e_n^2}}$, then, following Kirchhoff's law and neglecting the influence of C_{gd} for node d, we obtain

$$g_m V_{gs} = \sqrt{\overline{i_{ch}^2}} \tag{10.30}$$

Substituting the assumed value for V_{gs} in (10.30) we find

$$g_m^2 \overline{e_n^2} = \overline{i_{ch}^2} \tag{10.31}$$

Finally, substituting equation (10.27) for $\overline{i_{ch}^2}$ gives

$$g_m^2 \overline{e_n^2} = 4kT\tau g_m B \tag{10.32}$$

and rearranging to find $\overline{e_n^2}$ yields the equivalent noise voltage source,

$$\overline{e_n^2} = \frac{4kT\tau B}{g_m} \tag{10.33}$$

Note that the same expression can be found when the input terminals of both circuits are short-circuited with no load again.

To find the equivalent current source $\overline{i_n}$, V_{gs} for both circuits is calculated. For Fig. 10.6,

$$V_{gs}^2 = \overline{i_{gn}^2} Z_i^2 + \overline{e_{Ri}^2} \qquad (10.34)$$

Note here again that noise power densities are added and that it is assumed that there is no correlation of sources. For Fig. 10.7,

$$V_{gs}^2 = \overline{e_n^2} + \overline{i_n^2} Z_i^2 \qquad (10.35)$$

Equating equations (10.34) and (10.35) yields

$$\overline{i_{gn}^2} Z_i^2 + \overline{e_{Ri}^2} = \overline{e_n^2} + \overline{i_n^2} Z_i^2$$

substituting for $\overline{e_n}$ in equation (10.31) and rearranging to make $\overline{i_n}$ explicit we find

$$g_m^2 Z_i^2 \overline{{}_n^2} = g_m^2 \overline{e_{Ri}^2} + \overline{i_{ch}^2} + g_m^2 Z_i^2 \overline{i_{gn}^2} \qquad (10.37)$$

Note the change in sign for $\overline{i_{ch}^2}$ for reason of noise power densities. Dividing equation (10.37) by $g_m^2 Z_i^2$ gives

$$\overline{i_n^2} = \frac{\overline{e_{R1}^2}}{Z_i^2} + \frac{\overline{i_{ch}^2}}{g_m^2 Z_i^2} + i_{gn}^2 \qquad (10.38)$$

Finally substituting the respective expressions for the different noise sources (10.26), (10.27) and (10.29) delivers the equivalent noise current source at the input:

$$\overline{i_n^2} = \frac{4kTB}{Z_i^2}\left[R_i + \frac{\tau}{g_m}\right] + 2qI_G B \qquad (10.39)$$

where Z_i is the input impedance of the FET and is a frequency-dependent quantity.

In Figs 10.8 and 10.9, examples of the noise behavior characteristics versus frequency are presented for the bipolar and the field-effect transistor. The characteristics show the equivalent voltage noise and the equivalent current noise behavior versus frequency.

As said already, these noise measurements can be performed by measuring the output of the amplifier's stage with the input terminals short-circuited, delivering the noise voltage, or 'open-circuited', delivering the noise current and dividing both measured quantities by the gain of the amplifier. As can be seen from these graphs, flicker noise becomes important for lower frequencies.

10.5.3 Noise behavior of the diode

The equivalent noise circuit for a diode is drawn in Fig. 10.10. The following physical noise sources are to be found in a diode:

1 A thermal equivalent voltage noise source due to the presence of the physical resistance, r_s, of silicon,

$$\overline{v_s^2} = 4kTr_s B \qquad (10.40)$$

2 A shot noise current in the pn-region as a function of the DC current I_D,

$$\overline{i_d^2} = 2qI_D B \qquad (10.41)$$

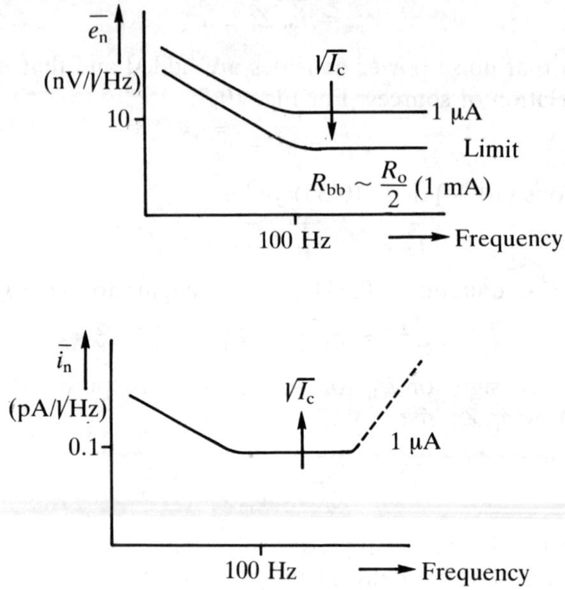

Figure 10.8 Noise versus frequency of the bipolar transistor

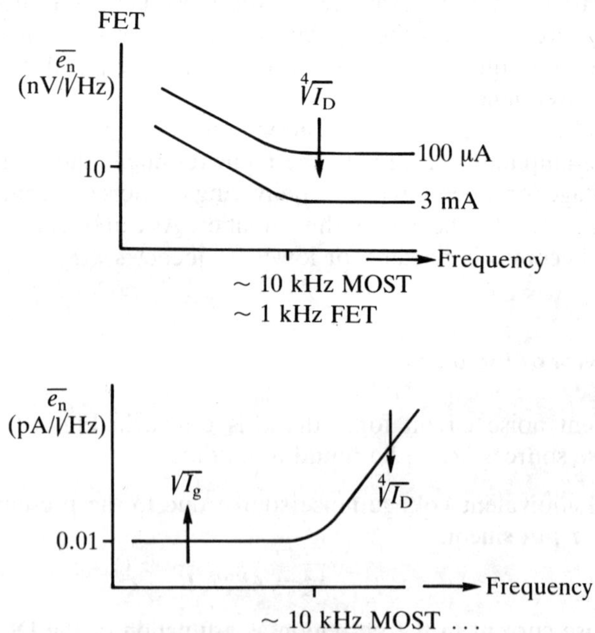

Figure 10.9 Noise versus frequency of the FET

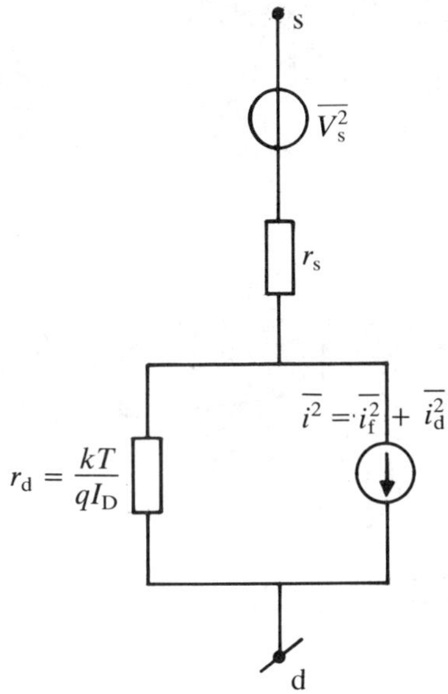

Figure 10.10 Diode equivalent small-signal circuit with noise sources

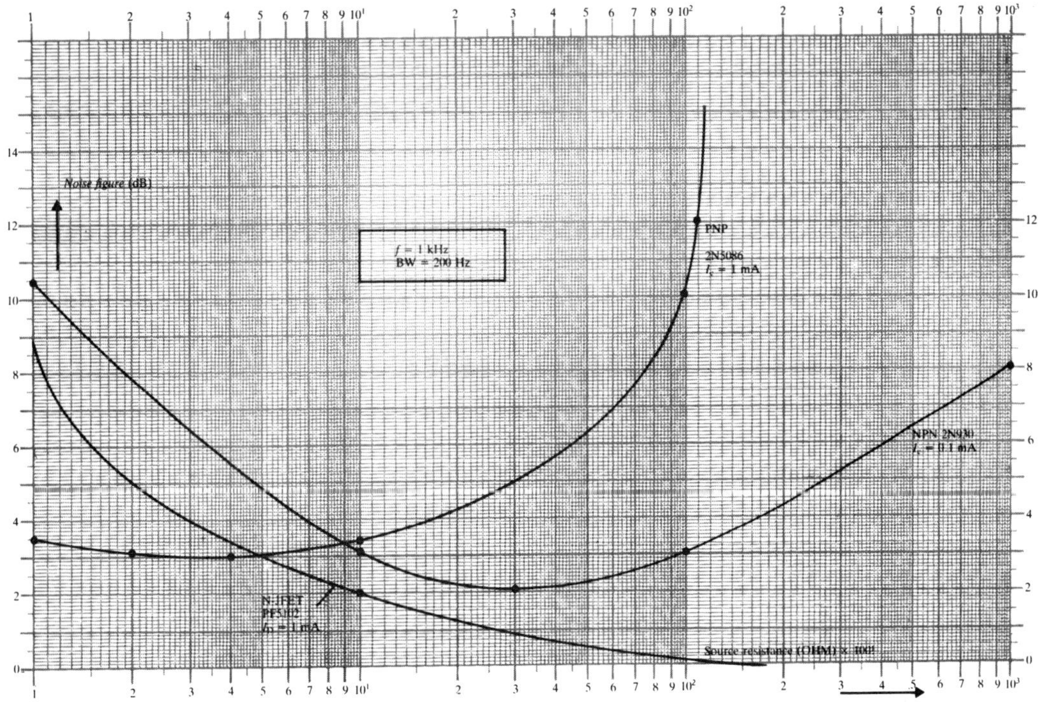

Figure 10.11 Noise comparison of bipolar transistor and junction-FET (courtesy National Semiconductors Ltd)

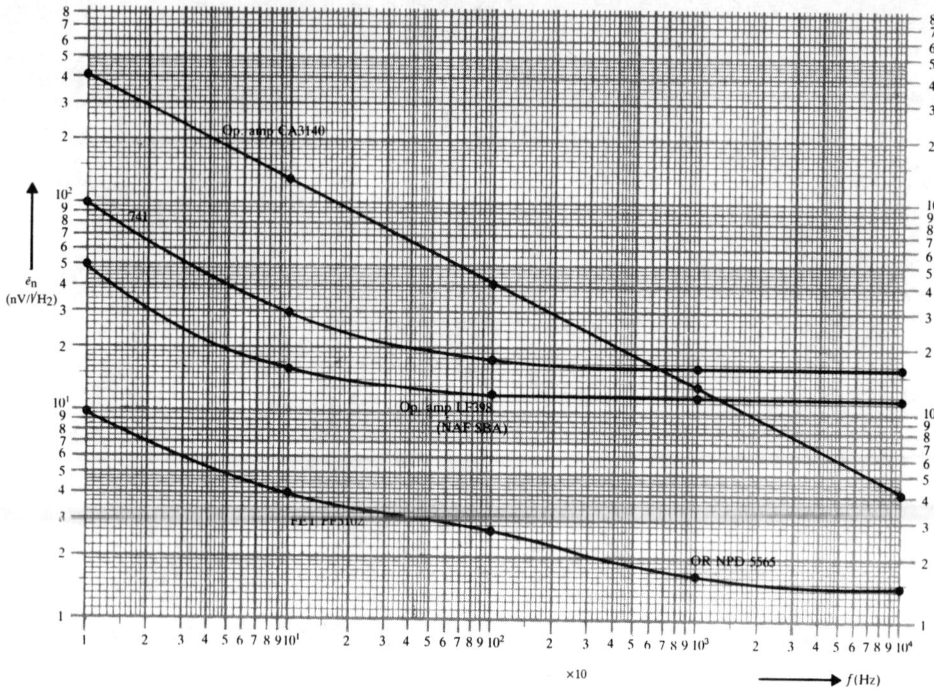

Figure 10.12 Noise comparison of junction-FET and op amp (courtesy National Semiconductors Ltd)

3 A flicker noise current component i_τ which is only of importance when a low frequency is present,

$$\overline{i_\tau^2} = K(I_D^a/f)B \qquad (10.42)$$

To be able to compare some other noise specifications of well known components two further figures are given. Figure 10.11 shows the noise behavior of a junction-FET and a bipolar transistor. Figure 10.12 shows the noise behavior of a discrete-junction FET and an op amp.

Eleven

ERGONOMICS OR HUMAN ENGINEERING

11.1 INTRODUCTION

This chapter is devoted to one of the most neglected disciplines in measuring techniques. The reason is easy to understand because it is the least possible to quantify and was considered for a long time to be of little importance. But equipment and systems have grown considerably in complexity and nowadays the relationship between humans and systems very often plays an important and vital role in efficiency and safety.

To understand what is meant by ergonomics the following definition will clarify the implications involved. Ergonomics is defined as the collection of knowledge about skills and limits of the human being and the science of applying this knowledge to equipment and systems to achieve a wellbeing and functioning of humans in which safety aspects and economical efficiency can play a vital role. The American equivalent for ergonomics is 'human engineering'.

The word 'ergonomics' was first introduced by K. F. H. Murrel in 1949 and is derived from the Greek word *ergon* = 'work' and *nomos* = 'knowledge'. The person involved in ergonomics is called a human engineer, for whom the following topics are a usual subject of investigation:

1 Creating good working conditions.
2 Creating efficient tools, equipment and systems in which a direct relationship with the human being is present.
3 The investigation of maximum load conditions for the human being performing various tasks under static and dynamic environmental conditions.
4 The mechanisms with which the required information is perceived and the relevant thresholds for the human being.
5 The way in which information be presented to allow unambiguous interpretation.

As stated already, in the inter-relationship between people and systems, safety aspects are normally also involved.

One of the best examples to illustrate the impact of these subjects is flying an aircraft. Here optimum conditions and circumstances have to be present to fulfill all requirements which can be put forward for the safe and good functioning of the pilot to fly the airplane. A pilot should be in good condition and be able to perform his tasks with a machine which is also maintained in good condition. The different flying tasks such as starting, flying, approaching and landing are tasks with different loads for the pilot with possible rapid changes in working circumstances. All equipment, systems and subsystems must be designed in such a fashion that optimum safety, attainability, and operability, are guaranteed and that the displayed information can be understood without ambiguity.

As will be recognized, in the field of ergonomics the human being is always at the center of the investigation; this means that the human being is the starting point and functions as the reference to whom all information must be adapted. The consequence of this is that we have to be well aware of the possibilities of our biological senses. This biological equipment will be discussed in the following section.

Ultimately, ergonomics can be defined as the science of finding optimum conditions in the interrelationship between the human being and his tools. In principle many disciplines are brought together in ergonomics, such as anatomy, physiology, cybernetics and information theory. Human engineering is not only related to tools such as machines, but may also be applied to the installation of offices, shops, schools, and so on.

As will be clear from the above, human engineering is a multidisciplinary area of science and for a long time it has been more art than science. Consequently, human engineering is difficult to quantify and is therefore, initially less attractive for the scientist. To get some feeling for ergonomics, we shall discuss some aspects related to electronic tools and equipment only. The chapter is not exhaustive, but its aim is to draw attention to those aspects involved in the design and application of electronic measurements and control systems. Due to their importance, we shall highlight aspects of display design.

11.2 BIOLOGICAL EQUIPMENT OF THE HUMAN BEING

Our biological equipment consists of a number of so-called windows, through which the different physical changes in the outer or inner world can be detected. Each window has a limited spectral width and a limited height of modality. Two of these windows are the most important from a technical point of view, i.e. the auditory and the visual windows.

11.2.1 The auditory window

The range or width of our auditory frequency window runs from 20 to 20 000 Hz and its modulation height goes from 0 to 140 dB. The resolution of the auditory organ involves 3 Hz below 1000 Hz and 0.3% of the frequency above 1000 Hz. This is the smallest step of distinction we can detect without technical means.

Within the auditory window it is curious that Nature has provided us with an optimum working area where the sensitivity is the highest. The optimal working area for a hearing organ lies between 300 and 6000 Hz and 40 and 80 dB, which roughly coincides with the area of speech and most music.

11.2.2 The visual window

The perceiveable wavelength of our visual window ranges from 380 to 720 nm and the modulation height varies from 10^{-9} to 10^{9} cd/m^2. The resolution for color changes varies from 1 to 20 nm and for light intensity changes the resolution must equal 100 cd/m^2 or $\delta I/I$ should be in the order of 5%. Within the visual window, the maximum sensitivity area lies between 450 up to 650 nm which is in the yellow to green range and for light intensities between 10 and 100 cd/m^2.

The candela (cd) is referred to as the standard intensity for a black radiator and proves to be a quite complicated definition. The candela is the light intensity in a perpendicular direction of a surface area of 1/600 000 m^2 of a black body at the congealing temperature of platina (2043 K) at a pressure of 101 325 Pa.

An illustration of the visual and the auditory windows is given in Fig. 11.1.

Beyond these limits we are not normally able to detect information, and transducers are required to transform the information within our windows. Besides this aspect, aging plays a role because with aging our perceptual windows and resolution capability tend to decrease. It is interesting to note that we transform all physical information just to these two types of detectable energies and never for instance to a type of smell or a temperature, because these senses are less developed in human beings and thus less efficient to detect and to present information.

Among these two perceptual senses, the visual perception seems to be the most important one. Therefore we will emphasize the way visual information is presented. There is still another reason why measuring instruments must help our biological perception. Perception takes place in the time domain. In order to be perceived the information should not only be equal to or greater than the minimum required size and step, but should be presented for some time also, to activate our natural transducers. A smaller change requires a longer duration of the actuating

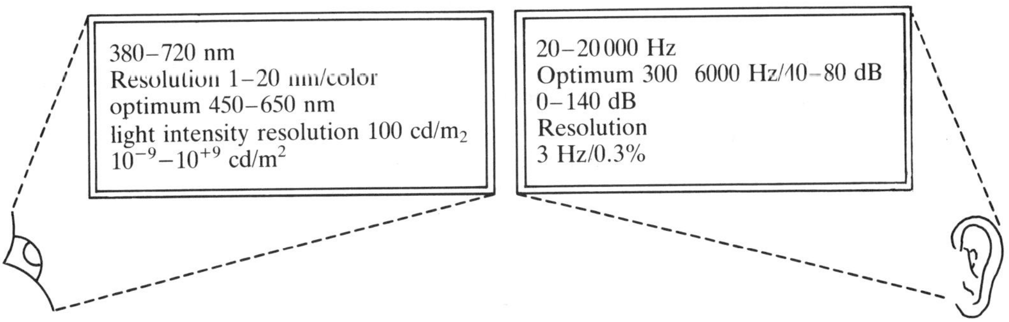

Figure 11.1 The visual and auditory windows of humans

process. The cones and rods of the human retina need some time to react and have a certain time constant.

The same reasoning applies for the process of transmitting the pulses across the nerve fibers. In many circumstances we not only need a transduction towards the perceptual area but also towards a perceptual time duration. In a lot of measuring instruments, both mechanisms are implied.

11.3 SOME OTHER ASPECTS OF HUMAN PERCEPTION OF INFORMATION

The artificial aids to natural perception mentioned already indicate the first essential functions of transduction: processing and displaying information. In general they represent a measuring instrument and bridge the gap between the real outer world and the observer to enable him to detect the physical changes in a broader, finer, or shorter fashion and on a more reliable and objective basis. This seems to be basic for our curiosity and need to control the outside world and our environment.

Three other essential functions must be added to the way we perceive information on higher levels. These functions are transmission, retrieval and processing of the acquired information.

11.3.1 Transmission of information

When transmission of information is involved a transmission channel is always present, in which barriers can be present and in which often noise is added to the signal. Besides these aspects, it will be clear that we are not able to observe synchronous phenomena occurring at different places without technical means. With well-equipped systems and instruments, it is possible to observe mechanisms at large distances, through walls and in places where we are not able to be physically present as for instance is the case in space travel.

11.3.2 Retrieval of information

The human brain can receive information in very limited amounts. As time passes the amount and accuracy of information that can be reproduced diminishes. During the first seconds the quantitive losses are huge and over longer periods the losses are also considerable. The human memory and/or brain, from which we can retreive information in normal circumstances, uses a process of selection, but also proves to deform or sometimes to complete information. So for instance we are able to read a sentence with some words missing. What is being perceived and retained differs strongly between individuals and circumstances in which expectation, prejudice and preoccupation can play an important role. This is the main reason why we want to record information.

11.3.3 Processing of information

As indicated already, the human brain is not a passive sensorial mechanism. In principle human perception involves a sequence of selection, deformation and completion in which billions of cells are busy coding the input and processing the received information. Information elements and their functional relations are actively structured into patterns which are easy to remember. A type of hierarchical order can often be recognized in these patterns. There is far from a one-to-one relationship between physical reality, direct perception and the patterns stored in the individual human being.

These aspects stress the fact that making objective statements about phenomena is very difficult, and make it necessary to de-individualize information as far as possible.

Here a more philosophical point of view can be made, which says that every observation will influence the process and the information which is retrieved, and is strongly related to the way the information is collected. But still, we constantly achieve a consensus about how information can be made more objective. For this purpose, intruments are very valuable: they can process, filter noise, weigh, transform, put into structures and handle information much faster and more reliably than we are able to do.

Throughout history all types of physical and technological principles have been applied to perform measurements. Today these measurements are based on an international system of standard units and measurements in order to have an invariable reference system on which everybody can rely.

For instance, the SI system is nowadays a well established system of measurements and units. This means that there are two systems or worlds of perception which exist together and can be compared. This is considered further in the next section.

11.4 PERCEPTION

Natural perception is human related and thus individualistic, inaccurate, not reliable, but with guessing capabilities and very inventive in a broad way of perception.

Instrumental perception is objective, very predictable, constant, highly accurate, reliable but very specific.

In a measurement chain the display at the output forms the plane of contact between these two worlds and bridges the gap in between. At this point all major aspects of perception come into scope because in displaying information the following requirements have to be considered:

(a) It must be possible to detect and scan the information under known conditions.
(b) Separate information elements have to be distinguished.
(c) It must be possible to decode the presented information.
(d) It must be possible to interpret the information involved without ambiguity.

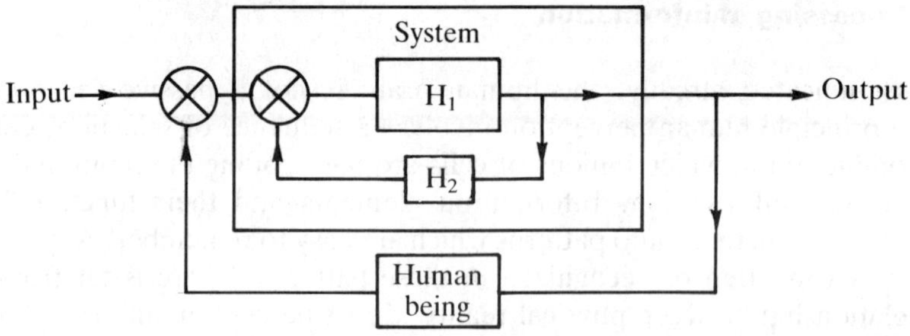

Figure 11.2 The concept of perception and feedback of a human being in a system

(e) It must be possible to remember the displayed information and to take the right actions if required.

Here it must be borne in mind that a human being is always in the feedback loop of the system and is part of the whole system to be controlled. Without this feed-back, the system is useless and not determined. Of course in a complex system, as for instance a power plant or an airplane, a lot of technical feedback loops will be present also, but ultimately the human being is always involved in the final feedback loop. This is schematically illustrated in Fig. 11.2.

The consequence of this is that the industrial designer who is involved in human engineering must himself be aware that he has to translate all the required information in some way on a display which must be interpreted by a human being. This means that the human engineer must be familiar with the laws of human perception.

A few examples can illustrate what is meant by these statements. In some automobiles the speedometer display shows a digital number only and no trend in speed can be detected. Any elementary law of ergonomics fails and in the design stage market trends have been important only. As another example, Fig. 11.3 shows a digital multimeter where for the same kind of information a combination is shown of an analog and a digital display.

If the display is only digital it is very accurate, but that is not always required. If it is only analog it shows a trend and is likely to be insufficiently accurate. Here only the right combination of analog and digital display techniques can offer the best solution.

Shortcomings with respect to display design, albeit visual or spoken, can be very costly in the sense of much loss of performance, dissatisfaction of operators in working circumstances, and wrong investments in equipment and manpower. We shall discuss some aspects and trends of human engineering in more detail in the following section.

Very recently there has been an increasing tendency to translate information into spoken commands in automobiles, aircraft and in an increasing variety of equipment. The voice-controlled computer and typing machine are already reality and these features can once again change the image of our society.

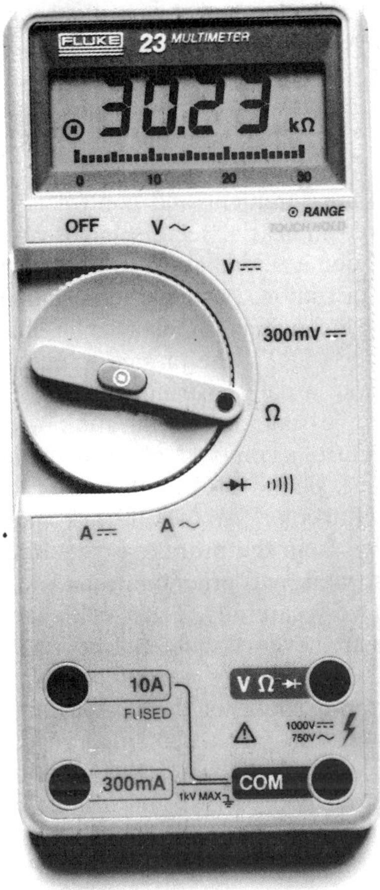

Figure 11.3 Example of a multimeter display with analog and digital display for the same kind of information (courtesy Fluke Inc. (Netherlands))

11.5 TRENDS IN THE DEVELOPMENT OF MEASURING SYSTEMS

In this section some trends shall be discussed in the development of measuring systems, not only from a viewpoint of human engineering but also to emphasize the impact of electronics on our society in general. In the September 1977 issue of *Scientfic American* a survey was presented of the then actual state of the art in microelectronics. One of the articles involved ended as follows: 'these decades will be recognized in the future as the beginning of the robotic revolution' (p. 190).

Within 10 years from that year we can recognize an already dramatic increase of the development of robots in manufacturing and control of a great variety of processes. This is mainly due to the dramatic increase in the application of microelectronics. The impact of this revolution can be treated in the following five points where the definition of a measurement system must be understood in a much broader sense than usual.

1 Miniaturization in electronics is still proceeding rapidly and has caused a qualitive change in information technology. It has become thousands of times smaller, faster, energy saving, more reliable, more complex and cheaper. One of the best examples is the development in memory chips; we are now talking about 1–4 Mbit DRAM chips and even 16 Mbit has already been announced. These chips contain over a million characters.
2 Decentralized measurement and control is now not only to be found in the professional and industrial sectors but is also being rapidly introduced in the consumer and domestic industry. Applications can be found in laboratories, in banking, administration, transport, entertainment, education, etc. The development in display technology for instance has made a big change too, especially the cathode-ray tube (CRT), the plasma screen and the liquid crystal display have become common in daily instrumentation.
3 Excellent examples can be found in the process industry where one person can monitor and control a complete plant by the applied principle of performing the management task via the CRT monitor which represents in real time all required information in full color. An overall view and detailed information by exception can be retreived from the process as required. The way the information is displayed is most vital, and ergonomics and working load for the operator must be considered very carefully. Even safety and health aspects play a role here because CRTs are never completely free of flickering and an uninterrupted work load of 4 hours for CRTs is a maximum allowed duration.
4 Another striking example of the impact of the electronic revolution is on the manufacture of newspapers. A complete revolution has occurred with a dramatic change in employment and working conditions. This change has caused in some countries a large labour unrest, but it must be stressed that here too a technological development has its own rules and cannot be wiped out or its existence denied.
5 A very important technological development is still occurring in the interface technique where the connections are made between the inner (systems) and outer world.

In the following section we shall discuss aspects of the observer display interaction in some detail.

11.6 THE OBSERVER–DISPLAY INTERACTION

In this section we will focus our attention on the observer–display interaction. This means that the process to be considered is at the front end of the measurement system where the human process of interpretation starts. In this process from the very start the human process involves the following steps:

- Scanning and detecting one or more quantities
- Identifying the different elements
- Combining different elements
- Interpreting the message
- Reacting or memorizing

For some tasks a passive registration can suffice but sometimes the tasks can be more vigilant and a qualitive interpretation is required. The steps to be investigated can be classified roughly according to the following four questions.

1 Which type of information is necessary?
To answer this question a system analysis concerning the understandability of the data and the way the data will be displayed has to be performed. A clear structure of the data must be developed which must result in a simple coding for the identification and the relationships of the displayed variables. So it can be necessary to display the data as a function of time, applying for instance the principle of displaying by exception.

2 For whom is this information destined?
A clear job description may be required defining task objectives and operating routines. The measurement and control procedures must be adapted to the least competent operator expected. Designing the procedures in case of occurring errors means that measurements, assessments and rectification of errors can be performed immediately and that if needed an analysis can be executed afterwards. In dangerous situations a display should not only present what is wrong, but preferably also the location and the right action to be executed.

3 Which general features characterize the required task?
Here one of the most disturbing characterizations must be recognized if involved in a process. A serious amount of monotony in a process can diminish the vigilance considerably, resulting in a loss of quality or even leading to dangerous situations. Diversity of tasks and performances can avoid these problems. The operator must be able to determine his own quality checking and his own work speed. At all times the quality and accuracy of performance must be stressed.

4 How is the information presented in the available workspace?
A lot of factors must be recognized which are impossible to quantify and which may only show in the long run, for example, when hours and hours must be spent by a pilot sitting in one place in a cockpit of an aircraft, or in such a case as an operator sitting in front of a console. Here every inconvenience will increase the workload and can cause undesired stress and fatigue especially in the eyes, hands, arms and neck. This is illustrated in the following example.

* * *

Example 11.1

An excellent example of a recently designed workspace for a helicopter pilot is shown in Fig. 11.4. This example shows how a multifunctional display and control-setting configuration can look. The cockpit is designed in flat-panel technology.

* * *

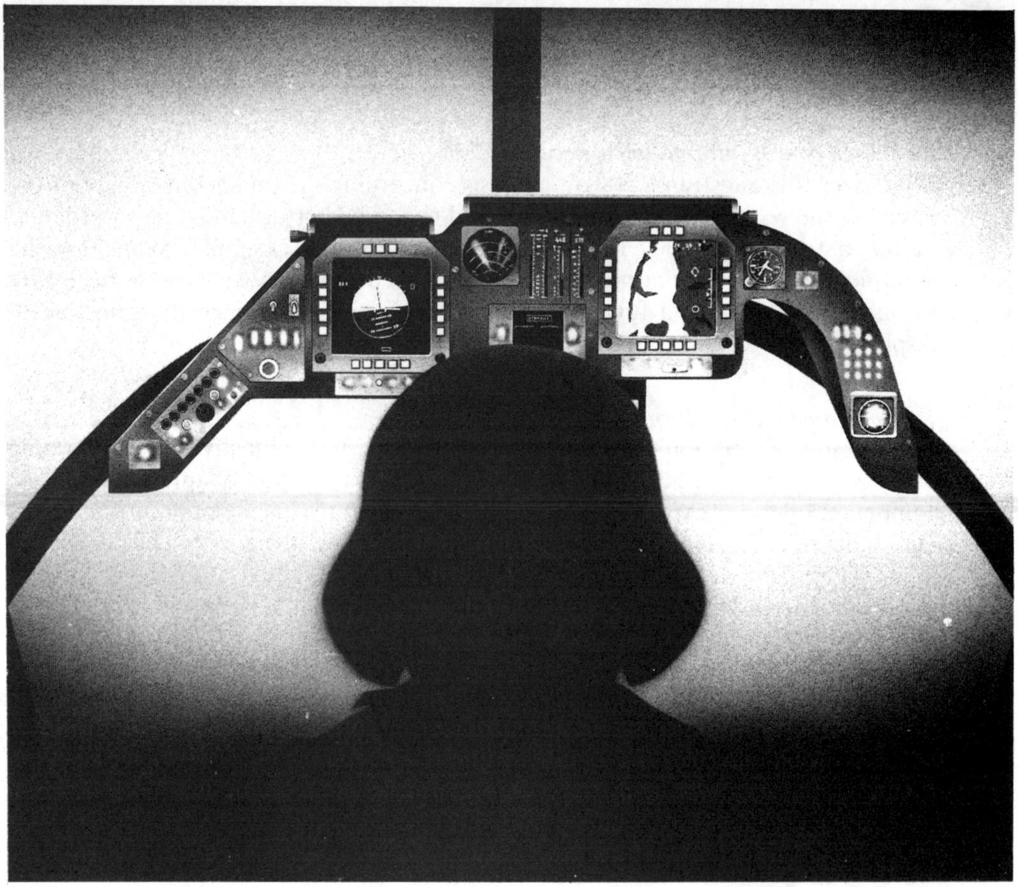

Figure 11.4 Design of a flat-panel cockpit display (courtesy of AEG GmbH)

The entire workspace must be designed starting from an acceptable position for the eyes and hands. The centre of the visual and manipulative field must be determined and all equipment and most frequently used instruments must be placed in the very middle of the workspace. The less important ones can be placed in the periphery. Any light source must be installed in a way that prevents glare. Very often, coated, tilted glasses and a hood are effective measurements.

In the following section we will present a number of basic rules of coding of information applicable in the design of displays.

11.7 CODING OF INFORMATION ON DISPLAYS

It is easily recognized that visual perception of information is still the most common way in which information is presented to us. Here, code symmetry plays an important role as the code used to present the information is, ultimately, the visual transducer from instrument to human. The code can be characterized as a system of agreements and, often, the user has to learn to understand these codes. Sometimes, this learning process is highly demanding and time consuming, requiring a long time to achieve complete comprehension. This is true for example of the CRT oscilloscope, the interpretation of the displayed information depends on different scaling factors and settings.

The human brain prefers the specific language code which already exists in the natural world. If the applied code differs substantially from the preferred code, the learning process is very difficult despite the logic of the applied code. In the design of coding it is always recommended to include some room for error, because in experienced users of instruments, a so-called half-word may be not enough. The designer of a code can make his choice from several so-called sensory windows, through which the information can be perceived.

A remark must be made here concerning the auditory sense. Auditive signals should be restricted to warnings supporting other functions as long as they do not involve human speech. The still-increasing possibilities of synthetic speech will undoubtedly offer possibilities which reduce the visual workload.

Other possibilities are tactile displays and keyboards originally designed for the blind, but which are still a relatively unknown area in ergonomic design. Up to now the visual display is the most common way of presenting detailed information in patterns. In the following table a review of coding rules are presented.

Table 11.1 Review of coding characteristics of humans (courtesy Delft University Press)

Code dimension		*Number of steps*	*Bits of information*	*Signal duration*
digits		≥100	≥6	short
letters		≥1000	≥10	short
figures	(standard)	≥15	≥3.9	short
color	(nm)	9	3.1	very short
size	(cm^2)	7	2.8	very short
grain	(n/cm^2)	7	2.8	short
brightness	(cd/m^2)	5	2.3	very short
number	(n)	6	2.7	long
length	(cm)	4	2.0	long
angle size	(deg.)	24	4.8	short
direction	(deg.rel)	15	3.9	very short
flicker	(c/sec)	4	2.0	very short
pitch	(Hz)	5	2.3	long
loudness	(dB)	5	2.3	short
smell	(?)	4	2	long

Table 11.1 lists the most effective information content of different visual coding dimensions for short duration of perception. The column ‘Number of steps’ shows the number of steps which can be distinguished in the time of signal duration.

The visual channel offers a wide opportunity for coding information according to different dimensions in form, color, size, resolution, texture, brightness, location, orientation and flicker. The number of steps that can be distinguished coming one ofter the other in regular order is not the same with all dimensions. Simultaneous signals can be distinguished more easily if they are next to each other than if they are separated in time.

Of course the coded information can contain a combination of two or more of the dimensions mentioned in Table 11.1. Examples of these combinations are a letters–color combination, or a digit–size–brightness combination. In such a way the information content can be enhanced but less than might be possible by a simple multiplication of the number of steps of the relevant dimensions which are chosen. A combination of dimensions can even improve the redundancy, for instance a velocity meter can depict all values above 180 km/hr in red as dangerous, and below that value in green as a safe velocity.

This table shows also that form, which involves digits, letters and standard figures, has the largest coding capacity, and alphanumeric characters have in their turn a larger capacity than pictograms. Color is very efficient for a quick search but cannot replace quantitive information. In the following an example is given of information displayed in two different ways for the same process monitored by an operator.

* * *

Example 11.2

Suppose the operator watches a machine of which the temperature is 189.75°C, the oil pressure is 5.78 Pa and 150 seconds ago the temperature was the same but the oil pressure was 4.78 Pa. This information may be interesting but the operator only wants to know if the machine is running smoothly, so if the temperature and the oil pressure needles are in the green part of the display the operator gets the same information in a more convenient fashion and with more efficiency.

* * *

A basic rule can be abstracted from this example: no more information should be displayed than strictly is required for the process under control.

* * *

Example 11.3

Another good example is a wristwatch. For a long time the display was completely analog. With the rise of modern electronics the wristwatch started to become digital, but it was soon clear that this was not satisfactory because too little information is available in a completely digital display. What we see now is a tendency to combine both possibilities.

* * *

Table 11.2 provides a survey of the parameters necessary to improve or to guarantee maximum legibility of displays, whether they be CRTs, paper or scale readings.

Table 11.2 Summary of parameters for maximum legibility and display design (courtesy Delft University Press)

normal reading distance	30–60 cm
sign height	≥15′ visual arc or 3.1 to 4.2 with an optimum of 3.7
width to height sign proportions	3:4 and 4:5
stroke width to height	1:8 to 1:6
spacing between signs	½ to ¼ of sign height
spacing between words	¾ to 1 of sign height
interline spacing	¼ to 1 of sign height
visual field for permanent attention	10–20 degrees of visual arc
color wavelength	550 nm (yellow to green)
background	dark
illumination level	80–160 cd/m^2
CRT screen illumination level	4–16 cd/m^2
keyboard illumination	500–700 lux
workspace illumination	≤300 lux
workspace ill. color temp.	3000 K

Following these guidelines excessive contrasts are avoided and eye fatigue can be prevented, so attention can be maintained for long periods as is often required. Up to a reading distance of 6 m the following formula can be applied for the sign height:

$$H = 0.0022D + K_1 + K_2$$

where

H is the sign height (cm)
D is the reading distance (cm)
K_1 is a correction factor ranging from 0.15–0.65 dependent on reading circumstances such as illumination
K_2 is a correction factor related to the importance of the message and ranges from 0.15–0.19

In many circumstances only capital letters are applied, which can be considered as a mistake. Lowercase letters can more easily be recognized than capitals, so lowercase letters should be preferred for short messages and for body text whenever possible. Also italic letters and letters with serifs are less easily distinguished and therefore should be avoided. If letters are made of a matrix of dots the matrix should preferably be composed of a rectangular pattern of 7 × 9 dots.

A special field of applications is the design of characters on a CRT display. If less than 10 lines are used for their height the so-called Kell effect can cause blurring and dotting.

Pointers with scales are a frequently ocurring combination for display design. The values of the scale should increase in a clockwise direction and from left to right or from bottom to top, dependent on the shape of the display involved. As said already, in many cases it is sufficient to have only different colored regions on a scale, such as green for a safe area and red for a dangerous area. Sometimes more quantitive information is required and a graduation can be applied consisting of three different sized markers: the headmarker for the decimal intervals, the intermediate markers for the middle of each interval and the minor markers for the smallest units.

These intervals should be fixed to whole units and simply to interpret series of divisions such as 1, 2 or 4, or 1, 10, 20, or 1, 5 and 10. A good example of these applied divisions can be found on a slide rule, still in use in some technical disciplines. The number of digits accompanying the markers can be limited to those most demanded. The designer should keep in mind that the human eye is able to interpolate very accurately the distances between two markers where the needle is in between. The pointer must just touch the baseline of the markers and parallax has to be avoided. The unit of a scale and the origin must be depicted very clearly without possible misunderstanding. On circular dials and scales it is recommended that the origin and the end of a scale are symmetrically divided with respect to 'six o'clock'. Sometimes a percentage scale ranging from 0–100% is very efficient and completely satisfactory. An example of a possible display design combining analog and digital information is shown in Fig. 11.3.

Graphical displays are also used to present information, for instance on a CRT or a printer output on paper. These possibilities offer excellent features to display structures and trends to which our natural perception is most sensitive. Processors are able to calculate a predictive path and so a type of a predictive display can be formed. Graphs need to be simple, for instance with vertical bars in a histogram ranging from left to right, whereas the number of curves in one graph should be limited to one or two. The identification of the process depicted in the graphs should be clear, with all the essential units and parameters shown on the graph. These provisions can make a graphical representation very attractive and easy to understand.

When designing a table of numbers a structure of columns proves to be very convenient especially when the decimal points are aligned and long numbers are divided into chunks of three digits. Besides these aspects, tables should be divided into blocks of rows and columns. But it should always be kept in mind that the information content of a graphical representation is always larger than a numerical representation because a graphical representation is in its very nature analog.

Last but not least there exists an increasing tendency to apply colors in all types and techniques of display design. This tendency can be seen in CRT design, in software applications and in all types of instruments. Although cost aspects have initially played an important role here, these are becoming less important due to the improved technology of color display devices.

11.8 PROBLEMS

11.1 Give a definition of ergonomics. What relationship exists between ergonomics and human engineering?

11.2 Which kind of topics are usual subjects of investigation in human engineering?

11.3 Give a definition of human perception.

11.4 What is meant by a natural transducer?

11.5 Which types of perception can be distinguished; explain these.

11.6 Which three functions are involved in human perception?

11.7 Discuss the pros and contras of displaying information in digital or analog form.

11.8 Which kinds of process steps have to be distinguished in the observer–display interaction?

11.9 What is meant by coding of information?

11.10 What is the disadvantage of displaying a digital voltage only?

Appendix 1

PERFORMANCE SPECIFICATIONS FOR CONVERTERS

A.1 ACCURACY

Accuracy is sometimes considered to be a non-specific term when applied to D/A and A/D converters. A linearity specification is generally considered as more descriptive. An accuracy specification describes the worst-case deviation of the DAC output voltage from a straight line drawn between zero and full scale; it includes all errors. A 12-bit DAC could not have a conversion accuracy better than $\pm\frac{1}{2}$ LSB or ± 1 part in 2^{12+1} ($\pm 0.0122\%$ of full scale due to finite resolution). This would be the case in Fig. A.1 if there were no errors.

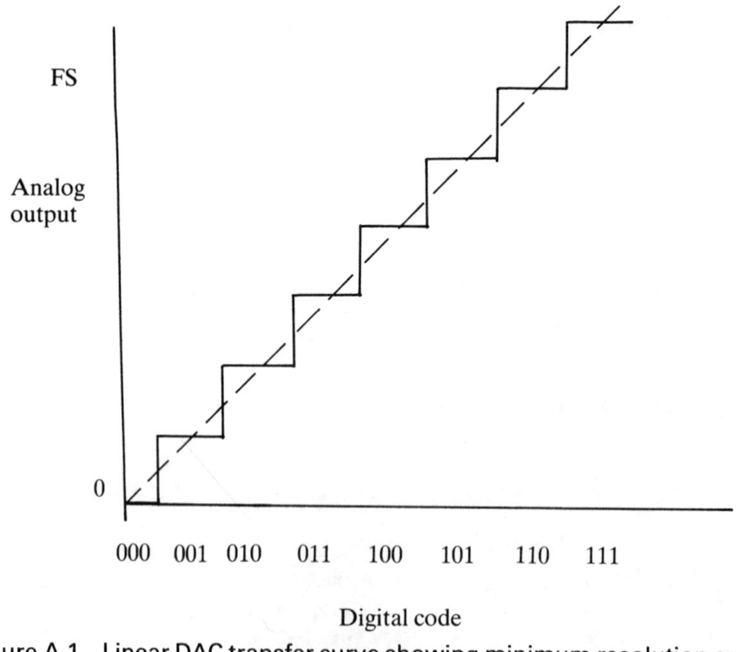

Figure A.1 Linear DAC transfer curve showing minimum resolution error and best possible accuracy

A.2 QUANTIZING ERROR

Quantizing error is the maximum deviation from a straight-line transfer function of a perfect ADC. As by its very nature, an ADC quantizes the analog input into a finite number of output codes; only an infinite resolution ADC would exhibit zero quantizing error. A perfect ADC, with suitable offset ½ LSB at zero scale as shown in Fig. A.2, exhibits only ±½ LSB maximum output error. If not offset, the error will be ±1 LSB as shown in Fig. A.3.

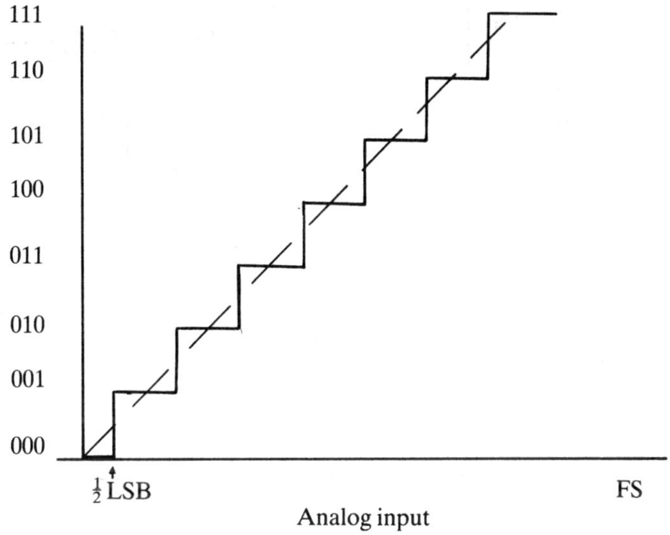

Figure A.2 ADC transfer curve, ½ LSB at zero

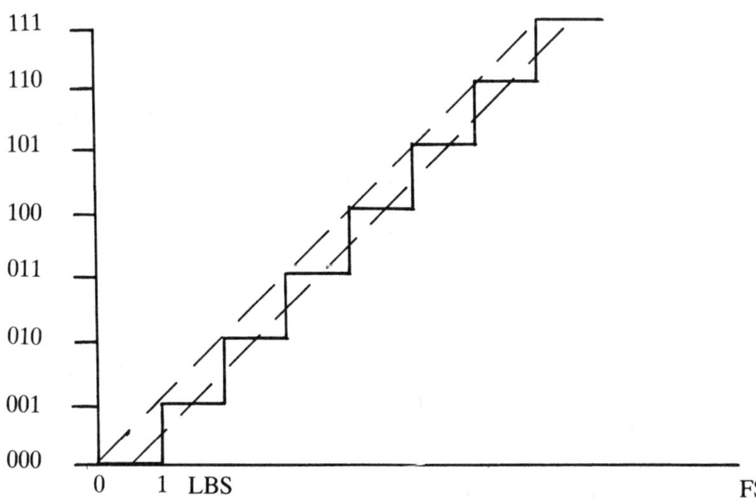

Figure A.3 ADC transfer curve, no offset

A.3 SCALE ERROR

Scale error (full-scale error) is the departure from design output voltage of a DAC for a given input code, usually full-scale code (see Fig. A.4). In an ADC it is the departure for a full-scale output code. Scale error can be caused by errors in reference voltage, ladder resistor values, or amplifier gain, etc. (see Section A.13).

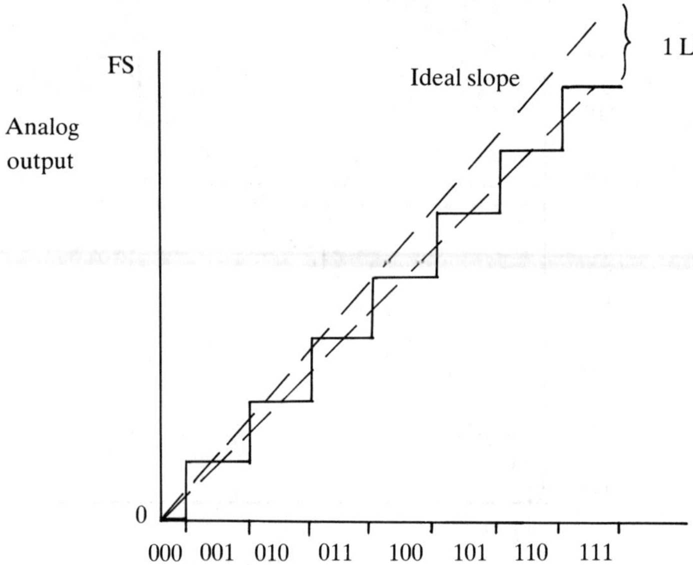

Figure A.4 Linear, 1 LSB scale error

A.4 GAIN ERROR

The gain error is essentially the same as scale error for an ADC. In the case of a DAC with current and voltage mode outputs, the current output could be to scale while the voltage output could exhibit a gain error. The amplifier feedback resistors would be trimmed to correct the gain error.

A.5 OFFSET ERROR

The offset error (or zero error) is the output voltage of a DAC with zero code input, or the required mean value of input voltage of an ADC to set zero code out. (See Fig. A.5.) Offset error is usually caused by amplifier or comparator input offset voltage or current. It can usually be trimmed to zero with an offset zero adjust potentiometer external to the DAC or ADC. Offset error may be expressed in % FS or in fractional LSB.

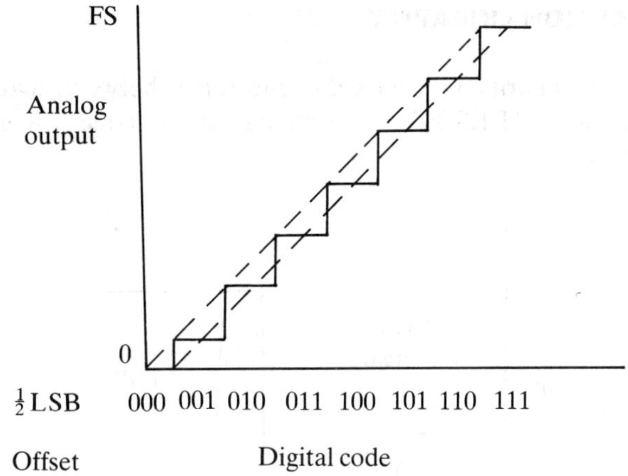

Figure A.5 Linear, ½ LSB offset error

A.6 HYSTERESIS ERROR

Hysteresis error in an ADC causes the voltage at which a code transition occurs to be dependent on the *direction* from which the transition is approached. This is usually caused by hysteresis in the comparator inside an ADC. Excessive hysteresis may be reduced by design; however some slight hysteresis is inevitable and may be objectionable in converters if hysteresis approaches ½ LSB.

A.7 LINEARITY

Linearity or more accurately, non-linearity specifications describe the deviation from a linear transfer curve for either an ADC or a DAC. Linearity error does not include quantizing, zero, or scale errors.

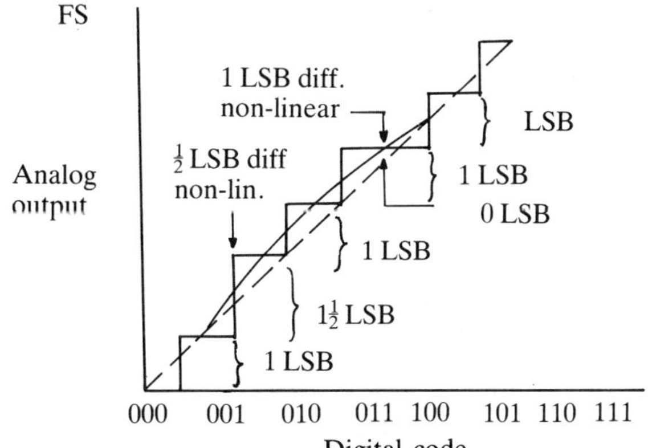

Figure A.6 ±½ LSB nonlinearity (implies 1 LSB possible error), 1 LSB differential nonlinearity (implies monotonicity)

A.8 DIFFERENTIAL NON-LINEARITY

Differential non-linearity indicates the difference between actual analog voltage change and the ideal (1 LSB) voltage change at any code change of a DAC. (See Figs A.6 and A.7.)

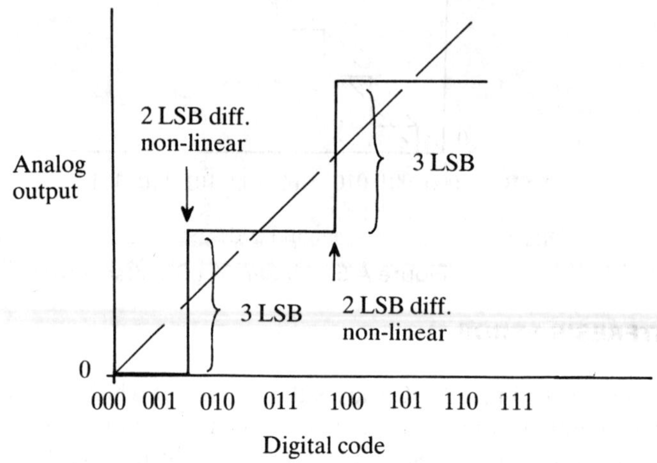

Figure A.7 ±1 LSB linear, ±2 LSB differential nonlinear

A.9 MONOTONICITY

A monotonic curve has no change in sign of the slope; thus all incremental elements of a monotonically increasing curve will have positive or zero, but never a negative slope. (See Fig. A.8.)

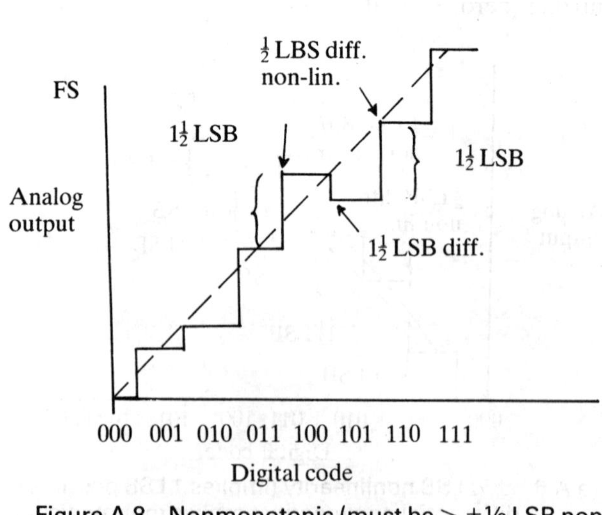

Figure A.8 Nonmonotonic (must be > ±½ LSB nonlinear)

A.10 SETTLING TIME

Settling time is the elapsed time after a code transition for DAC output to reach final value within specified limits, usually $\pm 1/2$ LSB. (See also Section A.16.) Settling time is often listed along with a slew rate specification; if so it may not include slew time.

A.11 SLEW RATE

Slew rate is an inherent limitation of the output amplifier in a DAC which limits the rate of change of output voltage after code transitions. Slew rate is usually anywhere from 0.2 to several hundred volts/μs. Delay in reaching final value of DAC output voltage is the sum of slew time and settling time.

A.12 OVERSHOOT AND GLITCHES

Overshoot and glitches occur whenever a code transition occurs in a DAC. There are two causes. The current output of a DAC contains switching glitches due to possible asynchronous switches of the bit currents (expected to be worst of half-scale transition when all bits are switched). These glitches are normally of extremely short duration but could be of half scale amplitude.

The current switching glitches are generally somewhat attenuated at the voltage output of the DAC because the output amplifier is unable to slew at a very high rate; they are, however, partially coupled around the amplifier via the amplifier feedback network and seen at the output.

A.13 TEMPERATURE COEFFICIENT

The temperature coefficient of the various components of a DAC or ADC can produce or increase any of the several errors as the operation temperature varies.

A.14 LONG-TERM DRIFT

Long-term drift is mainly due to resistor and semiconductor aging and can affect all those characteristics which temperature change can affect. Characteristics most commonly affected are linearity, monotonicity, scale, and offset. Scale change due to reference aging is usually the most important change.

A.15 SUPPLY REJECTION

This relates to the ability of a DAC or ADC to maintain scale, offset, temperature coefficient, slew rate and linearity when the supply voltage is varied. The reference must, of course, remain constant unless considering a multiplying DAC. Most affected are current sources (affecting linearity and scale) and amplifiers or comparators (affecting offset and slew rate). Supply rejection is usually specified only as a percentage FS change at or near full scale at 25°C.

A.16 CONVERSION RATE

Conversion rate is the speed at which an ADC or DAC can make repetitive data conversions. It is affected by propagation delay in counting circuits, ladder switches and comparators; ladder RC and amplifier settling times; amplifier and comparator slew rates, and integrating time of dual-slope converters.

Conversion rate is specified as a number of conversions per second, or conversion time is specified as a number of microseconds to complete one conversion (including the effects of settling time).

Appendix 2

STANDARD SYMBOLS FOR LOGIC ELEMENTS

Circuit	IEC norm	DIN norm 40700	American standard	Boolean function
AND	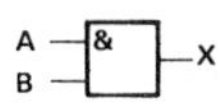	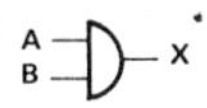	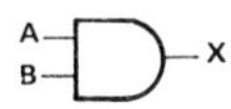	X = AB
OR	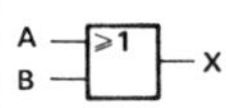	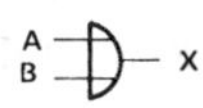		X = A+B
NAND	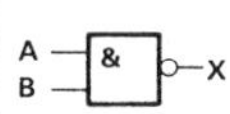	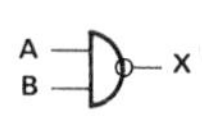	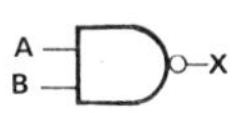	$X = \overline{AB}$
NOR	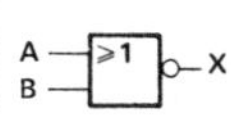	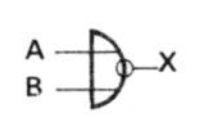	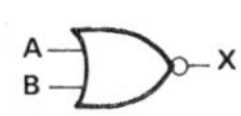	$X = \overline{A+B}$
NAND with one inverting input	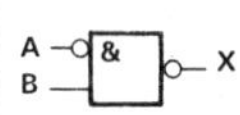	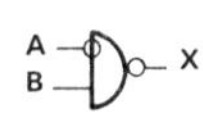	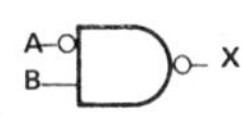	$X = \overline{\overline{A}B}$
NOR with one inverting input	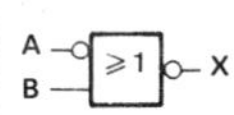	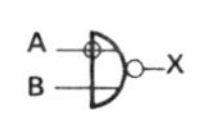	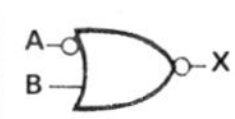	$X = \overline{\overline{A}+B}$
Inhibit gate	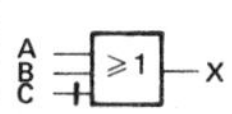		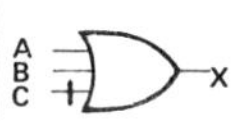	$X = (A+B)\overline{C}$
Exclusive OR	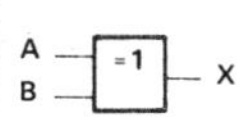	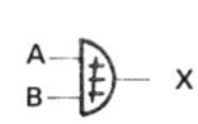	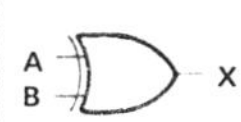	$X = A\overline{B} + \overline{A}B$ $A \oplus B$
Comperator	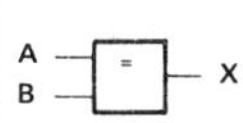	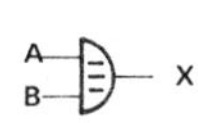	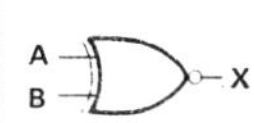	$X = AB + \overline{A}\overline{B}$ $A \equiv B$
Distributed AND			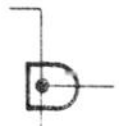	
Distributed OR	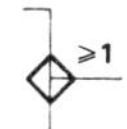			
Delay	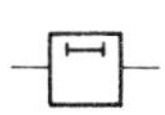	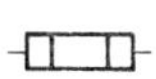		
Flip-flop	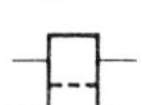	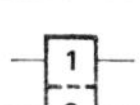		

BIBLIOGRAPHY

Albergi Quaranta, A., Sensors and transducers, *IEEE Trans. Ind. Electron. & Control Instruments*, **IECI-23**, 319, 1976.

Altman, L., The New LSI, *Electronics*, **48**, 81, 1975.

Berg, van de, H. A. M., Humphryes, R. F., Ruigrok, J. J. M. and Venema, A., Submicron I. D. T. fabrication, *Ultrasonics Symposium Proceedings*, IEEE, 1977.

Bergh, A. A. and Dean, P. J., Light emitting diodes, *Proc. IEEE*, **60**, 156, 1972.

Bergveld, P., Development, operation, and application of the ISFET as a tool for electrophysiology, *IEEE Trans. Biomedical Eng.*, **BME-19**, 342, 1972.

Biberman, L. M., *Perception of Displayed Information*, Plenum Press, New York, 1973.

Callagher, R. C. and Corah, W. S., A metal-oxide-semiconductor (MOS) Hall element, *Solid State Electron.*, **9**, 571, 1966.

Carroll, R. F., *Guidelines for the Design of Man/Machine Interface for Process Control*, Purdue Lab. for Applied Industrial Control, 1975.

Cole, B., Temperature transducer is monolithic, *Electronics* **48**, 148, 1975.

Dhillon, B. S., *Reliability Engineering in Systems Design and Operation*, Van Nostrand Reinhold Co., 1983.

Dirken, J. M., Alphanumeric symbols for mosaic printers and display tubes, *Icographic*, **6**, 12, 1973.

Dirken, J. M., Communication and display, *IBM-European Symposium, Stockholm*, 1977.

Department of Defense, *Reliability Stress and Failure Rate Data for Electronic Equipment*, Washington D.C., 1965 (4).

Fieret, J., Kwakernaak, A. and Middelhoek, S., Large linear area light spot position sensitive photodiode, *Electronic Letters*, **13**, 422, 1977.

Fry, P. W., Silicon photodiode arrays, *J. Physics. E.,* **8**, 337, 1975.

Gelder, van Z. and Matthey, M. M. P., Principle and technique in multi-colour DC gas discharge displays, *Proc. IEEE*, **61**, 1019, 1973.

Gieles, A. C. M., Subminiature silicon pressure transducer, *IEEE, TSSCC, Digest*, p. 108, 1969.

Gilbert, B., A new wide-band amplifier technique, *IEEE Journal of Solid State Circ.*, **3**, 353, 1968.

Hallis, J. E. L., Micro-electronic magnetotransducers, *Measurement and Control*, **6** (1), 38, 1978.

Horowitz, P. and Hill, W., *The Art of Electronics*, 2nd edn, Cambridge University Press, 1986.

Kmetz, A. R., Liquid crystal display prospects in perspective, *IEEE Trans. Electron. Devices*, **ED-20**, 954, 1973.

Maxwell, J. The low noise junction-Fet, the noise problem solver, *National Semiconductor Handbook*, 1976.

Microelectronics, *Scientific American*, New York, September, 1977.

Middelhoek, S., Angell, J. B. and Noorlag, D. J. W., Microprocessors get integrated sensors, *Spectrum IEEE*, p. 42 ff., February 1980.

Middelhoek, S., Broerse van Groenou, A. and Verwey, J. F., Vaste stof sensoren, *NNV Proc. Delft*, Kluwer Techn. Boeken bv., 1980.

Gray, P. R. and Meyer, R. G., *Analysis and Design of Analog Integrated Circuits*, John Wiley & Sons, 1980.

Hoogerwerf, A. C., Classification of measurement principles, MSc Thesis, University of Delft, 1986.

Middelhoek, S. and Noorlag, D., *Silicon Micro-transducers*, Delft University Press, 1978.

Norton, H. N., *Handbook of Transducers for Electronic Measurement Systems*, Prentice Hall Inc., Englewood Cliffs, 1969.

Oliver, B. M. and Cage, J. M., *Electronic Measurement and Instrumentation*, McGraw-Hill, New York, 1971.

Pennial, T. H., *Recommendations for the Design of Information Presentation on Visual Display Systems*, Human Factors Dept., Battersea Lab., British Steel Corp., 1979.

Plassche, van de, R. J., A wide band monolithic instrumentation amplifier, *IEEE Journal of Solid State Circ.*, **10**, 424, 1975.

Putten, van A. F. P., Temperature sensitive device for measuring the flow of a medium, Canadian Patent No. 10512119. 1979.

Putten, van A. F. P. and Middelhoek, S., Integrated silicon anemometer, *Electronic Letters*, **10**, 425, 1974.

Putten, van A. F. P., Ambient temperature compensated double bridge anemometer, US Patent No. 4,548,077, 1985.

Putten, van A. F. P., Device for measuring the flow of a medium, a double bridge anemometer, European Patent Application No. 83200379.2, 1983.

Putten, van A. F. P., Device for measuring the flow of a medium, US Patent No. 3,996,799, 1976.

Putten, van A. F. P., Ein Thermischer Gasströmumgsmesser, VDI Berichte 509, Tagung Bad Nauheim, 1984.

Putten, van A. F. P. An integrated double bridge configuration, *Conf. Proc. Transducer Tempcon 84*, Harrogate, UK, 1984.

Putten, van A. F. P., An integrated silicon double bridge anemometer, *Conf. Proc. of Solid State Sensors and Actuators*, Delft, 1983.

Regtien, P. P. L. and Makkink, H. K., A capacitive dew point sensor, *Delft Progr. Rep.*, **3** 107, 1978.

Regtien, P. P. L., *Modern Electronic Measurement Systems*, Delft Univerity Press, 1978.

Riet, van R. W. M. and Hui jsing, J. H., Integrated direction sensitive flowmeter, *Electronic Letters*, **12**, 647, 1976.

Riezenman, M. J. Integrated temperature transducers, *Electronics*, **47**, 130, 1974.

Sansen, W. and Spiegel, van der J., Solid state sensors and transducers, Summer Course, Heverlee, Belgium, 1982.

Sherwin, J., Noise specifications, confusing?, *National Semiconductor*, *Appl. Note 104*, May 1974.

Smith, D. J., *Reliability Engineering*, Pitman Publishing, Bath, UK.

Solomon, J. E. The monolithic op-amp, a tutorial study, *IEEE Journal of Solid State Circ.*, **9**, 314, 1974.

Steidle, H. G., Semiconductor magnetoresistors, *Siemens Review*, **15**, 177, 1973.

Sudar, T., *Digital Instrument Course*, Vol. 3, Philips Gloeilampen Fabrieken, 1980.

Sze, S. M., *Physics of Semiconductor Devices*, John Wiley & Sons, New York, 1969.

Trent, R. L., *Reliability/Availability Electronics Designers Handbook*, 2nd edn, section 28, McGraw-Hill Book Company, 1977.

Vanvari, N., An introduction to transducers and electrical transducers, *Electr.*, *India*, **16**, 1976.

Venema, A. and Humphryes, R. F., Inclusion of conductivity in silicon S.A.W. component design, *Electronic Letters* **13**, 594, 1977.

Widlar, R. J., Some circuit design techniques for linear integrated circuits, *IEEE Trans. on Circuit Theory*, **12**, 586, 1965.

Zias, A. R. and Hare, W. F. J., Integration brings a generation of low cost transducers, *Electronics*, **45**, 83, 1973.

Ziel, van der A., *Noise Source Characterization Measurement*, Prentice Hall Inc., Englewood Cliffs, 1971.

INDEX

absolute
  errors, 83
  temperature, 301
accuracy, 82
acousto-electro, 175
active
  guarding, 267
  on-line standby, 131
  physical quantity, 9
actuator, 161
A/D
  conversion, 19, 62, 67
  converter, 62, 67
ADC, 67
additive errors, 100
address (line) bus, 32
adjustable gain, 55
aging, 220
alias, 21
aliasing error, 21
alphanumeric display, 323
aluminum, 247
amount of information, 2
amplitude modulation, 19
  to digital converter, 62, 67
analog
  functions, 44
  multiplier, 25, 60
AND function, 127
angle size, 323
attenuation, 250
auditory
  resolution, 314
  window, 314
autocorrelation function, 23
automated electronic structures, 9
autozeroing, 69
availability, 123, 124

background, 325
bandwidth, 289 ff.
bath tub curve, 119
biasing
  circuits, 200
  current, 212
  voltage, 212
binary coding, 20, 62
biological
  equipment, 314
  perception, 315
  signal source, 275
  sources, 162
black
  body, 315
  radiator, 315
blurring, 326
Bode plot, 38
Boltzmann's constant, 280
Boolean algebra, 126 ff.
brain activities, 231
breakdown isolation, 252
bridge, 73 ff.
  amplifier configurations, 201
  circuitry, 238
  configurations, 196 ff.
brightness, 323
burn in period, 119
bus-line system, 31

cable resistance, 237
candela, 315
cardinal measurements, 3
cathode-ray tube, 320
CCDs, 186
cell membranes, 275
centralized data-acquisition
    system, 25, 27
characteristic frequencies, 37
charge coupled devices, 186
chemical
  dielectric, 177
  electric, 177
  energy, 163 ff.
  magnetic, 177
chromel–alumel, 262
CMRR, 89, 259
  and shielding, 26
code
  dimensions, 323
  symmetry, 323
coding, 62, 322
  capacity, 323
cold on-line standby, 131, 134
collector interference, 234
collisions, 210
color
  coding, 323
  resolution, 315
  wavelength, 315
common-mode
  rejection factor, 89, 229
  signal source, 232
  voltage, 88, 229
comparator, 57
compatible techniques, 184
compensation magnetic, 247
complex frequency, 35, 36, 40
components, 122
computer-operated systems, 31
computing power, 95, 107
conditional errors, 96
conductivity, 172
contact noise, 303
continuous stochastic variables,
    114
control bus, 32
controller-operated systems, 31
conversion, 64
  of shape, 166
convolution, 16
corona effects, 234
correlation, 22
  function, 22
Cotton–Mouton, 174, 177
  capacitive, 232
coupling

mechanisms, 249
cross correlation, 22, 97
cross-effect, 171
CRT, 320
screen illumination level, 323
cumulative distribution, 114, 116
Curie
and Néel, 177
temperature, 176
current
density, 172
limiting factor, 235
sink, 58
source, 58

DAC, 62
data bus, 32
data-acquisition, 25
part, 23
system, 23
data-distribution part, 23
data-processing part, 23
3 dB frequency, 37
De Morgan, 127
dead zone, 86, 98
decentralized data-acquisition system, 28
decoder, 70
decomposition, 140
definition of
equivalent open-circuit input RMS noise current, 276
equivalent short-circuit input RMS noise voltage, 276
ergonomics, 313
guarding, 229
interference, 229
noise, 300
current, 276
figure, 276
power, 276
voltage, 276
resolution, 81
sensitivity, 84
shielding, 229
deflection method, 74
Dember effect, 175
derating, 154
design aspects, 150 ff.
Destriau, 174
differential
amplifier, 56, 88
diode resistance, 302
mode signal source, 232
mode voltage, 88, 258
resistance, 302
sensitivity, 74
digital functions, 70
digital-to-analog conversion, 65
digits, 323
diode equation, 302
Dirac-pulse stimulus, 15
direct
compensation, 67
coupling, 249
effect, 171
discrete stochastic variables, 114
discrimination factor, 89, 259
display, 160, 317
distributed multiplexer data-acquisition system, 29
distributions
families of, 120
disturbance
additive, 100
dotting, 326
double-shielded cables, 267
downtime, 124
drift, 220
current, 220
voltage, 220
dual-slope integration ADC, 67
dynamic, 59, 91
range, 59, 91
state description, 172

early failure region, 119
earth, 236
economic reliability, 112
EEG, 231
effective
capacitance, 264
RMS noise voltage, 282
electrical
energy, 167
isolation, 252 ff.
electro-chemical, 176, 177
electrodes, 232
electroencephalogram, 231
electro-luminescence, 174
electro-magnetic, 177
electrostatic
fields, 229
interference, 229 ff.
electro-thermal, 176
encoder, 70
energy, 2
conversion, 160 ff.
space diagram, 166
types, 161
ENIAC, 181
environmental conditions, 96, 152, 155
equipment, 122
equipotential, 236
equivalent
current power, 281
noise power, 281
noise voltage source, 280, 281
open-circuit input RMS noise current, 276
RMS noise current, 280
short-circuit input RMS noise voltage, 276
ergonomics, 313
erratic currents, 237
errors, 83, 96
Ettinghausen effect, 177
excess noise, 303
exponential
expansion, 60
failure distribution, 120
extensive
physical quantity, 11, 170
quantity, 170
eye fatigue, 325

1/*f* noise, 303
failure
causes, 155 ff.
probability density, 116
rate, 116
Faraday, 174, 177
fault
potential, 262
value, 262
feedback
principle, 42, 98
resistors, 291
feedforward coupling, 101
figures, 323
flicker, 323
noise, 303
flickering, 320, 323
floating
guard, 263
shield, 267
Fourier
series, 17
transform, 18
fractional errors, 83
functional structure, 8 ff.
fundamental frequency, 17

galvanic separation, 255
galvano-electric, 177
gamma failure distribution, 120
gas discharge interference, 234
grain, 323
gravitational energy, 209
ground
definition of, 236
loop, 237
grounded systems, 237
guarding, 229, 235

Hall effect, 176, 192
Hall MOSFET, 193
hazard rate, 117
heart muscles, 232
height of modality, 314
high terminal, 235
hum and speech, 323

human
  behavior, 313
  brain, 323
  engineer, 313
  factors, 156
  perception, 316
  retina, 316
humidity, 220
  sensor, 194
hydrogen sensitive MOSFET, 195
hysteresis, 86, 98

identification unit, 161
ignition equipment, 234
illumination level, 325
impulse
  function, 15
  response, 15
inaccuracy, 82
incandenscence, 174, 176
infant mortality, 119
information technology, 320
injection laser, 188
input
  impedance, 103
  noise density, 278
  transduction, 12
instrumental perception, 317
integral transformation, 21
integrated
  flow sensor, 191
  magnetic transducer, 193
  resistive pressure transducer, 188
  transducer, 178
  transistor pressure transducer, 188
  Wheatstone bridge, 75
integrator, 57
intensive
  physical quantity, 9
  quantity, 9, 170
interactions, 210
interface, 27
interfacing, 196
  to sensors, 196
interference, 229
interfering source, 234
interline spacing, 323
interval measurements, 3
ion
  electrolytic conduction, 194
  sensitive field effect transistor, 195
ISFET, 195
isolating transformer, 252 ff.
isolation breakdown, 252

jamming, 19, 86
Johnson noise, 300

Kell effect, 326
Kerr
  electro-optic, 174
  magneto-optic, 174, 177
keyboard illumination, 325

Laplace transforms, 16
laser, 174
lateral
  photo-effect, 188
  photoelectric, 175
  photovoltaic, 175
lattice vibrations, 300
laws of human perception, 317
LCD, 187
least significant bit, 65
LED, 187
letters, 323
lifecycle, 113
lifetime, 113
light emitting diodes, 187
lightning, 240
linear
  amplification, 101
  system, 14
linearity, 85
liquid crystal displays, 187, 320
log dB scale, 36
logarithmic
  amplifier, 59
  compression, 59
  expansion, 60
Lossev, 174
loudness, 323
low terminal, 235
low-frequency noise, 303
lowpass filter, 39
LSB, 65
luminescence, 174

MAGFET, 193
Maggi–Righi–Leduc, 177
magnetic
  compensation, 248
  coupling, 251
  energy, 162
  fields, 246
  interference, 246
  loop, 241
  shielding, 246
magnetobridge, 193
magnetodiode, 192
magnetoresistance, 176
magnetoresistor, 192
magnetostriction, 175, 176
magnetotransistor, 193
manufacturing failures, 117, 118
Markov, 143
mean, 117
mean time between failure, 117, 118
  to failure, 118
  to first failure, 119
  to repair, 124
measurement, 3
measurements per unit of time, 94
measuring
  procedures, 8
  structure, 9
mechanical
  energy, 162, 174
  shocks, 234
median value, 283
metric information, 3
microprocessor system, 107
microwave coupling, 249
Miller indices, 167
MIL-standard classes, 151
minimal cut set, 142
mission availability, 125
modification unit, 161
modifier, 12, 166
modulating
  energy, 164
  transducers, 163
modulation, 19
  height, 314
moduli, 170
momentary availability, 125
most significant bit, 64
MSB, 64
MTBF, 118
MTTF, 118
MTTFF, 119
MTTR, 124
multichannel data-acquisition system, 25, 94
multiple
  instrument shielding, 241
  output, 24
multiplexer, 24, 70, 72
multiplexing, 72
multiplicative errors, 100
mumetal, 247
Murrel, K. F. H., 313

natural perception, 317
Néel temperature, 176
Nernst, 175, 176
nerve fibers, 316
network reduction, 137
NF, 277
Nichols chart, 40
noise, 85
  bandwidth, 282
  behavior of a diode, 309
  in bipolar transistors, 304
  calculating techniques, 295
  current, 276
  in FETS, 306
  figure, 277
    myth, 286

flicker current, 303
generator, 276
mechanisms, 300
power, 278
density, 278
in semiconductors, 304
sources, 97, 300
voltage, 276
nominal measurements, 3
non-inverting amplifier, 55
non-linearity, 85
non-repairable system, 123
normal
failure distribution, 120
reading distance, 325
nulling, 74
nullor, 44
Nyquist, 300
criterion, 43
frequency, 21

observer display interaction, 320
offset, 54, 86, 212
in bipolar transistors, 214
current, 212, 213
in differential amplifiers, 217
error, 54
in FETS, 215
voltage, 213
open windings, 253
operational amplifier, 44
optimum
signal-to-noise ratio, 289
source resistance, 287
opto-coupling, 255
OR function, 126
ordinal measurements, 3
output
impedance, 103
transduction, 11, 12
oxygen sensor, 195

parallel
connection, 130 ff.
system, 130
parameters
stochastic, 114
parasitic
capacitor, 230
current, 231
feedback, 239
voltage, 231
parts, 122
passive physical quantity, 11
path tracing, 139
patterns, 317
peak detector, 61
Peltier effect, 176
penetration depth, 246
perception, 317
periodic signal, 17
phase plot, 38
photochemical, 174
photoconductivity, 173, 186
photoconductor, 185
photodielectric, 174
photodiode, 185
arrays, 185
photoelastic, 175
photoelectric, 174
photoluminescence, 174
photointegrated circuit, 185
photomagnetic, 174
photomagneto electric, 173, 176
phototransistor, 185
photovoltaic, 173
effect, 186
physical effects in silicon, 183
physics of noise, 300
PI-FET, 189
pictograms, 324
piezoelectric, 174, 175
piezo-field effect transistor, 189
piezojunction, 188
piezo-optic, 175
piezo-resistance, 175
piezoresistor, 188, 189
piezotransistor, 188
pinch-off voltage, 216
pink noise, 303
pitch, 323
plasma screen, 320
pn-luminescence, 174
Pockel's effect, 174
Poisson, 115
polar plot, 36
poles, 40
pole-zero plot, 40
pollution, 236
power
delay product, 181
supply rejection ratio, 221
supply transformer, 252
supply variations, 221
preamplifier, 55
presentation unit, 161
price performance ratio, 182
primary winding, 252
probability distributions, 114
processing of information, 317
programmer-operated systems, 31
PSRR, 221
pulse
position modulation, 19
width modulation, 19

quality, 113
quality classes, 152

radiation
energy, 162, 173, 186
heating, 174
radio luminescence, 174
Raleigh failure distribution, 120
ramp function, 16
random factors, 156
range, 59
ratio
measurand, 13
measurements, 3
red noise, 303
redundancy, 131
rejection factor, 87, 89
relative sensitivity, 74
reliability, 112 ff., 181
repair rate, 124
repairable systems, 118, 123
resistivity, 172
resolution, 65, 81
response analysis, 14
rest impedance, 231
retrieval of information, 316
r.f. coupling, 249
Righi–Leduc, 177
rigidity, 170
root loci, 43

safety
aspects, 235
earth, 235
guard, 252
voltage, 235
sample and hold, 19
sampled signal, 19
sampling, 19
saturation, 86, 98
current, 302
SAW-filter, 190
secondary winding, 252
Seebeck effect, 175
self-generating transducers, 163, 164
semiconductor noise, 304 ff.
sensitivity, 84, 170
sensor, 161 ff.
sensorial mechanisms, 317
series
connection, 129
mode voltage error, 237
system, 129
Shannon, 21
shielded conductor, 245
shielding, 229, 239
cabinet, 239
shot noise, 301, 302
SI system, 3, 317
sign height, 325
average of, 107
signal
ground terminals, 237
structure, 9, 13
to-noise ratio, 284
silicon

complexity, 179
power consumption, 181
size, 179
technology, 179
size, 179
slew rate, 55, 86
smell, 323
S/N ratio, 284
soil resistance, 236
source resistance, 278
sources, 162
of interference, 234
spacing between signs, 325
spark-plug cables, 234
spatial structure, 9 ff., 23
specific
conductivity, 172
resistivity, 172
spectral width, 314
speedometer, 318
star structure, 31
starpoint, 252
state
description, 169
diagram, 144
steady-state availability, 125
steel, 247
step-down transformer, 252
step function, 16
stiffness, 170
stochastic
errors, 97, 98
signals, 21
variables, 114
variations in conductivity, 220, 300
stray
capacitance, 253
currents, 236
stroke width to height, 325
structural information, 2
Suhl effect, 176
superconductivity, 176
surface accoustic wave filters, 190
survival function, 115
synchronized detector, 19
system specifications, 80
systematic errors, 96
systems, 122

technical
reliability, 112
specification, 81
telemetry, 30
data-acquisition system, 30
temperature, 222
coefficient, 224
drift, 22 ff.
in bipolar transistors, 223
in FET, 225
effect, 234
integrated sensor, 190
texture, 324
thermal
biasing, 77
channel noise, 307
energy, 162, 175, 190
equivalent RMS noise voltage, 281
feedback, 78
noise, 300
thermochemical, 176, 177
thermoconductivity, 176
thermodielectric, 175
thermodiode, 190
thermoelastic, 175, 176
thermoelectric, 175
thermoluminescence, 176
thermopile voltage, 234
thermoresistor, 190
thermotransistor, 190
threshold voltage, 216
thyristors, 234
time
analysis, 14
sharing, 19
tools, 34
tolerances, 81
total noise
power, 278
voltage, 279
transducers, 160 ff.
transduction, 12
transfer function, 35
transformer coupling, 251
transient
response, 15
signal, 15
transition impedance, 144
transmission of information, 316
tribo-electro, 175
twisted pairs, 248
two-wire cable, 242
types of errors, 96

unreliability, 116
uptime, 124
useful life, 119

variance, 114
vibration, 220
virtually earth, 54
visual
coding, 323
field for permanent attention, 323
window, 315
volta effect, 177
voltage matching, 92
voltage-to-frequency conversion, 67
voltage-to-time conversion, 67

wavelength, 315
wear-out period, 119
Weibull failure distribution, 120
Westcott theorem, 103
Wheatstone bridge, 73, 196
configuration, 73
white noise, 22, 301
width to height sign proportion, 325
windows, 315
working area, 315
workspace, 321
illumination, 325
color temperature, 325

zero signal reference potential, 239, 244
zero error, 98
zeros, 40